Geology of the c
Buxton, Leek ar

The countryside whose geology is described in this memoir includes some of the most beautiful and varied in the Peak District National Park. It extends from the high gritstone moorlands of North Staffordshire in the west, across the central limestone plateau deeply dissected by wooded valleys, such as that of the River Wye that changes its name several times along its length, to the lush dales and wooded gritstone scarps in the east around Bakewell and Stanton.

The Carboniferous and Triassic rocks that form this landscape are described and interpreted to reveal something of the history and geography of the district some 340 to 200 million years ago when it was successively transformed from clear warm ocean with sporadic volcanic eruptions to dark muddy sea, turbulent delta slope, floodplains of powerful rivers, steamy peat swamps and arid desert.

This succession of environments was a consequence of the northward drift of this part of the earth's crust across the equatorial and tropical zones, movements that eventually raised, folded and faulted the rocks in a manner described in the chapter on structure. Mineral-bearing fluids were also generated and mobilised by these movements, the minerals being subsequently precipitated in the veins and orebodies described in the mineralisation chapter. Interpretations are also given of the variations in the earth's gravitational and magnetic fields found in the region, and the deposits left during and since the Quaternary glaciations are described and explained.

As well as being of great beauty and geological interest, the district continues to have considerable mineral and agricultural resources and, being within easy access of several densely populated conurbations, is of high recreational value. The memoir, especially when used in conjunction with the geological maps, provides essential reference material to help plan and maintain the balance between these conflicting interests.

Plate 1 Ramshaw Rocks: escarpment of the Roaches Grit (R_{2b}) on the east side of the Goyt Syncline near the Leek–Buxton A53 road.

The spectacular crags consist of fluviatile coarse-grained sandstone showing large-scale cross-bedding cut by masses of similar sandstone with faint internal lamination and strongly erosive bases (L 1221).

BRITISH GEOLOGICAL SURVEY

N. AITKENHEAD,
J. I. CHISHOLM AND
I. P. STEVENSON

CONTRIBUTORS

Palaeontology
M. Mitchell and A. R. E. Strank

Petrography
N. G. Berridge

Geophysics
J. D. Cornwell

Geology of the country around Buxton, Leek and Bakewell

Memoir for 1:50 000 geological sheet 111
(England and Wales)

1835   Geological Survey of Great Britain

150 Years of Service to the Nation

1985   British Geological Survey

Natural Environment Research Council

LONDON   HER MAJESTY'S STATIONERY OFFICE   1985

First published 1985

Bibliographical reference

AITKENHEAD, N., CHISHOLM, J. I. and STEVENSON, I. P. 1985. Geology of the country around Buxton, Leek and Bakewell. *Mem. Br. Geol. Surv.*, Sheet 111.

Authors

N. AITKENHEAD, BSc, PhD and I. P. STEVENSON, MSc
British Geological Survey, Keyworth, Nottingham NG12 5GG
J. I. CHISHOLM, MA
British Geological Survey, Murchison House, West Mains Road, Edinburgh EH9 3LA

Contributors

N. G. Berridge, BSc, PhD, J. D. Cornwell, MSc, PhD and M. Mitchell, MA
British Geological Survey, Keyworth, Nottingham NG12 5GG
A. R. E. Strank, BSc, PhD
formerly *British Geological Survey*
now *British Petroleum Company plc, BP Research Centre, Chertsey Road, Sunbury-on-Thames, Middlesex TW16 7LN*

Other publication of the Survey dealing with this and adjoining districts

BOOKS

British Regional Geology
The Pennines and adjacent areas, 3rd Edition 1954

Memoirs
Stockport (98) 1963
Chapel on le Frith (99) 1971
Sheffield (100) 1957
Chesterfield (112) 1967
Derby (125) 1979
Ashbourne (124) *In preparation*
Macclesfield (110) 1968

Reports
Hydrogeochemistry of groundwaters in the Derbyshire Dome with special reference to trace constituents, 71/7, 1971
A standard nomenclature for the Dinantian formations of the Peak District of Derbyshire and Staffordshire, 82/8, 1982
A revision of the stratigraphy of the Asbian and Brigantian limestones of the area west of Matlock, Derbyshire, 83/10, 1983

Mineral Assessment Reports   Limestone and dolomite resources of 1:25 000 sheets
No. 26   Monyash SK 16   1977
No. 47   Wirksworth SK 25 and part of SK 35   1980
No. 77   Buxton SK 07 and parts of SK 06 and 08   1981
No. 79   Bakewell SK 26 and part of SK 27   1982
No. 98   Tideswell SK 17 and parts of SK 18 and 27   1982
No. 129 Ashbourne SK 15 and parts of SK 04, 05 and 14   983
No. 144 Peak District (parts of 1:50 000 sheets 99, 111, 112, 124 and 125)   1985

MAPS

1:625 000
Great Britain, South Solid geology
Great Britain, South Quaternary geology
Sheet 2 Aeromagnetic

1:250 000
Liverpool Bay   1978
Humber–Trent   1983
Bouguer gravity anomaly, Liverpool Bay   1977
Aeromagnetic anomaly, Liverpool Bay   1978
Bouguer gravity anomaly, Humber-Trent   1977
Aeromagnetic anomaly, Humber-Trent   1977

1:50 000 or 1:63 360
Sheet 98   (Stockport)   1976
Sheet 99   (Chapel en le Frith)   1975
Sheet 100   (Sheffield)   1974
Sheet 112   (Chesterfield)   1963
Sheet 125   (Derby)   1972
Sheet 124   (Ashbourne)   1983
Sheet 110   (Macclesfield)   1968

1:25 000
SK 06   Roaches and upper Dove valley   1976
SK 07   Buxton   1975
SK 16   Monyash   1977
SK 17   Millersdale   1976
SK 18   Castleton   1976
Parts SK 25, 26, 35 and 36 Matlock   1978

ISBN 0 11 884389 3

Printed for **HMSO** by Commercial Colour Press
Dd 737397 c.20 9/85

CONTENTS

PLATES

FIGURES

TABLES

NOTES

Throughout the memoir the word 'district' refers to the area covered by the 1:50 000 geological sheet 111 (Buxton) except where used in the name Peak District.

National Grid References are given in square brackets; those beginning with the figure 9 lie in the 100 km square SJ and those beginning with the figures 0, 1 or 2 lie in the 100 km square SK.

Numbers preceded by the letter E refer to the sliced rock collection of the British Geological Survey.

The authorship of fossil species is given in the Index of Fossils.

PREFACE

The district covered by the Buxton (111) Sheet of the 1:50 000 geological map of England and Wales was originally surveyed on the one-inch scale as Old Series sheets 81 SW, by A. H. Green, published in 1864, and 81 SE, by J. Phillips and W. W. Smyth, published in 1852. Near its southern limit, the district includes small proportions of sheets 72 NW (1857) by W. W. Smyth and E. Hull, and 72 NE (1853) by J. Phillips, A. C. Ramsay and E. Hull.

Small marginal areas of the Buxton Sheet were surveyed during work on adjoining sheets between 1939 and 1958. Systematic fieldwork was commenced in 1964 by I. P. Stevenson, joined later by E. A. Francis, D. Price, Dr N. Aitkenhead and J. I. Chisholm; the Wye valley area survey incorporates work by R. A. Eden in 1953–1954. The mapping of the 1:50 000 sheet was completed in 1972 and six-inch maps by the various surveyors are listed on p. 148. Solid and drift editions of the map were published in 1978. In addition, parts of the district are covered by the following 1:25 000 sheets: SK 06 (The Roaches and Upper Dovedale), SK 07 (Buxton), SK 15 (Dovedale), SK 16 (Monyash), SK 17 (Miller's Dale), SK 25 (Wirksworth and Matlock), and SK 26 (Bakewell).

The Dinantian and Namurian chapters which make up most of the present memoir were written mainly by Dr N. Aitkenhead; the remainder is largely by J. I. Chisholm and I. P. Stevenson. The memoir was edited by I. P. Stevenson. M. Mitchell, who identified the macrofauna from the Dinantian, and Dr A. R. E. Strank, who identified the foraminifera, both contributed to the memoir; Dr D. J. C. Mundy also identified many of the reef macrofossils, and M. J. Reynolds the conodonts. Identifications of fossils from the Namurian and Westphalian are by Dr W. H. C. Ramsbottom and Dr M. A. Calver respectively, with assistance from Dr N. J. Riley. Petrography of the sedimentary and igneous rocks is by Dr N. G. Berridge, who also contributed to the relevant chapters. The chapter on geophysics is by Dr J. D. Cornwell. The photographs, a list of which is given in Appendix 4, were taken by K. E. Thornton, J. M. Pulsford and J. Rhodes.

Grateful acknowledgement is given to the many quarrying and mining concerns in the district for their cooperation, which in many cases involved the release of information. In particular, Laporte Industries Ltd released information and provided details of the Longstone Edge area and Raper Mine, while Dresser Minerals International Incorporated supplied and released information regarding Long Rake.

SIR MALCOLM BROWN, DSc, FRS
Director

British Geological Survey
Keyworth
Nottingham NG12 5GG

12 July 1985

SUMMARY OF GEOLOGICAL SEQUENCE

SUPERFICIAL DEPOSITS (DRIFT)

Recent and Pleistocene

Landslips	Calcareous tufa
Peat	Scree
Alluvium	Head
Alluvial Fan	Glacial Sand and Gravel
River Terrace, undifferentiated	Boulder Clay

		Generalised thickness
SOLID ROCKS		
Pliocene and earlier		m
Pocket Deposits	pebbly sand, silt, clay and mudstone	—
Triassic		
Sherwood Sandstone Group		
Hawksmoor Formation	red pebbly sandstones	up to 184
Carboniferous		
Westphalian		
Westphalian A	mudstones and sandstones with thin coals and seatearths; the *Gastrioceras subcrenatum* Marine Band at base	155
Namurian		
Yeadonian (G$_1$)	mudstones with Rough Rock at top and *Gastrioceras cancellatum* Marine Band at base	85
Marsdenian (R$_2$)	mudstones, with sandstones including Chatsworth Grit, Roaches Grit, Corbar Grit and Ashover Grit; two thin coals at top; *Reticuloceras gracile* Band at base	up to 540
Kinderscoutian (R$_1$)	mudstones and mudstones-with-sandstones, with Kinderscout Grit, Longnor Sandstones, Kniveden Sandstones and Blackstone Edge Sandstones; *Hodsonites magistrorum* Band at base	95–215
Alportian (H$_2$) and Chokierian (H$_1$)	mudstones and mudstones-with-sandstones, with Lum Edge Sandstones; lowest *Homoceras subglobosum* Band at base	18–50
Arnsbergian (E$_2$) and Pendleian (E$_1$)	mudstones and mudstones-with-sandstones, with Hurdlow Sandstones and Minn Sandstones; *Cravenoceras leion* Band at base	10–400
Dinantian: shelf province		
Brigantian (D$_2$)		
Longstone Mudstones	dark mudstones	
Eyam Limestones	dark limestones with knoll-reefs and flat-reefs	
Monsal Dale Limestones	pale and dark limestones with lavas and tuffs	240
Asbian (D$_1$)		
Bee Low Limestones	pale limestones with apron-reefs and units of lava and tuff	180
Holkerian to ?Tournaisian (S$_2$–C$_1$)		
Woo Dale Limestones	dark and pale limestones and dolomites	400
Dinantian: off-shelf province		
Brigantian (D$_2$) and Asbian (D$_1$)		
Mixon Limestone-Shales	thinly interbedded dark limestones and mudstones with subordinate sandstones	
Hopedale Limestones	pale and dark thin-bedded limestones with knoll-reefs	190
Asbian (D$_1$) ?Holkerian (S$_2$) and Arundian (S$_1$)		
Ecton Limestones	dark thin-bedded limestones	200
Arundian (S$_1$) to Tournaisian (C$_1$)		
Milldale Limestones	pale and dark thin-bedded limestones with knoll-reefs, lavas and tuffs: base not proved	220–790
Pre-Carboniferous	altered acid tuffs and lavas	proved to 38

CHAPTER 1

Introduction

AREA AND LOCATION

The district covered by this memoir lies mainly in the counties of Derbyshire and Staffordshire, and includes a small area of Cheshire in the north-west. The higher ground in the north, centre and east, forms part of the Peak District, and most of it lies within the Peak District National Park.

Much of the central and eastern part of the district is occupied by the Dinantian limestone outcrop (Figure 1), a largely upland area mostly lying between 200 and 400 m above sea-level. It is dissected locally by valleys of the rivers Wye, Lathkill, Dove and Manifold (Figure 2); the smaller valleys are normally dry, though intermittent streams are present in places. The area shows typical karst features such as sink-holes, caves and limestone scars. In the eastern parts of the limestone outcrop, lead mining has left spoil heaps, shafts and open workings along the lines of veins and in areas of flats; active mining is mainly in the Longstone Edge and Youlgreave areas (see pp. 118 and 120). Drainage levels or soughs de-water extensive areas of the limestone away from the main river valleys.

On the west of the district the outcrop of sandstones and mudstones of Namurian and early Westphalian age forms moorlands with well marked sandstone scarps and dip-slopes around the Goyt Syncline. The highest ground is on Axe Edge, at 551 m. The outcrop of the lower, predominantly mudstone, part of the succession shows a more subdued topography south of Longnor and east of Leek. This area is drained by the rivers Churnet, Manifold and Dove. In a smaller area of Namurian rocks on the east of the district, gently dipping sandstones produce plateau-like features; drainage is by the Wye and by the Derwent which runs just beyond the eastern margin.

A small area of low ground at Leek is underlain by Triassic sandstones.

OUTLINE OF GEOLOGICAL HISTORY

The known geological history of the region, on the evidence from the Eyam Borehole (p. 3), begins in the Ordovician (probably Llanvirn), when the area probably formed a marine platform (Anderton and others, 1979, p. 89) flanking the northern edge of the major continent of Gondwanaland, possibly then situated in high southerly latitudes (Cocks and Fortey, 1982, p. 472).

Following the Caledonian Orogeny, when the Iapetus Ocean to the north of Gondwanaland closed and was replaced by mountain ranges of the Old Red Sandstone or Laurasian continent (Johnson, 1981, p. 221), the district was probably in a tropical area of eroding hills and intermontane basins. At the beginning of Carboniferous times, these continental conditions gave way to a marine environment of mainly carbonate deposition, with repeated minor transgressions and regressions in an unevenly subsiding arm of the sea that lay between the main landmass of Laurasia to the north, and the Wales–Brabant Island to the south. Later in the Carboniferous, this sea gradually filled with sediment brought in by rivers from these land areas until eventually fluvial and swamp conditions predominated, interrupted by only a few brief marine transgressions. Crustal tension – thought to have been caused by southward creep of the deeper crustal layers towards a subduction zone far to the south – and partial melting of the mantle are thought to have led to volcanic eruptions during the Carboniferous (Leeder, 1982). It is also known, from palaeomagnetic evidence (p. 124), that the region drifted northwards across the equator during the Devonian to Permian time interval (literature summarised by Johnson, 1981). During the Hercynian or Armorican orogeny at the end of Carboniferous times, the rocks were folded, faulted and uplifted and, by the end of the Permian, erosion had produced a widespread peneplain. Renewed deposition during the Triassic was of a continental nature and gave rise to the red pebbly sandstones of the Leek Basin. Cycles of erosion and deposition during the interval from the Triassic to the Tertiary have left no known representatives in the south Pennines. This gap in the record ends with the sands and conformably overlying Pliocene clays of the Brassington Formation, which have been preserved as scattered Pocket Deposits filling solution cavities in the Carboniferous Limestone. The current Quaternary Period has been marked by successive glacial advances and retreats. A few scattered patches of till on the limestone outcrop probably represent the penultimate (Wolstonian) glaciation but only the latest (Devensian) glaciation is represented by extensive deposits, confined mainly to the western part of the district, indicating that the higher ground to the east was essentially free of any thick mobile ice cover at this time. Of the milder interglacial intervals, only the last (Ipswichian) is evident from cave deposits. Post-glacial erosional and depositional processes have continued up to the present day, either adding to or eroding the Recent deposits of head, peat, alluvium, calcareous tufa, and landslip.

Figure 1 Sketch map showing the general geological setting of the district.

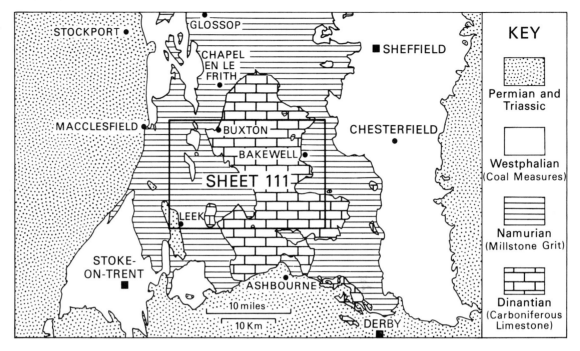

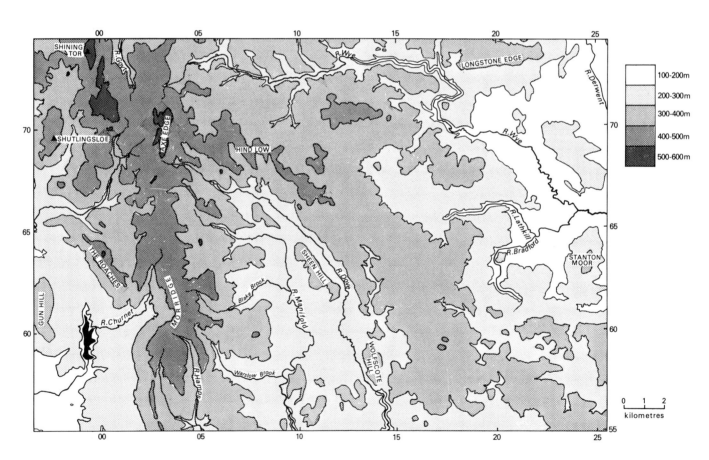

Figure 2 Principal physical features and drainage.

CHAPTER 2

Pre-Carboniferous

The existence of a stable block of pre-Carboniferous rocks beneath the Dinantian limestone outcrop of the Derbyshire Dome (p. 109) has long been inferred from variations in lithofacies of both Dinantian and Namurian strata, and from the contrast in tectonic style between the dome and some adjacent areas (*see*, for example, Kent, 1966, p. 335; Stevenson and Gaunt, 1971, p. 320).

Studies of gravity data (Maroof, 1976; *see also* pp. 125 – 129) confirm the existence of a structural high, corresponding to the stable block and roughly coincident with the present limestone outcrop. The upper surface of the block is inferred to have a gentle easterly inclination, with high points in the north and south. The nature of the rocks involved is known from only three boreholes: at Woo Dale which lies in the present district, Eyam situated just beyond the northern edge and Caldon Low, 7 km to the south of the district.

The Woo Dale Borehole [SK 07 SE/24] (*see* Appendix 2) at the junction of Woo Dale and the Wye valley, entered pre-Carboniferous rocks at 273.6 m and continued in red and green 'altered lava, volcanic breccias and pyroclastic rocks' (Cope, 1973, p. 30) to the bottom of the borehole at 312 m.

The lavas show flow banding and approach soda-dacites or soda-rhyolites in composition. Potassium/argon dating gives a maximum age of 382 ± 6 Ma indicative of a Devonian or older age (Cope, 1979, pp. 319 – 320).

The Eyam Borehole [SK 27 NW/15] proved mudstones of probable Llanvirn age between 1803.25 and the bottom of the hole at 1851.05 m (Dunham, 1973). These strata have yielded a microflora and a shelly fauna including sponge spicules, orthocones, brachiopods (*Conotreta?*, *Lingulella*, orthide) and trilobites including *Platycalymene* cf. *tasgarensis*, *Pliomeroides?*, an ogygiocarinid and a cybeloid fragment.

The lowest beds in Caldon Low Borehole [0804 4822] comprised a 170.30 m-formation, the Redhouse Sandstones, consisting of reddish-brown sandstone, siltstone and mudstone with some concretionary dolomite (Institute of Geological Sciences, 1978a, p. 11; Aitkenhead and Chisholm, 1982, pp. 7 – 9). The uppermost part of this formation probably belongs to the earliest Dinantian (Hastarian) Stage since the conformably overlying formation, the Rue Hill Dolomites, belongs mainly to this stage (Welsh and Owens, 1983), but the remainder may be of Devonian age.   IPS, NA

CHAPTER 3

Dinantian

The Dinantian rocks of the district consist mainly of marine limestones, traditionally referred to as 'Carboniferous Limestone'; they also include a varying proportion of inter-bedded mudstone in the highest part of the sequence (Aitkenhead and Chisholm, 1982, p. 1). The outcrop extends over half the area of the district and is largely continuous (Figure 3). Elsewhere the Dinantian rocks are concealed beneath Upper Carboniferous strata, but have been proved in boreholes at Gun Hill (p. 46) and in the Stanton Syncline (p. 45).

Dinantian rocks are estimated to have a minimum total thickness around Buxton of about 660 m, the estimate being based partly on the Woo Dale Borehole results (Cope, 1973). Boreholes at Eyam (Dunham, 1973) and Caldon Low (Institute of Geological Sciences, 1978a) close to the district (Figure 3) have also proved the base of the Dinantian. From these borehole data, the total thickness is inferred to be about 1900 m in the north-east part of the district (nearly three times that near Buxton) and about 1000 m in the south-west.

PREVIOUS RESEARCH

The combination of mineral wealth, great natural beauty and good rock exposure has encouraged geological research in the Dinantian outcrop of the Peak District for at least two centuries. Useful summaries of early research work have been published by Smith and others (1967), Morris (1969), Stevenson and Gaunt (1971) and Ford (1977); the few additional observations given below are intended to supplement these accounts. Although Whitehurst (1778) first recognised and described the succession of alternating limestone and 'toadstone' (basaltic lava) strata in the Matlock area just outside the eastern boundary of the district, he evidently failed to understand that this represented a succession of depositional and eruptive events, and concluded that the toadstone 'is actual lava' but that it 'did not approach the open air, but disgorged its fiery contents between the strata in all directions'.

Watson (1811) illustrated the general succession and structure of the Derbyshire Dome with diagrammatic cross-sections from Combs Moss to Chatsworth, Grange Mill to Bakewell, Bakewell to Longstone Edge, and Bakewell to Hartington (Ford, 1960). Farey (1811) published the first geological map showing the outcrop of the Carboniferous limestone in the south Pennine region. He subdivided the outcrop into two parts, one in the east with interbedded limestone and toadstone, and another in the west in which his lowest stratigraphic unit (Fourth Limestone) cropped out. He also showed two outcrops of 'shale-limestone' strata, one around Butterton and Mixon Hay and the other between Blore and Kniveton in the Ashbourne district. Following the publication of the Old Series geological map, Green and others (1887) gave some details of the limestone succes-

sion, particularly in the railway-cuttings along the Wye valley where an upper thinly bedded cherty sequence was underlain by 'a great thickness of exceedingly massive pure limestones'. They also recognised that the toadstones were mainly the result of 'contemporaneous volcanic eruptions'. The classical account of the igneous rocks of the region was provided many years later by Arnold-Bemrose (1894 and 1907).

Sibly (1908) first applied the coral-brachiopod subdivisions, established in the Avon Gorge by Vaughan (1905), to the Derbyshire area, the limestones of which he correlated with the upper part of the Bristol succession. The dolomites and dolomitic limestones that have extensive outcrops in the south-eastern part of the Dinantian tract were mapped and described by Parsons (1922) whose account remains the standard work on the subject.

A series of papers by Cope (1933, 1936, 1937 and 1939) on the succession in the Wye valley marked the start of the modern period of detailed stratigraphical work. The publication of the results of the deep boreholes at Gun Hill (Hudson and Cotton, 1945) and Woo Dale (Cope, 1949a) added greatly to the knowledge of the succession and showed the marked contrast between rocks of approximately the same age in the shelf ('massif') and off-shelf ('basin') provinces (p. 6). This work was followed by studies in the south-west by Parkinson (1950), Prentice (1951 and 1952) and Morris (1969); around Earl Sterndale by Wolfenden (1958); around Hartington by Sadler and Wyatt (1966); in the east by Shirley (1959); and in the north by Butcher and Ford (1973) and by Walkden (1977).

Other works of local specialised significance are mentioned in appropriate sections of the following account. Many data on the chemical and physical properties of the limestones and their petrography are given in the series of BGS Mineral Assessment Reports covering the limestone outcrop (p. 139).

CLASSIFICATION

Chronostratigraphy

The chronostratigraphical subdivision used, particularly the stages, are those proposed by George and others (1976) and modified by Ramsbottom and Mitchell (1980). The stages, based on precisely defined stratotype sections, replace the coral-brachiopod zonal subdivisions of Vaughan (1905) previously in common use, and were devised as a standard for correlation. The new stages and their approximate coral-brachiopod zonal equivalents are (from oldest to youngest): Hastarian (K and Z), Ivorian (γ), Chadian (C_1 and early C_2S_1), Arundian (late C_2S_1), Holkerian (S_2), Asbian (D_1), and Brigantian (D_2). The two oldest stages together comprise the Tournaisian Series, and the remainder the Viséan

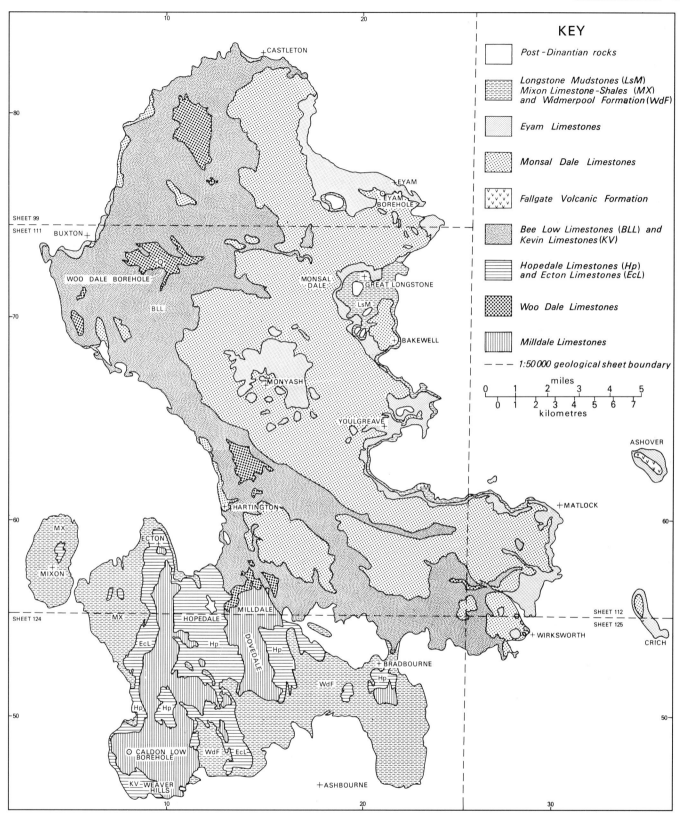

Figure 3 Sketch map of Dinantian formations in the Peak District.

Series. Recognition of these stages depends largely on the identification of characteristic fossil assemblages (p. 55–58).

In the shelf province the Holkerian–Asbian and Asbian–Brigantian stage boundaries are approximately coincident with the Woo Dale Limestones–Bee Low Limestones and Bee Low Limestones–Monsal Dale Limestones formational boundaries respectively (*see below*). In the off-shelf province, stage-diagnostic macrofossils are scarce and, although some progress has been made using conodonts and foraminifera, much work still needs to be done, particularly on the latter forms, before the stage boundaries can be delineated. Further subdivision into goniatite-bivalve zones and subzones (Earp and others, 1961, p. 174) is made where diagnostic goniatites are present, mainly in the Asbian apron-reef facies and the late Asbian to Brigantian off-shelf formations.

Lithostratigraphy

The Dinantian rocks of the district, and of the whole Peak District, lie in two main areas of differing stratigraphy – the shelf province (Stevenson and Gaunt, 1971, p. 11) and the off-shelf province – which were clearly differentiated during the Asbian and Brigantian but less so in the earlier Dinantian stages (*see below*).

Since the publication of the Buxton 1:50 000 geological sheet, the formations in the two provinces have been defined and described by Aitkenhead and Chisholm (1982). Figure 4 shows this classification and the main local and earlier classifications which have been incorporated in the scheme; the outcrops are shown in Figure 3. Most of the standard formation names in the shelf province have been adapted from those of Stevenson and Gaunt (1971); in the off-shelf province the majority are derived from Hudson (*in* Hudson and Cotton, 1945) and Parkinson (1950). Two new formations have been defined, the Fallgate Volcanic Formation, present largely at depth in the eastern part of the district, and the Hopedale Limestones in the off-shelf province in the south-west. Where a formation contains local sub-units of member status, the names given by the original authors have been retained where possible; this applies particularly in the shelf province. Not all members are named, however.

Shelf and off-shelf provinces

The shelf province is characterised by mainly shallow water deposits. It includes the bulk of the Dinantian outcrop and extends eastwards beneath later cover rocks. The most typical formation, the Bee Low Limestones, includes a narrow discontinuous belt of outward-dipping apron-reef facies

SHELF PROVINCE							OFF-SHELF PROVINCE		
REGIONAL FORMATION NAMES	LOCAL AND EARLIER CLASSIFICATIONS						REGIONAL FORMATION NAMES	LOCAL AND EARLIER CLASSIFICATIONS	
	Wye valley *(Cope 1933, 1937 & 1958)*	Matlock area *(Smith and others 1967)*	Wirksworth area *(Frost and Smart 1979)*	Monyash and Wirksworth *(Shirley 1959)*	North-east of Hartington *(Sadler 1966)*	Wolfscote Dale & Alsop Moor *(Parkinson 1950)*		Mixon & Manifold Valley *(Hudson in Hudson and Cotton 1945)*	Manifold Valley *(Prentice 1951 & 1952)*
LONGSTONE MUDSTONES		Cawdor Group	Cawdor Shale				MIXON LIMESTONE-SHALES	Mixon Limestone-Shales	Posidonomya Beds
EYAM LIMESTONES	Ashford Beds		Cawdor Limestone	Cawdor Limestone					
MONSAL DALE LIMESTONES	Monsal Dale Beds	Matlock Group	Matlock Limestone	Lathkill Limestones			HOPEDALE LIMESTONES and ECTON LIMESTONES	Mixon Limestones and Ecton Limestones	Brownlow Mudstones
	Priestcliffe Beds / Upper Lava / Station Quarry Beds					Alsop Moor Limestone			Apestor & Warslow limestones / Waterhouses Limestone and Crassiventer Beds
BEE LOW LIMESTONES	Miller's Dale Beds	Hoptonwood Group	Hoptonwood Limestone	Via Gellia Limestones					
	Lower Lava				Upper Limestones	Wolfscote Dale Limestone			
	Chee Tor Rock				Lean Low Beds				
WOO DALE LIMESTONES	Daviesiella Beds	Griffe Grange Bed			Hand Dale Beds	Iron Tors Limestone	MILLDALE LIMESTONES	Manifold Limestones with Shales	Manifold Limestones -with-Shales
					Vincent House Beds			Calton Limestones	Massive Series
									Cementstone Series

The formations include the following named members:-
MONSAL DALE LIMESTONES: Shothouse Spring Tuff, Winstermoor Lava, Lathkill Lodge Lava, Conksbury Bridge Lava, Shacklow Wood Lava, Lees Bottom Lava, Litton Tuff, Lower Matlock Lava, Upper Miller's Dale Lava and Station Quarry Beds.
BEE LOW LIMESTONES: Miller's Dale Limestones, Lower Miller's Dale Lava and Chee Tor Rock.
WOO DALE LIMESTONES: Griffe Grange Member, Hand Dale Member, Vincent House Member and Iron Tors Limestones.

Figure 4 Dinantian formation and member names in the district compared with earlier classifications.

along the western and southern margins of the province. In the off-shelf province, which lies to the west and south of the reef belt, rocks of roughly equivalent age are largely of deeper-water facies. Pre-Holkerian rocks of the two parts of the district do not show the same contrasts in lithology and consequently there is little evidence for the existence of clearly differentiated shelf and off-shelf provinces during the earlier part of the Dinantian. However, for convenience, the older formations are described as part of whichever province is appropriate to the overlying formations.

Reef-limestones

Reef-limestones occur in formations belonging to both shelf and off-shelf provinces; none of the occurrences has been given a member name. The terminology used here follows that of Stevenson and Gaunt (1971, pp. 17–19) in the district to the north, knoll-reefs and apron-reefs being the commonest forms. This usage does not necessarily imply the presence of a rigid organic framework as it does in most modern definitions of reef (see, for instance, Wilson, 1975, pp. 20–22). Knoll-reefs, apron-reefs, and flat reefs (Stevenson and Gaunt, 1971, p. 19) would all be included in the term 'buildup' (Heckel, 1974), now in common use. It should be noted, however, that Heckel (1974, p. 110) regarded the whole of the Derbyshire Dinantian limestone mass as a 'buildup', and his more restricted term 'lime-mud buildup' would be more appropriate for the reef masses.

The different types of reef have several lithological features in common. They are composed largely of poorly bedded pale to medium grey micritic limestones; they show dips reflecting the original slopes of the reef flanks at progressive growth stages, quaquaversal in the case of knoll-reefs and sub-parallel to the shelf margin palaeoslope in the case of the apron-reef; they all contain a varied, distinctive and, in places, abundant fauna, mainly of brachiopods. They are of varied origin, however, as has been shown by the few detailed petrographical and palaeoecological studies that have been published (Wolfenden, 1958; Eden and others, 1964; Orme, 1970; Broadhurst and Simpson, 1973; Adams, 1980; Miller and Grayson, 1982).

The micritic limestone forms either the whole mound, as in some of the knoll-reefs in the Milldale Limestones, or a core surrounded by crinoidal flank beds. The junction between core and flank, or between core and inter-reef limestones, is generally sharp, and for mapping purposes is taken as the boundary of the knoll-reefs. In a few reefs in the Eyam Limestones, however, the flank beds have been included in the reef.

The apron-reefs that fringe much of the shelf margin in the Bee Low Limestones have a discontinuous wall-like core of micritic algal biolithite; this passes into a transitional back-reef on the shelf side, and into an outward-dipping fore reef on the off-shelf side.

The origin of the lime mud forming the reef micrite is still problematical, though it is generally assumed that algae played a significant part in mud production and at times also acted as sediment binders. The mechanism by which the mud banks were built up relative to the surrounding sediments also remains uncertain.

PALAEOGEOGRAPHY AND DEPOSITIONAL HISTORY

After a long period of erosion and local continental red-bed sedimentation in the Devonian, the earliest Dinantian rocks of the district were probably deposited during intermittent marine transgressions 'across a geologically diverse land surface of some relief' (Aitkenhead and Chisholm, 1982). The evidence for this comes from a few deep boreholes, of which one, at Woo Dale (Cope, 1973; 1979), lies in the present district (pp. 8 and 10).

Pre-Asbian

The late Tournaisian and early Viséan outcrop in the southwest of the district is characterised by large composite knoll-reefs which probably formed in relatively deep water after the initial marine transgressions were complete (p. 45). The reefs reached their maximum development in Chadian times. Growth may have been halted in the Arundian, when a period of erosion is indicated by the presence of a large block of reef-limestone in beds of this age in the Manifold valley (p. 5). The relatively thin Arundian succession in the Woo Dale Borehole (157.7 m, p. 10) and the thick Chadian–Arundian one in the Eyam Borehole (about 1050 m) were extensively dolomitised, probably during one or more restricted lagoonal phases in pre-Asbian, possibly late Holkerian, times. Local volcanic activity in the off-shelf province is represented by basaltic lavas proved in the Gun Hill Borehole (p. 10) and a thin tuff in the Milldale Limestones near Ecton.

Holkerian palaeogeography remains uncertain, with insufficient work yet done on the foraminifera to confirm the distribution pattern of rocks of this age. However, the sections exposed in the Woo Dale Limestones of the type area have now been firmly established as Holkerian, and the same age can be inferred for the remaining exposed parts of the formation elsewhere. Some of the micritic beds in the uppermost part of the sequence include 'birdseye' structures and rare thin coals, indicating a late Holkerian regressive phase when intertidal and supratidal conditions prevailed.

Over much of the off-shelf province the presence of Holkerian strata is still in doubt. Limestones, for which a Holkerian age would be expected as they lie between proven Asbian and Chadian or Arundian strata, tend to contain non-diagnostic faunas. However, probable late Holkerian to early Asbian foraminifera have been found in one section of the Ecton Limestones in the Manifold valley (p. 51).

Asbian

During Asbian times a clear differentiation evolved between an upstanding shelf province fringed by apron-reefs and an off-shelf province beyond the reef-belt. On the shelf thickly bedded pale grey calcarenites (Bee Low Limestones) were deposited in shallow water, and there were frequent periods of subaerial emergence. Intermittent volcanic activity became widespread at this time. At the shelf margins the fore-reef limestones, the main subfacies of the apron-reefs, consist of pale micritic limestone with abundant and diverse shelly faunas (Wolfenden, 1958). These beds have a depositional dip component away from the shelf into deeper water. In the off-shelf province, two main sets of sediments were laid down. Limestone turbidites (Ecton Limestones) accumulated in the deeper troughs, whereas a more diverse set of inter-reef sediments (Hopedale Limestones) was laid down in water of variable depth between the upstanding masses of older and contemporaneous knoll-reefs. The Hopedale Limestones contain conglomerates and well laminated beds, indicating that high current speeds were reached, perhaps in constricted channels between the reef masses. Evidence of emergence is lacking, but the water depth may not have been great and so the term 'off-shelf province' has been introduced to allow for the presence of both shallow- and deep-water sediments in the areas previously described as 'basinal'. The Ecton Limestones (together with the mainly Brigantian Mixon Limestone-Shales) are the only off-shelf deposits here regarded as genuinely 'basinal'. The distribution of shelf and off-shelf facies towards the end of Asbian times has been illustrated by Aitkenhead and Chisholm (1982, fig. 8).

Brigantian

The shelf province established during Asbian times persisted into the Brigantian, but the growth of fringing apron-reefs diminished or ceased. The Monsal Dale Limestones contain many pale grey shallow-water limestones with emergent surfaces, but bodies of dark argillaceous limestone, deposited in deeper water, are also included. In the eastern part of the shelf, interbedded volcanic rocks from an important proportion of the sequence and locally predominate over limestones (p. 59). Many of the volcanic rocks were erupted subaerially, but there are also hyaloclastites of undoubted subaqueous origin. During the later part of the Brigantian (at the top of the Monsal Dale Limestones), a well marked emergence and resubmergence of the shelf was followed by a short-lived phase of knoll-reef development unrelated to the shelf margin (knoll-reefs in Eyam Limestones).

The shelf margins were the site of erosion during late Asbian or early Brigantian times, for in places the lowest of the Monsal Dale Limestones rest with disconformity on the remnants of the Asbian apron-reefs, as between Hartington and Pilsbury (Aitkenhead and Chisholm, 1979). The beds above the disconformity are not reef-limestones, but are variably shelly limestones of shelf facies; reefs of proven Brigantian age are rare in this belt. The shelf margins were affected by further erosion in late Brigantian or early Namurian times, with resultant overstep of the basal Namurian mudstones on limestones of various Asbian and Brigantian ages.

The off-shelf province was the site of deeper-water deposition of mudstones interbedded with limestone turbidites (Ecton Limestones, Mixon Limestone-Shales), and also with sandstone turbidites in places (Onecote Sandstones).

The final late Brigantian phase of submergence was persistent, affecting shelf and off-shelf provinces alike and leading to the period of basinal mudstone deposition that prevailed over the whole south Pennine region during early Namurian times.   NA

STRATIGRAPHY OF SHELF PROVINCE

WOO DALE LIMESTONES

The Woo Dale Limestones were defined formally by Aitkenhead and Chisholm (1982), using the name introduced by Stevenson and Gaunt (1971, p. 21) 'for the S_2 beds of the Wye valley' synonymous with the 'Daviesiella Beds' of Cope (1933, p. 129). For convenience, the pre-Woo Dale Limestones sequence in the Eyam Borehole is dealt with in this part of the account.

The outcrop in the present district is restricted to inliers (*see* Figure 5). The most extensive of these lies east of Buxton and includes the type area around Woo Dale [097 726] where the uppermost 132 m* of the formation are well exposed; the lower part, totalling 270.26 m, is proved in the Woo Dale Borehole [SK 07 SE/24]. Thus the formation here extends from the base of the Dinantian succession to that of the Bee Low Limestones. The dolomites and dolomitised limestones at the base of the exposed sequence and in the borehole (Cope, 1973), are known informally as the Woo Dale Dolomites. The formation as a whole is inferred to underlie the younger Dinantian rocks over the entire shelf province, and the dolomitic facies is probably also widespread as it has been proved in all three boreholes that reach the lower parts of the formation: Woo Dale Borehole; Eyam Borehole [SK 27 NW/15]; and Ryder Point No. 3 Borehole [SK 25 NE/48] in the Via Gellia valley (Chisholm and Butcher, 1981). The limits of this facies are, however, likely to be highly irregular and diachronous.

In some of the other inliers (Figure 5), distinctive parts of the sequence named by previous authors have now been assigned member status (Aitkenhead and Chisholm, 1982). They include the Hand Dale Member and the Vincent House Member in the two inliers north-east of Hartington, the Iron Tors Limestones in the Wolfscote Dale – Iron Tors outcrop (Parkinson, 1950) and the Griffe Grange Member ('Bed') in the Via Gellia inlier (Smith and others, 1967). However, it has been found practicable to delineate only the Iron Tors Limestones on the map.

Lithology and facies variations

In the Woo Dale Borehole (Cope, 1973; 1979) there is a 2.44 m basal breccia containing fragments of lava and pyroclastic rocks presumably derived from the underlying pre-Carboniferous volcanics. This bed is overlain by 27.96 m of grey and dark grey limestones including thin calcilutites and shaly partings; this sequence is given a ?Tournaisian age by Cope (1979). The succeeding grey, dark grey and brown mainly dolomitised limestones comprising the remaining 243.20 m of the borehole sequence are assigned to the Woo Dale Dolomites. The sparse faunas recorded by Cope were taken by George and others (1976) to indicate a Chadian–Arundian age for the basal 133.00 m and a Holkerian age for the remaining 110.20 m. However,

---

* This figure supersedes the estimate of 100 m given by Aitkenhead, p. 8, *in* Harrison, 1981.

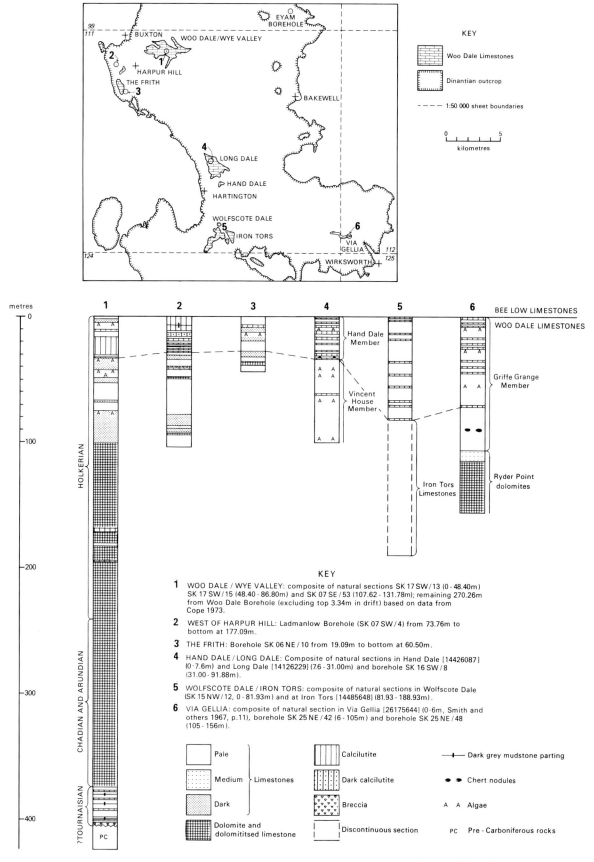

Figure 5 Comparative composite vertical sections and outcrop locations for the Woo Dale Limestones.

the evidence from newly discovered foraminiferal and algal assemblages (Strank, 1985) indicates that the basal 157.7 m should be assigned to the Arundian: older Dinantian strata are apparently absent.

The Eyam Borehole (Dunham, 1973, pp. 84–85), a little beyond the northern boundary of the district, provides a record of Dinantian beds underlying the Woo Dale Limestones. The junction with the underlying Ordovician at 1803.25 m has been taken at the base of a 0.66 m grey-green and red sandstone. This was separated by 7.96 m of dolomite with some mudstones, from 64.99 m of banded anhydrites with thin bands of mudstone and dolomite extending up to a depth of 1729.64 m; the evaporite sequence yielded mio-spores suggesting a late Tournaisian age. Above, the bore-hole proved a 1324.52 m sequence of Woo Dale Limestones. The lowest 1048.44 m of this contained much doloite with some grey-brown and dark grey limestones; it carried a coral and foraminiferal fauna indicative of the Chadian and Arun-dian stages. The upper part of the formation comprised 276.08 m of dark and pale grey limestones with subsidiary dolomites, up to the base of the Bee Low Limestones at 405.12 m; it contained a coal 0.1 m thick, at 508.92 m.

The exposed part of the formation (which complements the Woo Dale Borehole section), approximately 132 m thick, is exposed largely in various incomplete sections from which a general description can be derived (see also Cope, 1933). The basal 32 m comprise the uppermost part of the Woo Dale Dolomites, and consist mainly of brown granular dolomite with minor interbedded limestone. These are overlain by some 67 m of limestones characterised by their predominantly dark grey colour and irregular partial dolomitisation. The uppermost part of the sequence is distinguished by the presence of numerous irregular beds of mainly medium to pale and grey-brown micritic limestone or calcilutite, the calcite mudstones of Cope (1933). These rocks are interbedded with peloidal calcarenitic limestones (pelsparites and biopelsparites). A similar facies occurs in the uppermost part of the formation in all the other inliers in the district (Figure 5). The immediately underlying rocks in these inliers are of more uniform lithology, and nowhere else do dark limestones and dolomites predominate as they do in the type area. Instead, pale or medium grey calcarenites, particularly crinoidal biosparites and pelsparites, form the main rock type, and are characteristic of the Vincent House and Iron Tors members and the lower part of the Griffe Grange Member.

Non-carbonate rocks, though forming only a very small proportion of the total thickness (less than 0.5 per cent), are a significant part of the formation. Black or dark grey mudstone partings are the most common, and are normally carbonaceous or bituminous. They occur mainly in the dark limestone sequence in the Woo Dale – Wye Valley inlier and are mostly less than 2 cm thick. At least four coals up to 3 cm thick have also been recorded from these strata; it has not been established that they represent plant remains coalified in-situ.

In contrast with the overlying Bee Low Limestones, very few rocks of volcanic origin have been recorded, the most notable being a 1.66 m bed of tuff in a borehole [SK 07 SE/39] at Tunstead Quarry (Harrison, 1981).

With the exception of some nodular chert noted in a borehole [SK 25 NE/42] in the Via Gellia between depths of 80.50 and 87.10 m (Cox and Harrison, 1980), the formation is chert-free.

Environment of deposition

The main deduction about the basal breccia in the Woo Dale Borehole sequence is that it is probably a marine shoreline deposit representing the first transgression phase in Dinan-tian sedimentation at that locality. As there were probably several minor cycles of transgression, and possibly regression (Ramsbottom, 1973), over an undulating landscape, the deposit does not necessarily correlate with the lowest beds elsewhere. These comprise an evaporite-bearing sequence in the Eyam Borehole (see above), that probably represents a sabkha environment of wide extent (Dunham, 1973).

Following the transgressive phase, a shallow marine shelf environment was established, but the predominance of stratigraphically controlled dolomites in the sequence in-dicates that restricted water circulation and intense evapora-tion tended to produce the hypersaline conditions favourable for dolomitisation*. Comparison with the Eyam Borehole record (Dunham, 1973) shows, in addition, that the se-quence is significantly condensed in the Wye valley (Cope, 1973). The presence of 42 m of dolomite in the concealed basal part of the Via Gellia succession (Figure 5) indicates similar restricted shelf conditions there. However, the con-figuration of the shelf area probably had little resemblance to that which later became so clearly delineated in Asbian times (p. 7).

In the upper part of the formation the beds become thin-ner and more variable in their lithology with calcilutites and micrites interbedded with the pelsparites and biopelsparites. In places, the coarser lithologies show parallel and cross-lamination, and irregular or lenticular bedforms; taken together these indicate strong current action, either in tidal channels or in areas subjected to wave action. Algae and algal-encrusted clasts are common and small vughs filled with clear spar, known as 'birdseyes' (Shinn, 1968), and possible rootlet structures occur in some of the micritic beds, indicating a supratidal environment. This interpretation is supported by the local occurrence of a few thin coals, sug-gesting the presence of a low island or islands with a dense vegetation cover in the Wye valley area. Since there are many thin alternations of these lithologies, the shelf area must have been fluctuating just above and below sea level many times during the late Holkerian. Ramsbottom (1973) believes that the late-Holkerian marks the acme of a major regression of the sea that is evident in both the United States and Europe at this time.

NA, JIC, IPS

---

* A fundamentally different view of the formation of the Woo Dale Dolomites has recently been propounded by Schofield (unpublished PhD thesis, 1982). He proposes that dolomitisation was effected during deep burial in Upper Carboniferous times by the influx of magnesium- and iron-rich fluids from surrounding basinal sediments and underlying decompos-ing volcanic rocks.

DETAILS

Wye valley and adjacent areas

The Woo Dale Dolomites crop out in two closely linked areas in the Wye valley. Unexposed strata were proved in the Woo Dale Borehole [SK 07 SE/24] (p. 8), which commenced at a horizon close to that of the lowest exposed beds in the vicinity.

The most important outcrop lies around and to the west of the confluence of Woo Dale and Wye Dale, and a section [SK 07 SE/53] shows 24.16 m of brown granular dolomite with subsidiary bands of pale grey and grey-brown limestone; the top lies some 7.62 m below the top of the Woo Dale Dolomites.

On the southern side of the valley, dolomites are well exposed in cuttings in the disused railway [0971 7244 to 0999 7242]. The western exposed area lies around the confluence of Cunning Dale and the Wye valley; here, exposures [such as 0826 7277] are generally small.

In Cunning Dale the Woo Dale Limestones are estimated to be about 100 m thick; elsewhere their measurement is complicated by the presence of faulting. A 25.5 m section in a quarry, [SK 07 SE/60], shows variable dark and paler grey limestones and calcilutites with a coal up to 0.03 m in the lower part; *Daviesiella sp.* and *Delepinea?* occur near the top.

The main sections are summarised in Figure 6. Detailed correlation between them is rendered difficult by rapid lateral variation, particularly in the calcilutites, which are in places highly lenticular. The most extensive section [SK 17 SW/65] is near Meadow; it shows Chee Tor Rock on dark grey, grey and brown limestones with an algal band 9.7 m, pale grey to grey-brown calcilutite 20.5 m, grey-brown limestone with 0.4 m of calcilutite at base 1.95 m, dark limestone with algal bands 9.8 m, grey-brown and dark grey limestones 15.3 m.

Sections in the lower part of the formation are not common, the best [SK 17 SW/66] being in the railway-cutting north of Topley Pike: well bedded mainly dark grey fine-grained limestone 18.1 m; on grey-brown and pale grey-brown fine-grained limestones, with *Daviesiella* in their lower part, 16.27 m; dark fine-grained well bedded limestone with *Daviesiella* 4.0 m. This overlaps with the preceding section; together these span much of the upper part of the formation.

The road section on Topley Pike (Cope, 1933, p. 127–130) shows 31.7 m of Woo Dale Limestones; pale and dark grey limestones with two thick calcilutite bands. Elsewhere, particularly in the lower part of the formation, thin coals are present. An example is in an old railway-cutting [1028 7252] north-east of Topley Pike which shows 12.45 m of thin-bedded limestone, mainly dark grey, with a 10 mm coal at 1.4 m above the base of the section. In Cunning Dale a 25 m section [SK 07 SE/60] shows a 30 mm coal overlain by a thin clay parting at 7.43 m above the base.

At the time of survey (1968) the uppermost 40 m or so of the Woo Dale Limestones were visible at the south end [099 734] of Tunstead Quarry, and were distinguishable from the overlying Chee Tor Rock by their darker colour and more closely spaced bedding. Topley Pike Quarry [100 722] showed some 38 m of the formation in the same year. To the south of the latter quarry, the uppermost part of the Woo Dale Limestones is exposed in Deep Dale [SK 07 SE/56] (Figure 6) and Back Dale.

In the Wye valley the Woo Dale Limestones are exposed [1229 7319] in an inlier beneath Chee Tor. This shows a small thickness of dark limestone, the exposure lying close below the base of the Chee Tor Rock.

Harpur Hill, The Frith, and Hind Low

Just to the east of Countess Cliff, the uppermost part of the Woo Dale Limestones is exposed in a small inlier. A typical section [0565 7097] shows 21.2 m of pale limestone, part finely crinoidal, with calcilutite bands.

Ladmanlow Borehole [SK 07 SW/4] proved Bee Low Limestones to 73.76 m, and continued to 114.91 m in pale and dark grey limestones with bands of calcilutite and a 0.76 m algal band at 76.66 m; from thence to the bottom at 177.09 m the hole passed through pale grey limestone with dark bands.   IPS

Discontinuous exposures, in approximately the top 50 m of the formation, occur in small disused quarries in the inlier of The Frith. The limestones are mainly mid-grey well sorted calcarenite, which appear oolitic in hand specimens but are pelsparites (Harrison, 1981). At one locality [0554 6911], a 1.52 m bed has yielded *Axophyllum vaughani* and *Daviesiella sp.. A. vaughani* is also recorded within a 1.37 m bed of micrite with birdseye structures in a borehole [SK 06 NE/10] west of High Edge which penetrated the topmost 41.41 m of the formation beneath 18.09 m of the Bee Low Limestones (Harrison, 1981): the junction between the formations is taken here at the top of the highest medium to dark grey bed in the sequence. The highest beds are also exposed north-east of Dalehead in a section [SK 06 NE/33] that continues up into the Bee Low Limestones (Harrison, 1981, pp. 32–33). They include a number of pale grey-brown calcilutites, thinly interbedded with coarser peloidal calcarenites. The mapped boundary with the Bee Low Limestones is taken at the top of the highest calcilutite.

Two boreholes, at quarries in the Hind Low area [SK 96 NE/3 and SK 06 NE/15] on the Bee Low Limestones outcrop, have reached the Woo Dale Limestones. The log of the former is unreliable to a depth of 183.18 m. Below this depth to the bottom at 210.92 m, mainly grey finely crystalline and 'oolitic' (peloidal) limestones predominate with *Linoprotonia corrugatohemispherica* (at 184.40 m) and *Davidsonina carbonaria* (at 185.62 m), indicating a Holkerian age. After penetrating 122.81 m of Bee Low Limestones, the second borehole [SK 06 NE/15] passed into 12 m of thinly interbedded mid to pale grey calcilutites, calcisiltites and peloidal calcarenites, underlain by pale finely bioclastic calcarenites to the base at 141.33 m.   NA

Hartington

About 100 m of Woo Dale Limestones are exposed or proved by drilling in the Hartington Anticline. The sequence comprises the Vincent House Member below, overlain by the Hand Dale Member (Aitkenhead and Chisholm, 1982, p. 3). Both consist of pale well bedded calcarenites, but the latter is distinguished by the additional presence of thin calcilutite bands. Detailed descriptions are given by Sadler and Wyatt (1966) and Sadler (1966). Additional petrographic details, together with chemical data, have been published by Cox and Bridge (1977). In areas of poor exposure the calcilutite bands are not necessarily visible, and so the outcrops of the two units cannot be reliably distinguished.

The top of the formation appears to be an erosion surface at the railway-cutting [144 616] north-west of Hartington Station, and the existence of this erosional episode was used by Sadler and Wyatt (1966, p. 59) to explain apparent thickness variations in the upper member. Such thickness variations are, however, as easily explained by lateral facies variation within the Woo Dale Limestones or by poor exposure of the calcilutite bands.

The Vincent House Member consists mainly of pale crinoidal spar-cemented calcarenites with shelly bands; Sadler and Wyatt (1966, p. 58) recorded typical Holkerian fossil assemblages from several localities. A borehole [SK 16 SW/8], near Vincent House, started close to the top of the member and proved 61 m of pale mainly crinoidal algal biosparite (Cox and Bridge, 1977) without reaching the base. The top 30 m or so are exposed in numerous crags and small quarries in Long Dale. The best section is a quarry [1398 6197] where about 26 m of thickly bedded limestone with shelly bands can be seen. As in the borehole, the limestones here are

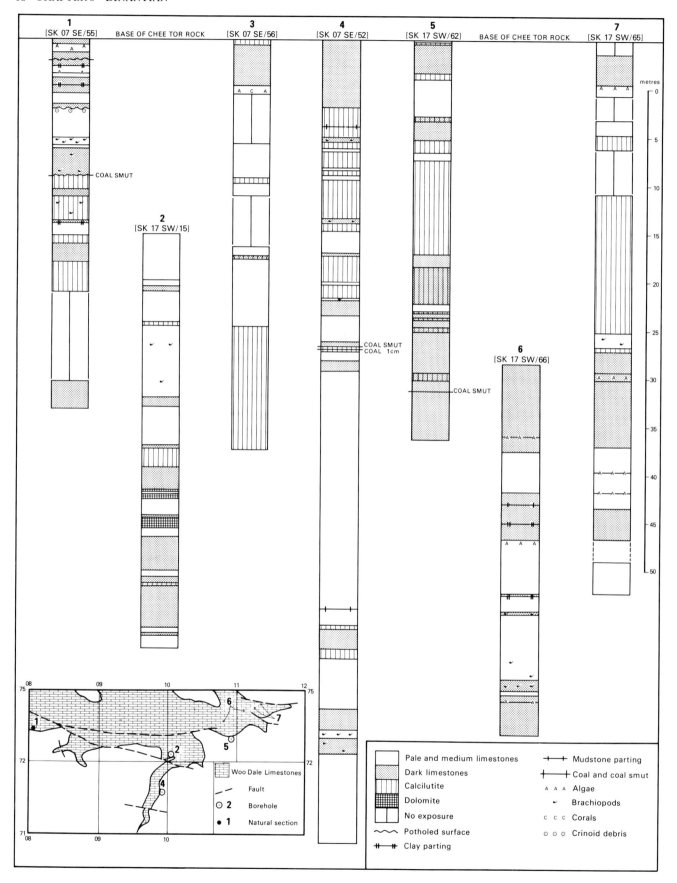

Figure 6   Sections of Woo Dale Limestones in the Wye valley.

mainly crinoidal algal biosparites (Cox and Bridge, 1977, p. 66). An unusually deeply weathered section in the wall of a silica-sand pocket [1410 6259] shows that the limestones are well laminated, partly in parallel beds and partly in low-angle cross-beds.

The Hand Dale Member is a well bedded alternation of pale calcarenite, calcisiltite and calcilutite about 35 m thick. Typical Holkerian faunas were recorded by Sadler and Wyatt (1966, p. 59) and collected during the present survey. The base is gradational, being drawn at the that of the lowest calcilutite band. In a daleside section [1413 6224] north of Hartington-moor Farm, the basal calcilutite bed is medium grey in colour and overlain by about 30 m of incompletely exposed well bedded pale limestones, mainly biomicrites and biopelsparites (Cox and Bridge, 1977, p. 66). A crag [1375 6170] in Long Dale, a few metres below the top of the member, has yielded stromatolitic algae, *A. vaughani*, *Lithostrotion martini*, *Daviesiella derbiensis*, *Linoprotonia corrugatohemispherica* and *L. hemisphaerica*. The top of the member is well exposed in Hand Dale, the best section [1442 6087] showing 7.6 m of well bedded pale calcisiltite and calcilutite with calcarenite bands, separated from the overlying poorly bedded Bee Low Limestones by a prominent bedding plane: there is no evidence of erosion at the contact. The top is also exposed over a distance of 80 m in the railway-cutting [144 616] north-west of the former Hartington Station where about 4.5 m of poorly bedded calcisiltite and subordinate calcarenite and calcilutite contain algal nodules, *Axophyllum vaughani*, *Davidsonina carbonaria* and *L.* cf. *corrugatohemispherica*. The top of this bed is an undulating surface, with about 2.5 m of relief, that probably represents an interval of erosion (Sadler and Wyatt, 1966, pp. 59–60); above lie calcarenites and calcisiltites assigned to the Bee Low Limestones. JIC

WOLFSCOTE DALE TO IRON TORS

The lower part of the sequence comprises the Iron Tors Limestones Member which is estimated to be about 107 m thick. The limestones are mainly exposed in a few isolated crags [1439 5647, 1448 5648 and 1458 5643] up to 12 m high at Iron Tors, and comprise massive pale grey to grey-brown peloidal calcarenite. In the bottom of the dale [around 1432 5654] they pass laterally into darker thinner bedded limestones assigned to the Milldale Limestones. No macrofossils have been found.

Good sections of the upper part of the formation are seen near the confluence of Wolfscote Dale and Biggin Dale, where the crags of Drabber Tor [1390 5702] form a striking feature (Plate 2). The best composite section [SK 15 NW/12] (Figure 5) shows 81.93 m of grey to pale grey calcarenites, commonly laminated, and subordinate thinly interbedded calcilutites. The calcarenites include a proportion of texturally distinctive biopelsparites containing prominent micrite-coated intraclasts as well as lumps and peloids. Macrofossils are sparse and not diagnostic of age.

There is no trace of this facies in Hopedale, just beyond the southern boundary of the district, and lateral passage of these uppermost beds into the Milldale Limestones is inferred to take place immediately to the south of Alstonefield.

In Biggin Dale, a section [SK 15 NW/13] including a 15.6 m sequence of thinly interbedded peloidal calcarenites and calcilutites lies within an area shown on the 1978 edition of the 1:50 000 map as Bee Low Limestones. These beds have yielded foraminifera indicating a Holkerian age and are now included in the Woo Dale Limestones. The map has also been revised in the light of new information from the area south-east of Coldeaton. Here, 69.90 m of limestone of a similar facies to that of Wolfscote Dale [SK 15 NW/12] (*see above*), and containing a Holkerian to early Asbian foraminiferal assemblage (p. 57), have been proved in a borehole [SK 15 NE/5] at Lees Barn (Bridge and Kneebone, 1983, p. 45). They are also exposed in a valley [156 566] nearby. NA

GRANGEMILL, BALLIDON AND PARWICH

The Woo Dale Limestones are present beneath the Bee Low Limestones in the Via Gellia valley [254 565], where an upper division, now named the Griffe Grange Member, can be distinguished from a lower division, the Ryder Point dolomites (Smith and others, 1967, pp. 8–11; Chisholm and Butcher, 1981).

The Ryder Point dolomites are mainly medium grey fine-grained quartzose dolomites, 51 m of which have been proved in Ryder Point No. 3 Borehole [SK 25 NE/48], a short distance to the east of the district (Chisholm and Butcher, 1981). Their regional extent is unknown.

The Griffe Grange Member is 105 m thick at Ryder Point (Chisholm and Butcher, 1981). Lithological details have been published by Cox and Harrison (1980, p. 56–57). The uppermost 75 m closely resemble the Hand Dale Member of the Hartington area (*see above*), consisting of pale and medium grey interbedded calcilutite, calcisiltite and calcarenite, with fenestral textures ('birdseye structures') indicating deposition in very shallow water. The lowest 30 m consist predominantly of crinoidal biosparite and include rare chert nodules. West from Ryder Point the outcrop of the upper part of the member extends into the present district for about 1 km until cut off by a fault [246 568]. There are small exposures along the sides of the Via Gellia valley, the best [SK 25 NW/38] showing about 12 m of buff interbedded porcellanous and granular limestone with an apparently gradational passage into the overlying Bee Low Limestones. The lowest beds contain tuffaceous material. At Prospect Quarry [SK 25 NW/37] a similar section contains several greenish clay partings of probable volcanic origin.

South and west of Grangemill the member has been proved beneath the Bee Low Limestones in several boreholes and in mine workings. At Longcliffe Quarry and near Longcliffe the top 20 m were penetrated in boreholes [SK 25 NW/19 and 18], and near Lowmoor Farm a borehole [SK 15 NE/6] proved the top 35 m. In each case the top of the member has been drawn at the top of the highest significant calcilutite bed in the sequence. The top few metres of the unit have also been encountered in inclined tunnels below the bottom of the shafts at Golconda Mine (Worley, in preparation). JIC, IPS

BEE LOW LIMESTONES

The Bee Low Limestones formation was defined by Aitkenhead and Chisholm (1982). It is synonymous with the Bee Low Group as used in the Chapel en le Frith memoir (Stevenson and Gaunt, 1971). The limestones are characterised by lateral and vertical homogeneity, in contrast to the variability of the adjacent formations. The thickness varies from a maximum of about 213 m in the west to a minimum of 68 m near Middleton, in the south-east, just beyond the boundary of the district (Figure 7).

In the present district the formation includes two volcanic members, the Ravensdale Tuff and the Lower Miller's Dale Lava (pp. 61–62). Where the latter is present around Buxton and the valley of the River Wye, the limestone sequence comprises the Chee Tor Rock below the lava and the Miller's Dale Limestones above; elsewhere it is undivided. Underground provings in the eastern part of the district have shown the presence of volcanic rocks in the upper part of the formation beneath Haddon Fields (p. 60) and in Millclose Mine (p. 60).

Plate 2 Drabber Tor overlooking the River Dove in Wolfscote Dale; crags comprising the uppermost part of the Woo Dale Limestones consisting of laminated peloidal calcarenites and subordinate calcilutites of probable Holkerian age (L 1152).

The name Chee Tor Rock was introduced by Green and others (1869) after the 47 m cliff of massive limestone at the type locality (Plate 3) [123 733]. Subsequently the unit was defined and described by Cope (1933). As its top is taken at the base of the Lower Miller's Dale Lava, its outcrop is confined to approximately the same area as that member, around Buxton and the valley of the River Wye. The Chee Tor Rock consists of thickly bedded pale grey to grey calcarenite of very uniform appearance. Beds, whose boundaries are generally defined by either stylolitic or clay partings, range mostly from 0.5 to 5.0 m, up to a maximum of 10.2 m. The total estimated thickness is generally about 120 m. Rare calcilutites or very fine calcarenites also occur, particularly in the basal 13 m, where beds tend to be thinner, marking a vertical transition to the facies characterising the highest part of the underlying Woo Dale Limestones. The Upper *Davidsonina septosa* Band is recognised locally in

Ashwood Dale, and helps in correlating the sequence with that in the adjacent district (Stevenson and Gaunt, 1971, pp. 23–25).

The Lower Miller's Dale Lava (Stevenson and Gaunt, 1971, p. 23) crops out extensively within the northern part of the district (Figure 7) at a level some 40 m below the top of the Bee Low Limestones. The slightly lower Ravensdale Tuff is restricted to a small area around Ravensdale Cottages [172 736].

The Miller's Dale Limestones are approximately equivalent to those referred to as Miller's Dale rock by Hull and Green (1869) and Miller's Dale Beds (Cope, 1933, 1937, 1939; Stevenson and Gaunt, 1971, p. 24). They lie between the Lower Miller's Dale Lava and the overlying Monsal Dale Limestones. The member consists of about 40 m of limestone and is of broadly similar facies to the Chee Tor Rock, except that the beds are rather thinner and richer in fossils, particularly corals, brachiopods and crinoid debris. Rare cherts occur locally in the middle part of the sequence. Faunal horizons include a widely developed bed with *D. septosa*, 0.61 to 0.76 m thick, just above the base, as well as more localised coral–brachiopod bands higher in the succession.

Outside the limited area where the Lower Miller's Dale Lava crops out, the Bee Low Limestones are undivided, except for areas of dolomitisation and lateral passage into apron-reef (Figure 7). Faunal marker bands are rare though several bands with *Davidsonina septosa* occur locally. Wolfenden (1958) believed he could trace the 'Cyrtina' (Davidsonina) Band in the Earl Sterndale area, and Sadler (1964) mapped it in the general area east and south of Buxton. Subsequently, work in both the Chapel en le Frith and the present district has shown the presence of up to four

similar bands. Conclusions based on the assumption of only one band are, therefore, unreliable. Tentative correlations between major sections, using clay wayboards (p. 59), were proposed by Walkden (1972, fig. 3; 1974, fig. 2). This author, together with J. R. Berry (written communication, 1982), using a combination of palaeontology, sedimentology and emergence characteristics, has built up a complete correlation for all major sections between Bee Low, Hartington and Miller's Dale. The succession contains some forty correlatable clay-bearing horizons, many of which are now stylolitised and each of which probably represents a period of subaerial exposure (Walkden, 1974). Of these, fifteen are well developed palaeokarsts, including one beneath each of the Miller's Dale lavas, and this multiplicity of emergent surfaces evidently represents a complex interplay between eustatic and tectonic controls on sedimentation.

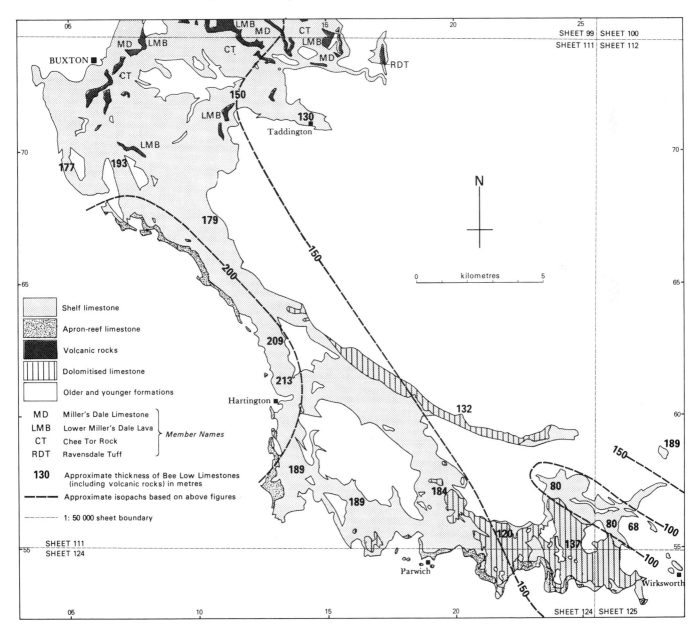

Figure 7 Outcrop distribution and thickness of the Bee Low Limestones.

Plate 3 Chee Tor: cliff overlooking the River Wye, in massive very thick beds of calcarenitic limestone of the Chee Tor Rock (L 2262).

Lithology and facies variation; main part of shelf province

The Bee Low Limestones (including the Chee Tor Rock and Miller's Dale Limestones) consist predominantly of pale grey, pale brownish grey and grey, fine to medium-grained calcarenites with scattered crinoid debris and some comminuted brachiopod and coral fragments. These rocks (for example, Harrison, 1981) are mainly biosparites, though biopelsparites and pelsparites also occur, particularly beneath palaeokarstic or pedogenic crusts (*see below*). Peloidal calcarenites also tend to occur in the basal 60 m of the formation and in the back reef facies; the spergenites reported by Wolfenden (1958) probably belong to this category. Some micritic limestones, biomicrites and calcilutites are also present. Biomicrites are also less rare in the highest part, but the texture of these tends to be mixed, with micrite mainly enveloping fossil fragments and intraclasts in an otherwise sparry rock.

The calcarenites generally have a uniform even-grained appearance and lamination is rare, except in the well sorted peloidal limestones which locally show parallel or low amplitude ripple-lamination. However, vague traces of bioturbation are commonly present and this process may have destroyed pre-existing current-induced lamination.

Pale to dark brown laminated micritic crusts (Walkden, 1974) occur at certain horizons, usually beneath clay wayboards. Walkden established that the crusts 'were produced by contemporaneous in situ subaerial alteration' of the limestone, and that 'this alteration took place beneath soil cover concomitant with karstic solution'.

Environment of deposition

The faunas are dominated (Wolfenden, 1958; Stevenson and Gaunt, 1971, pp. 394–399) by crinoids and relatively few species of simple and compound corals and brachiopods, together with some foraminifera, ostracods and gastropods. The latter are common in only a few very fine calcarenite or calcilutite beds where they may indicate very shallow water or high-salinity conditions. Shallow-water conditions are also indicated by the presence of the dasycladacean alga *Koninckopora*. The work of Wolfenden (1958), Sadler (1964; 1966), Walkden (1974) and Cox and Bridge (1977) indicate deposition in a warm mostly clear shelf sea whose level fluctuated repeatedly, with emergence and soil formation at times. Water conditions varied from fairly calm, when the crinoid–coral–brachiopod fauna flourished, to wave-agitated when the remains of these animals were finely comminuted and redeposited.

Minor cyclicity has been widely recognised in Carboniferous rocks in general and in the Asbian in particular (for instance, Somerville, 1979); Ramsbottom (1973) has attributed it mainly to eustatic changes in sea level. However, various other factors, including tectonism, local volcanicity and fault movement, presumably affected the sedimentation pattern in this region (Walkden, 1977, p. 365).

Lithology, facies variation and environment of deposition; margin of shelf province

Apron-reef limestones occur in a discontinuous belt on the western and southern margins of the Bee Low Limestones outcrop. Wolfenden (1958) referred to the 'reef complex', which he subdivided into fore-reef, reef, and back-reef. Following Stevenson and Gaunt (1971) the term 'apron-reef' as used here is synonymous with 'reef complex', and the 'reef' subdivision is referred to as the 'algal reef'. The apron-reef outcrop shown on the map does not include back-reef as this is difficult to delineate with any precision.

The following summary of the facies change that occurs at the shelf margin is based mainly on the work of Wolfenden (1958) and largely confirmed during the present survey. Within about 0.4 km from the reef there is an increase in the proportion of well rounded grains, including peloids and bioclasts with oolitic coatings, in the well bedded shelf calcarenites. Conglomeratic limestones occur locally, presumably resulting from intense wave action. Nearer the reef, the limestones become poorly bedded, increasingly micritic, and the number and variety of fossil species, particularly brachiopods, increases.

The algal reefs are the only part of the complex which is an organic framework reef as defined by Wilson (1975): that is, 'a build-up formed in part by a wave-resistant framework; constructed by organisms'. Here, these are mainly stromatolitic algae and encrusting bryozoans which form wall-like masses of pale micritic limestone up to 24 m high and 9 m wide at two levels (pp. 20 and 22).

The algal reef passes laterally into irregularly bedded micritic fore-reef limestones with a steep and largely original depositional dip. The fauna of these limestones is both varied and abundant in places: it includes corals, brachiopods, bivalves, gastropods, nautiloids, goniatites, trilobites, ostracods, bryozoans, crinoids, foraminifera and algae. This abundance and variety, with benthonic, planktonic and nektonic forms, reflects the well oxygenated and nutrient-rich conditions that must have prevailed around the fore-reef slopes. At Castleton, Broadhurst and Simpson (1973) have

demonstrated that, at certain times, particular genera occupied restricted levels on the fore-reef slope that were related to depth of water. Timms (1978) has shown that there is a similar depth-related distribution of distinctive brachiopod communities both at Castleton and, in the present district, at Chrome Hill and Dowel Farm near Earl Sterndale.

Another feature common to both areas is the presence of a boulder bed composed of disorientated blocks of fore-reef limestone, on the slopes of Treak Cliff (Simpson and Broadhurst, 1969) and Chrome Hill (Timms, 1978). Its significance is discussed on p. 24.

The apron-reef in the Earl Sterndale area probably commenced as a linear feature in late Asbian (early B_2) times, but developed a highly irregular form with several deep embayments and knoll-like promontories (Figure 9). A possible explanation is that north-west and south-east of the Dowel Farm embayment, channels carried sediment from the shelf through gaps in the earlier algal reef to form the promontories on which the later algal reef subsequently grew.

NA, JIC, IPS

DETAILS

WYE VALLEY

With the presence in this area of the Lower Miller's Dale Lava, the Bee Low Limestones are subdivided into the Chee Tor Rock and the Miller's Dale Limestones.

In Ashwood Dale the Chee Tor Rock forms cliffs on the valley sides. The Upper *D. septosa* Band (of Stevenson and Gaunt, 1971, pp. 22–23), 0.61 m thick, occurs [SK 07 SE/57] some 8 m below the Lower Lava near the gasworks. In Ashwood Dale Quarry [079 726] the Chee Tor Rock is faulted against the Woo Dale Limestones.

Below Cowdale Quarry the lowest part of the unit is seen in a section 12.88 m thick [SK 07 SE/55] overlying Woo Dale Limestones. These beds contain a thin calcilutite band, sporadic *Daviesiella* and an algal band.

Cowdale Quarry [SK 07 SE/54] (Figure 8) shows a section 50 m thick in massive limestones with several clay partings and potholed surfaces; scattered *Daviesiella* occur close to the base. In the vicinity, algal bands are commonly present at or close to the base of the Chee Tor Rock; one such band, 0.53 m thick and about 1.37 m above the base, is exposed [0854 7223] on the northern side of Cow Dale.

Deep Dale, Back Dale and Horseshoe Dale provide conspicuous sections of the Chee Tor Rock, one [SK 07 SE/56] showing 6.71 m of pale massive limestone overlying Woo Dale Limestones. At the junction of Back Dale and Horseshoe Dale [0961 7073], a section 19.25 m thick contains a calcilutite band 0.36 m thick, 9.14 m above the Woo Dale Limestones. Several other calcilutite bands are present, at a slightly higher level, just beyond the eastern side of the dale near the above confluence.

South of Cow Dale, a conspicuous band of calcilutite, with a maximum recorded thickness of 3.8 m [0824 7170], is traceable for 90 m; the band lies about 25 m above the base of the Chee Tor Rock. Some 15 m higher, an oolitic band with a maximum of 1.52 m [0854 7152] is present locally. Both bands are absent in Cowdale Quarry to the north and in a borehole [SK 07 SE/51] on the south side of the outcrop (Harrison, 1981, pp. 88–90).

Tunstead Quarry [around 099 740] showed an extensive face of Chee Tor Rock overlying Woo Dale Limestones. The section extends northwards beyond the district (Stevenson and Gaunt, 1971, p. 43). Detailed borehole sections from the quarry area have been published by Harrison (1981, pp. 74–80; *see also* Figure 8).

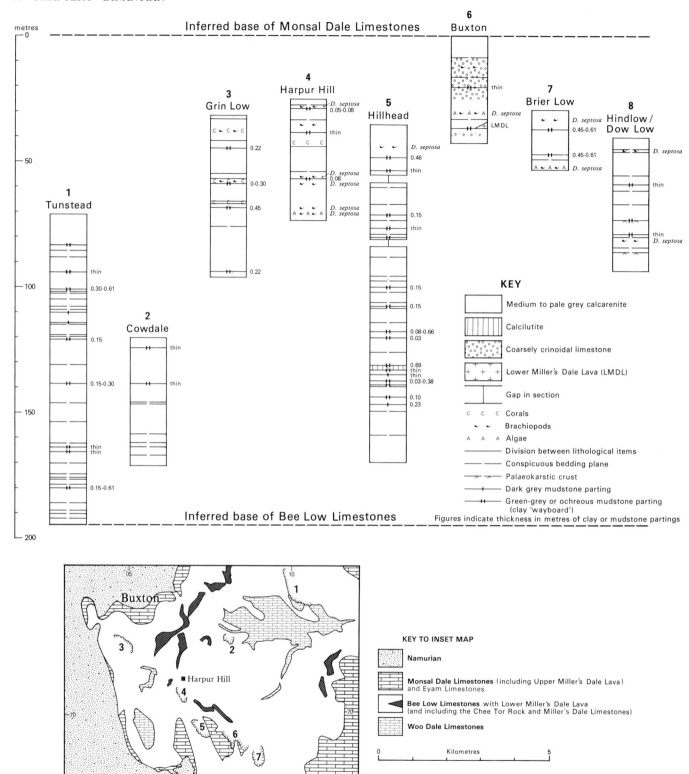

Figure 8 Comparative sections of major quarries in the Bee Low Limestones around Buxton, showing the vertical extent of workings within the formation.

At Chee Tor [1229 7331] the unit is seen in a 47 m cliff of massive pale limestone overlying Woo Dale Limestones (Plate 3).

On the east and south-east side of Buxton the Lower Miller's Dale Lava, up to 15 m thick, forms a nearly continuous outcrop. In Ashwood Dale, a section in the railway-cutting shows 7.62 m of vesicular lava [SK 07 SE/57]. A short distance to the north-west, however, the lava dies out locally. To the south-west, an exposure in a road-cutting [0640 7229] shows pale green clay, 3.35 m, on vesicular lava, 2.59 m. The main outcrop terminates north of Harpur Hill, though an isolated lens occurs [062 709] to the west. The lava reappears north of Hillhead Quarry, where it is seen in a railway-cutting [0710 7023].

The lava forms isolated outcrops east of Calton Hill and in Blackwell Dale, a section [1290 7223] in the latter showing 3.3 m of mainly non-vesicular rock. At Wormhill, exposures continuous with the more extensive outcrops in the Chapel en le Frith district extend southwards to Miller's Dale, the lava dying out in the vicinity of Chee Tor. To the east, a small inlier is present in the dale immediately west of Ravenstor.

In Tideswell Dale the lava is intruded by dolerite (p. 97), which, in a quarry on the eastern side of the dale [1547 7378], overlies 0.61 m of purple prismatic clay-rock. The latter clay is normally present beneath the lava, and is here baked by the intrusion. The prismatic band was formerly more extensively exposed, its thickness being given as 1.52–3.65 m; the columns were 2.44–2.74 m in length and up to 0.13 m wide (Arnold-Bemrose, 1899, pp. 242–245).

The Miller's Dale Limestones are some 38 m thick around Buxton. The lowest beds are seen in a section above the railway in Ashwood Dale [SK 07 SE/57] where they contain a band 0.61 m thick with productoids and *D. septosa* 1.83 m above the base. Higher strata are exposed in a railway-cutting in Higher Buxton [SK 07 SE/59], which shows 8.20 m of pale massive limestone with a 0.30 to 0.38 m shell band 0.46 m above the base; the junction with the Monsal Dale Limestones is a potholed surface.

The lowest part of the Miller's Dale Limestones is well exposed in Blackwell Dale. A discontinuous section [1311 7245 to 1317 7247], starting some 4 m above the Lower Lava, shows 19.8 m of the subdivision; the uppermost 2 m are a band with a varied coral fauna, *Gigantoproductus sp.* and *Linoprotonia hemisphaerica*, which is traceable for a short distance on both sides of the dale. Cope (1972) has suggested that a knoll-reef (or patch reef) is present at this locality, but the present work has failed to confirm this. A section [1315 7252] on the western side of the dale shows 2.2 m of cherty limestone with its base 6.2 m below the coral band. Nearby, these limestones yield a fauna including *Dibunophyllum bourtonense*.

South of Wormhill, a 2.41 m-section [1243 7380] near the base of the member contains a shell bed 0.76 m thick, with corals including *D. bourtonense*, *Gigantoproductus semiglobosus* and common *D. septosa*.

Station Quarry, Miller's Dale [SK 17 SW/67], shows on its east side a section of Miller's Dale Limestones 18.11 m thick overlain by 'Station Quarry Beds' (p. 27). The section in the western part of the quarry [1302 7357] differs from this in the presence of a thick clay parting (0.05–0.60 m) 19 m below the top of the member, and in the occurrence of chert between 17.5 and 23.67 m below the same horizon.

In Monk's Dale, the full thickness of about 60 m of Miller's Dale Limestones is more or less continuously exposed. A coral band, 1.1 m thick, is present locally [1395 7360] some 25 m from the top; it contains a varied fauna including *D. bipartitum*. At Ravenstor the subdivision is seen in a section [1503 7328] of pale massive limestone 30 m thick resting on a strongly potholed surface of Lower Lava; the potholes, up to 1.5 m deep, are filled by limestone.

In the Cressbrook Dale area the Lower Miller's Dale Lava is absent, but the Ravensdale Tuff, up to 20 m thick, occurs some 80 m from the top of the formation in the lower reaches of the dale, near Ravensdale Cottages. An exposure in the stream [1724 7393] shows

2.44 m of banded tuff with both cognate and accidental lapilli.

EYAM AND LONGSTONE EDGE

The Eyam Borehole [SK 27 NW/15] (Dunham, 1973, pp. 84–85) proved the formation, 161.94 m thick, with 18.58 m of fragmented lava and some tuff in the upper part; the limestones are predominantly pale though dark intercalations occur towards the base.

At Longstone Edge, a borehole [SK 27 SW/20] proved beds equivalent to the Bee Low Limestones and uppermost Woo Dale Limestones in an abnormal dark facies below the base of the Monsal Dale Limestones at 358.98 m. The lowest strata between 447.38 m and the bottom of the hole at 611.46 m are dark cherty limestones, with some paler and locally bioclastic beds in their lower part, which show graded bedding in places; thin dolomites (up to 1.5 m at 486.61 m) are spaced throughout this part of the succession. Between 447.38 m and the base of the Brigantian at 358.9 m the borehole passed through predominantly dark limestones with rare chert; bioturbation or burrows are common and there are some slightly coarser beds with graded bedding. Paler, commonly bioclastic, intercalations also occur. For the lower part a probable early Asbian age is indicated by a fauna from 416.13–425.04 m (with *Pojarkovella nibelis* and *Archaediscus krestovonikovi*), while the beds between 601.90 and 611.48 m yield a typical Holkerian assemblage of foraminifera. The higher beds contain Asbian microfaunas in the upper part (with *Gigasbia sp.* in an assemblage between 364.92 and 374.55 m).

HARPUR HILL AND STANLEY MOOR

In this area the Bee Low Limestones are some 180 m thick. To the west of Harpur Hill, the lowest part of the formation is exposed [0578 7109] near Ferny Bottom, where a 0.66 m band with algae and corals lies some 9.6 m above the base. Beds in the upper part of the formation, all above the horizon of the Lower Lava, are exposed in Harpur Hill Quarry [SK 07 SE/58] in a section 49 m thick (*see also* Harrison, 1981, pp. 98–99). At the base, a band with *D. septosa* has been traced throughout the area of the quarry.

On the south side of Grin Low, three fossiliferous beds, just above the Lower Lava, bear algae, corals and brachiopods. The Ladmanlow Borehole [SK 07 SW/4] proved 73.76 m of Bee Low Limestones before reaching the Woo Dale Limestones; an oolitic band 0.38 m thick was present at 39.32 m, and a calcilutite band 0.3 m thick at 44.04 m.

To the south of Stanley Moor Reservoir, beds near the top of the Bee Low Limestones are exposed on Anthony Hill [0474 7043] where a section 105 m thick of pale massive limestone contains a shell bed 3.96 m thick about 25 m below the top of the unit; the band yields *Gigantoproductus sp.* and *Linoprotonia hemisphaerica*. Farther south, the shell bed is also present [048 701] near Turncliff.

At Burbage, the Bee Low Limestones crop out in the core of an anticline which extends northwards to the vicinity of Edgemoor. IPS

DALEHEAD TO EARL STERNDALE

North-west of Earl Sterndale the Bee Low Limestones are best seen in a section [SK 06 NE/24] at High Edge. The total thickness of the section is about 127 m, but within this are several unexposed parts amounting in total to about 51 m. Its base lies about 29 m above that of the Bee Low Limestones, proved in a nearby borehole [SK 06 NE/10] (Harrison, 1981, p. 34). The calcarenitic limestones are predominantly pale grey biosparites in thick well defined posts, but a few thinner finer-grained peloidal beds are present, conspicuous by their off-white weathering. The only useful marker is a *D. septosa* band exposed near the summit of High Edge [0627 6882], some 12.79 to 14.61 m below the top of the section. The strong palaeocurrent activity typically associated with these bands (Sadler,

1964) is shown by fine parallel and cross-lamination, and by up to 23 cm of conglomeratic limestone at the base of the overlying bed. The band probably correlates with the lower of two similar bands which have been traced nearby to the south-west. These are estimated to lie about 8 and 49 m below the top of the formation. The higher band can be traced as far as Thirkelow Rocks [0501 6891] where it is exposed on the steep (45°) dip-slope near the margin of the Dinantian limestone outcrop.

At Hillhead Quarry the section [SK 06 NE/25] (Figure 8), starting about 25 m above the base of the Bee Low Limestones, shows a total of 133 m of these beds. The top lies about 35 m below that of the formation, here about 193 m thick. The limestones are very thickly bedded, with bedding planes defined by stylolites, clay partings and thicker 'wayboards'; bed thicknesses range up to 10.67 m and average 3.57 m. The rocks are classified by Harrison (1981) as mainly biosparites or biopelsparites of fine to coarse arenite grade. Ten clay wayboards are present: they exceed 3 cm in thickness, ranging up to a maximum of 66 cm, with a total thickness of 2.75 m or 2% of the sequence (*see also* Walkden, 1972, fig. 3). A *D. septosa* band is present about 44 m below the top of the formation; *D. septosa* was also noted in a fossiliferous band lying about 25.43 m below this band.

Buxton Quarry exposes a section [SK 06 NE/26] in the highest 44 m of the formation and about 8 m of the overlying Monsal Dale Limestones. The section also includes a probable correlative of the Lower Miller's Dale Lava (see p. 62).

Brierlow Quarry exposes about 25 m of beds, possibly lying about 35 m to 60 m below the top of the formation. The sequence [SK 06 NE/27] contains two *D. septosa* bands, the lower one yielding *Chaetetes depressus, Palaeosmilia murchisoni, Lithostrotion portlocki, D. septosa, Delepinea sp.* and *Linoprotonia sp.* [*hemisphaerica* group]: the higher band is inaccessible. These two bands are the highest of four in this vicinity, the other two being exposed in old quarries alongside the A 515 road [0896 6966 and 0935 6950].

Hindlow [095 677] and Dow Low [100 675] quarries are separated by only a narrow wall of limestone. The total thickness of the Bee Low Limestones in this vicinity is estimated to be about 179 m, the two quarries working a sequence 70 m thick [SK 06 NE/28], the top of which lies approximately 110 m below that of the formation. Bed thicknesses range up to 9.5 m and, apart from thin clay partings less than 2 cm thick, only one conspicuous clay wayboard was noted. This varies from 0 to 15 cm thick and overlies a prominent bed 1.22 m thick with a dark brown laminated crust and irregular peloidal laminae in the top 15 cm, similar to those associated with palaeokarstic surfaces (Walkden, 1974). The bed lies in the middle of the sequence and about 5.94 m above a *D. septosa* band which probably correlates with that exposed in the roadside quarry [0898 6750] near Jericho (Cope, 1958), about 700 m to the west.

The lower B_2 algal reef is exposed in the dry gorges around Dowel Farm [0760 6771] and Glutton [0839 6726 and 0857 6714] (Figure 9). The age was given by Wolfenden (1958, p. 875) and is accepted here, though no diagnostic faunas have since been found at these localities. However, scattered small exposures of reef-limestone near Abbot's Grove [0977 6663], not previously recorded, have yielded a specimen of *Goniatites sp.* [*hudsoni* group] of lower B_2 age. The most accessible exposure is on the west side of Dowel Dale [0760 6771]. There, on the weathered surface of the algal-reef, D. J. C. Mundy (personal communication) has noted that the algal stromatolites contain former growth cavities, with faunas including small spherical *Emmonsia parasitica* and encrusting *Tabulipora*, similar to the algal reef community on Stebden Hill in the Cracoe reef-belt of Yorkshire (Mundy *in* McKerrow, 1978, p. 158). The overstep of the algal reef by well bedded shelf limestones is also well seen at this point. The algal reef appears to pass into fore-reef limestones about 30 m to the south, beyond which [0752 6755] they form a steep (28–30°) dip-slope descending to the floor of a large

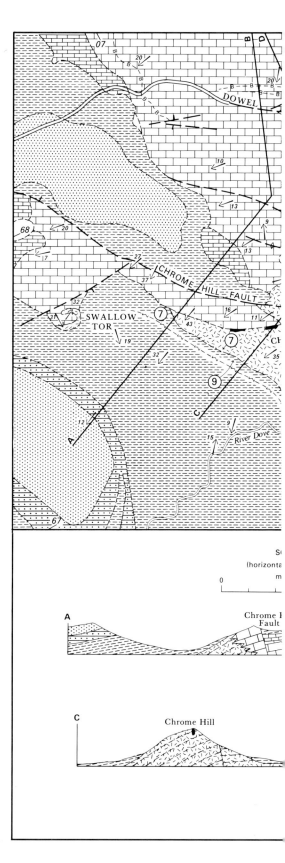

Figure 9 Sketch map and cross-sections of the margin of the Dinantian limestones outcrop near Earl Sterndale.

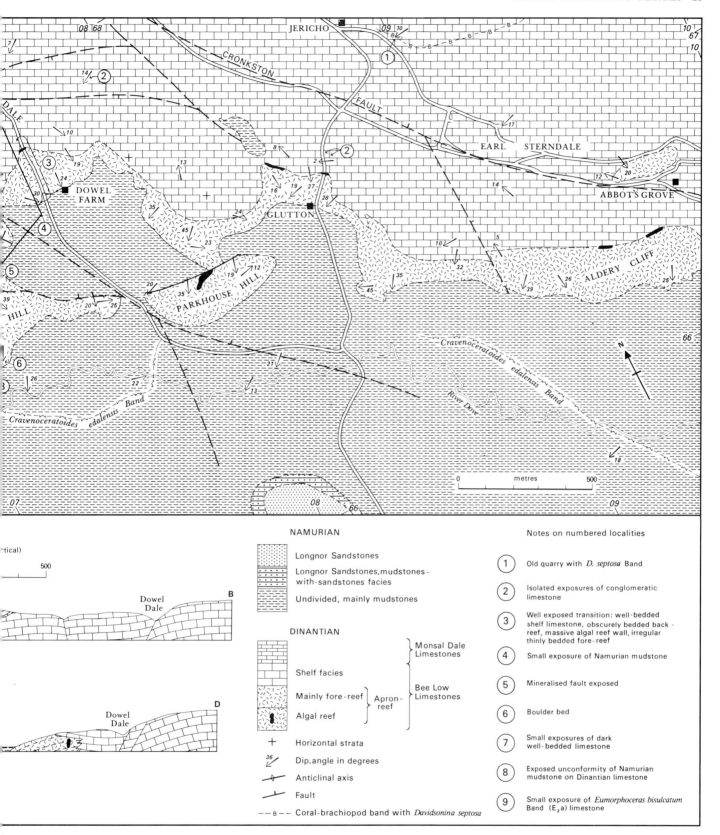

NAMURIAN

Longnor Sandstones

Longnor Sandstones, mudstones-
with-sandstones facies

Undivided, mainly mudstones

DINANTIAN

Monsal Dale
Limestones

Shelf facies

Mainly fore-reef ⎫ Apron-   ⎫ Bee Low
                   ⎬ reef   ⎬ Limestones
Algal reef   ⎭   ⎭

+ Horizontal strata

³⁶ ↙ Dip, angle in degrees

Anticlinal axis

Fault

– – B – – Coral-brachiopod band with *Davidsonina septosa*

Notes on numbered localities

① Old quarry with *D. septosa* Band

② Isolated exposures of conglomeratic
   limestone

③ Well exposed transition: well-bedded
   shelf limestone, obscurely bedded back-
   reef, massive algal reef wall, irregular
   thinly bedded fore-reef

④ Small exposure of Namurian mudstone

⑤ Mineralised fault exposed

⑥ Boulder bed

⑦ Small exposures of dark
   well-bedded limestone

⑧ Exposed unconformity of Namurian
   mudstone on Dinantian limestone

⑨ Small exposure of *Eumorphoceras bisulcatum*
   Band (E₂a) limestone

(·tical)

500

Dowel
Dale   B

Dowel
Dale   D

embayment of Namurian mudstone. A study of the geopetal infillings of shells by Timms (1978) has shown this dip-slope to be largely an original palaeoslope. The limestones have yielded *Goniatites moorei*, indicating an upper B$_2$ age.

The upper B$_2$ algal reef is well exposed on the summit ridges of Chrome Hill [0713 6729], Parkhouse Hill [0796 6696] and Aldery Cliff [0960 6641] (Wolfenden, 1958; 1959). Within the limits of exposure, these wall-like masses of fossiliferous micritic limestone do not appear to extend more than 160 m laterally and 21 m vertically. Elsewhere, the relatively gently dipping shelf and back-reef limestones appear to pass abruptly into the steeply dipping fore-reef limestones without any intervening algal reef.

The spectacular topography in the Earl Sterndale area (Plates 4 and 5) was thought by Green and others (1869) to be wholly fault-controlled. Hudson (1931) described Chrome Hill and Parkhouse Hill as reef knolls and recognised that the limestone topography had

been exhumed from beneath the unconformably overlying mudstones, but concluded that 'such topography can only be the result of subaerial limestone erosion accompanied by considerable foundering and subsequently modified by marine erosion'. Following Wolfenden (1958), it is now believed that the present fore-reef limestone slopes are largely original. They form the side of the large embayments floored by Namurian mudstones around Dowel Farm and Glutton, and so these embayments must also be original submarine features of upper B$_2$ age. Recent work by Timms (1978) on the upper B$_2$ brachiopod communities from the fore-reef limestone slopes has demonstrated a bathymetric zonation which tends to follow the present contours on the south and south-west slopes of Chrome Hill.

However, it is clear that other factors complicate this general picture. A fault plane with associated mineralised breccias is exposed in places along the foot of the north-east face of both Chrome and

Plate 4   Asbian apron-reef scenery at the western margin of the Derbyshire Shelf near Earl Sterndale.

Apron-reef limestones form the foreground slope and the two steep-sided hills of Parkhouse Hill and Wheeldon Hill beyond, the former with a fault-line scarp marking the line of the Chrome Hill Fault, on its northern side. The more rounded hills to the left are of shelf limestones and the embayment in the middle ground and the Dove valley to the right are eroded in the unconformably overlying Namurian shales (L 1164).

Plate 5 Asbian apron-reef scenery at the western margin of the Derbyshire Shelf near Earl Sterndale viewed from the opposite direction to Plate 4.

Shelf limestones form the foreground and the rounded hill in the centre while the peaks to the left (Parkhouse Hill), and to the right (Chrome Hill) are of apron-reef limestones, faulted on their north faces (L 1164).

Parkhouse hills (Figure 9, Plate 5). Named the Chrome Hill Fault (p. 114) this has a downthrow to the north-east of up to 90 m. A minor anticline with the same structural trend, 200 m to the north-east of the fault, may have formed in association with the early fault movement (Figure 9). The south-westerly dips steepen in the fore-reef limestones towards the north-western part of Chrome Hill, and probably include a tectonic component accounting for the westerly disappearance of the apron-reef beneath the mudstone cover. The most north-westerly inlier, which forms a craggy knoll at Swallow Tor [0645 6765], shows quaquaversal dips and resembles a small knoll-reef. Both Chrome and Parkhouse hills also show a partial knoll form, but the main component is the fore-reef limestones dipping between south-west and south from the shelf towards the basin area to the south-west beneath the Namurian cover.

Between Chrome Hill and High Edge [067 680] an outlier of the Longnor Sandstones lies in a hollow in the underlying Dinantian; it was penetrated by the Stoop Farm Borehole [SK 06 NE/20], which proved Monsal Dale Limestones in the basal 1.82 m at 107.69 m, beneath Namurian mudstones (p. 74). The hollow was probably formed mainly by the down-flexing of the Greensides Syncline, cut off at its southern end by the Chrome Hill Fault. Another mudstone-filled hollow occurs north of Glutton [0837 6739] and coincides with a faulted synclinal structure in the Bee Low Limestones, modified by late Dinantian erosion.

Rocks of possible post-upper B_2 age occur in three places on the lowest parts of the fore-reef slopes on the south and south-west sides of Chrome Hill (Figure 9). Timms (1978) recognised a brachiopod community characterised by the presence of *Buxtonia spp.*, *Echinochonchus subelegans*, *Productus productus* and *Striatifera striata*, comprising the lowest, and by inference the deepest-water, of four different communities that he distinguished at successively lower levels down the slope. However, both the age and water depth postulated for this community are in doubt as a similar association of P_{1a} age is closely associated with shallow-water algal stromatolites in the Cracoe reef-belt of Yorkshire (Mundy *in* McKerrow, 1978, p. 166). Thus the community may have been a shallow-water one, post-dating the formation of the reef, and formed during a marine regression. If the basal part of the fore-reef was under shallow water, the upper part would be emergent; this is confirmed by the presence of a boulder bed exposed in a small area at the foot of the fore-reef slope [0719 6706; Figure 9]. Goniatites indicating a P_{1a} age have been recorded in close association with the boulder bed, but their precise derivation is unclear (Aitkenhead and Holdsworth, 1974; Timms, 1978). Dark rubbly limestones with chert appear to be banked up against fore-reef limestones at two small exposures [0682 6750 and 0692 6729] (Figure 9), but have not yielded age-diagnostic faunas.

EARL STERNDALE TO GRATTON HILL

South of Aldery Cliff no algal reef has been detected, the shelf limestones apparently passing directly into back-reef and fore-reef. At the shelf margin, near Wheeldon Trees [1014 6614] and east of Crowdecote [1061 6502], hillside exposures show pale thickly bedded calcarenites dipping south-west at angles between 5° and 30° near the apron-reef. At Crowdecote, a small inlier of undivided Bee Low Limestones extends westwards from the foot of the fore-reef slopes and is surrounded on three sides by Namurian mudstones. As in the area to the north, the highly indented contact is interpreted as an unconformity. An exposure [1003 6520] shows 2.7 m of thickly bedded grey bioclastic calcarenite with a brachiopod fauna including *Davidsonina septosa transversa*, indicative of late Asbian age. This facies contrasts with the more steeply dipping micritic fore-reef limestones exposed higher up the dip-slope to the south-east [1025 6506]; these have yielded an abundant and varied fauna including the upper B_2 goniatites *Bollandites castletonensis* and *Bollandoceras micronotum*. Small exposures of dark grey limestone occur at the foot of the fore-reef slope.

Farther south, between Pilsbury and Beresford Dale, the apron-reef becomes discontinuous at outcrop, though it may well be present beneath the Namurian cover. At Pilsbury Castle Hills [1147 6389] a 10.8 m 'stack' of fore-reef limestone is probably entirely surrounded by unconformable Namurian mudstones. In outcrops between this locality and Beresford Dale, Monsal Dale Limestones appear to rest directly on both the apron-reef and shelf facies of the Bee Low Limestones (p. 36; Aitkenhead and Chisholm, 1979).

In Beresford Dale the River Dove has cut a gorge in obscurely bedded micritic shelly apron-reef limestones. These pass laterally into massive very fine-grained biosparites, with a few laminae of dispersed crinoid debris and ?gigantoproductoid shells in concave-upwards position, seen in a 20 m crag high on the east side of the gorge [1291 5894] and around Frank i' th' Rocks Cave [1312 5842]; though not typical, these beds are assigned to the marginal shelf or back-reef facies. On the opposite side of the gorge, a section clearly shows the lateral passage and interdigitation from shelf to fore-reef limestones; no algal reef is present.

From Beresford Dale, the apron-reef outcrop continues SSW for about 1 km, before merging with the knoll-reef which forms Gratton Hill. This is poorly exposed and recognised mainly by its knoll-like topographic expression. Beyond Gratton Hill, the apron-reef is absent and the Bee Low Limestones either pass laterally into the Hopedale Limestones or are faulted against older rocks.   NA

HARTINGTON AREA

The formation is estimated to be about 210 m thick hereabouts. The base is exposed in an old railway-cutting [1440 6165], where 12.5 m of pale calcarenite rest on a markedly undulating, possibly erosional, surface of finer-grained Woo Dale Limestones (Sadler and Wyatt, 1966, pp. 59–60). The lowest 3 m of the Bee Low Limestones contain calcisiltite bands and show a well developed undulating bedding partly parallel to the basal surface. Smaller sections through the contact can be seen by the roadside in Hand Dale [1442 6086 and 1483 6116] and at the entrance [1495 6125] to the former railway station. The sections suggest a gradual transition, rather than erosion, at the formation boundary.

Some 20 to 30 m higher in the sequence lies a slightly darker band of poorly bedded calcarenite about 6 m thick. It forms prominent crags [137 610 to 139 615] in Long Dale, and has been recognised at Caskin Low [148 617], Lean Low [149 622] and in the old railway-cutting [147 630] south of Parsley Hay.

At Hartington Station Quarry [SK 16 SE/15] the darker bed is overlain by 106 m of thickly bedded pale limestone with at least 13 clay wayboards, some of which overlie potholed palaeokarstic surfaces (Walkden, 1974, fig. 2). The limestones are mainly biosparites and pelsparites (Cox and Bridge, 1977, pp. 74–76).

Near Mosey Low [128 642] is exposed a *D. septosa* band about 0.2 m thick in a small quarry [SK 16 SW/11]. It lies in the upper part of the formation but the exact level cannot be estimated, due to faulting.

The highest beds are exposed in Hartington Dale, especially at the east end [SK 16 SW/12], where about 70 m of pale thickly bedded calcarenite are discontinuously exposed beneath the basal dark bed of the Monsal Dale Limestones. This sharp contact can also be seen in Hartington village [SK 16 SW/13], while at Banktop [129 615], not far to the north, the contact is an undoubted erosion surface (p. 37) with up to 11 m of relief. Just west of Banktop two small inliers of reef-limestone have been mapped [1258 6158 and 1258 6133]; they are interpreted as upstanding parts of the apron-reef, mainly buried under a disconformable cover of the Monsal Dale Limestones and Namurian mudstones. South of Hartington, at Bullock Low [1290 6006], the apron-reef projects through the later deposits and is in contact with the shelf limestones.   JIC

WOLFSCOTE HILL TO ALSOP EN LE DALE

The Bee Low Limestones are about 189 m thick in this area (Figure 7). Biggin Dale contains exposures in the lowest 43 m of the formation which are recorded in a composite section 156 m thick [SK 15 NW/13]. The underlying beds are now included in the Woo Dale Limestones (p. 13). Approximately the upper 36 m of the Biggin Dale succession are probably represented by beds proved in the lowest 36 m of the borehole 141 m deep [SK 15 NW/8] near the top of Wolfscote Hill. The hole started about 30 m below the top of the formation, and just below a *D. septosa* band exposed nearby [1368 5832]. A number of pedogenic crusts, probably representing palaeokarstic surfaces, have been recognised (Bridge and Kneebone, 1983, p. 6). They occur at intervals, mainly in the range 10 to 15 m, throughout the sequence; this repetition is thought to be cyclical as other lithologies are also repeated.

Following the drilling of a borehole [SK 15 NE/5] at Lees Barn (Bridge and Kneebone, 1983, p. 45) that proved Bee Low Limestones to 39.30 m on Woo Dale Limestones, the outcrop of the formation is now known to be less extensive than was previously thought (p. 13). The base of the Hindlip Quarry section [SK 15 NE/12], to the east, is estimated to lie about 60 m above the top of the borehole section, but there are no exposures of the intervening limestones. In the quarry are exposed about 44 m of pale thickly bedded bioclastic calcarenite with several clay wayboards, some of which rest on potholed surfaces. About 25 m above the base of the section is a *D. septosa* band that can be traced to near the base of the adjoining Coldeaton railway-cutting section [SK 15 NE/11]. In the cutting are a further 68 m of limestone similar to that in the quarry, but without significant fossil bands, overlain sharply [1606 5762] by darker Monsal Dale Limestones with a clay wayboard on a potholed surface at the junction.   NA, JIC

BALLIDON AND GRATTON DALE

The base of the formation has been proved in a borehole near Roystone Rocks [SK 15 NE/6]. Here 64 m of pale thickly bedded calcarenite, with a *D. septosa* band near the top, rest on porcellaneous limestones taken to be at the top of the Woo Dale Limestones. Above lie an estimated 120 m of poorly exposed pale grey and cream calcarenite, partly dolomitised, as at Roystone Rocks [197 568], and overlain on a spur of higher ground [1997 5725] north of Roystone Grange by darker shelly beds at the base of the Monsal Dale Limestones. The formation boundary has been penetrated by two boreholes in this general area [SK 15 NE/2 and 3] (Bridge and Kneebone, 1983, pp. 42–43), and in both it proved to be a sharp contact of medium or dark grey limestone on pale.

The top of the formation was proved in boreholes near Gratton Dale [SK 16 SE/6, SK 26 SW/18 and 19]. A thickness of 132 m of pale calcarenite was encountered above a thin dark grey bed which is thought to lie near or below the base of the formation. At outcrop in the dale [201 598] the Bee Low Limestones are all dolomitised.

A working quarry at Ballidon [202 554] displays 85 m of pale calcarenite, and a borehole at Twodale Barn [SK 15 NE/7], not far away, has provided a detailed record of 101 m of the formation (Bridge and Kneebone, 1983, pp. 47–48). The exact stratigraphical level of both sections is in doubt, but the base of the formation cannot be far beneath, as it was proved in a borehole [SK 15 NE/6] not far to the north (*see above*) and beneath the dolomitised zone in a borehole [SK 25 NW/18] about 1.5 km to the east. An occurrence of *D. septosa* has been recorded [2057 5567] near the quarry.

The apron-reef reappears in places along the southern margin of the Bee Low Limestones outcrop but mostly lies in the adjacent Ashbourne district. Exposures up to 3.6 m thick near Dale End Farm [1803 5542] and on the western slope of Parwich Hill [1861 5521] have yielded an abundant and varied fauna including, at the former locality, the goniatite *G.* cf. *hudsoni* suggesting a lower B$_2$ age.

LONGCLIFFE AND GRANGEMILL

Around Longcliffe and Harboro Rocks the Bee Low Limestones are extensively dolomitised. A small quarry [2348 5589] near Curzon Lodge, in an undolomitised patch close below the top of the formation, has yielded a typical Asbian fauna, including *Dibunophyllum* cf. *bourtonense*, *Lithostrotion aranea*, *L. martini*, *L. pauciradiale*, *Palaeosmilia murchisoni*, *Gigantoproductus sp.* and *Linoprotonia* cf. *hemisphaerica*. North of here, around Grangemill [243 576], the limestone is worked in three quarries. At Prospect Quarry [SK 25 NW/37], 46 m of mainly pale calcarenite with several clay wayboards overlie porcellaneous beds assigned to the Woo Dale Limestones; a tuffaceous mass recorded at the contact has been interpreted as part of a tuff mound (Walkden, 1972, p. 156). The base of the formation has also been recorded at 35 m depth in a borehole [SK 25 NW/19] in the floor of Longcliffe Quarry, and is exposed in a natural section in the Griffe Grange Valley [SK 25 NW/38]. The section in Longcliffe Quarry [SK 25 NW/35] shows 28 m of limestone with palaeokarstic surfaces (see also Walkden, 1974, loc. 57). A similar section [SK 25 NW/36] in the nearby Grangemill Quarry includes a *D. septosa* band. JIC, IPS

MILLCLOSE MINE

The main orebody lay along a north–south line beneath the Namurian rocks east of Stanton Moor Plantation [248 632] and close to the district boundary. The Dinantian sequence encountered in the workings has been described by Traill (1939; 1940), Shirley (1950) and by Smith, Rhys and Eden (1967, pp. 27–32). A recent recorrelation (Chisholm and others, 1983) suggests that the top of the Bee Low Limestones should be drawn at that of the 10-foot

Limestone rather than at the top of the 129 Limestone as was done by Shirley (1950, p. 354). The section at No. 2 Winze [2549 6460] showed 10.5 m of pale limestone (the 10-foot Limestone) resting on the steeply dipping front of a lava, the 144 Pump Station Toadstone. The lava contained some pillow structures and reached a thickness of 14.9 m nearby (Traill, 1940, pp. 210–211, 219, plate XI). Shirley (1950, p. 354) obtained D$_1$ coral faunas both from the limestone just above, and 15 m below, the lava; the age of these faunas has recently been confirmed (Chisholm and others, 1983).
JIC

MONSAL DALE LIMESTONES

The Monsal Dale Limestones (Aitkenhead and Chisholm, 1982) occupy a broad tract to the east of the Bee Low Limestones outcrop (Figure 10). There is also a large outlier around Biggin, and smaller ones at or near the western margin of the Dinantian outcrop; in the east there is an inlier north-west of Bakewell. The general stratigraphy is summarised on p. 28. The limestones at outcrop in the southern part of the district (Figure 37) are widely affected by secondary dolomitisation (p. 102).

Within the district the thickness of the formation is generally between 100 and 200 m, though this increases to at least 375 m at Longstone Edge (Figure 13). The isopachyte lines shown in Figure 10 give little support to the contentions of Shirley (1959, p. 421) and Butcher and Ford (1973, p. 191) that areas of thick and thin deposition were related to the sites of structural synclines and anticlines. Indeed, in the thickest area the relationship appears to be inverse, for it lies beneath the Longstone Edge Anticline. It is more likely that the local changes of facies and thickness that take place within particular stratigraphical units are related to localised tectonism associated with penecontemporaneous volcanic activity and/or changes in sea level. There is, however, a reasonable agreement between the trend of isopachytes in the depositional basin near Longstone Edge and the south-eastward dipping palaeoslope inferred at Ashford by Adams and Cossey (1978) from slump structures in the Rosewood Marble.

The base of the formation is generally sharp, with darker limestones conformably overlying the pale Bee Low Limestones. Locally, as at Litton Tunnel (Cope, 1933, p. 132; Walkden, 1977, p. 350) there is evidence of erosion at the contact; this is particularly evident at the western shelf margin as at Banktop, where the base rests on an erosion surface with a relief of several metres. In those areas where exposure is poor, or where pale limestones are present at the base, the formation boundary is drawn on faunal evidence. At Buxton, on the western edge of the shelf area, a marked break with overlap has resulted in the absence of the lower part of the formation.

The top of the formation is sharp, and in places lies below a clay wayboard (Figure 17). In the Wirksworth area, Shirley (1959, p. 418) and Smith and others (1967, p. 12) have shown that the top of the Monsal Dale Limestones was eroded before deposition of the Eyam Limestones, but in the present district there is no evidence for a significant amount of erosion at this level. Across a wide area between Bradford Dale and Deep Dale, for example (Figures 14, 15), a persistent marker horizon, the Lathkill Shell Bed, (p. 36) can be recognised at a constant level below the top of the formation.

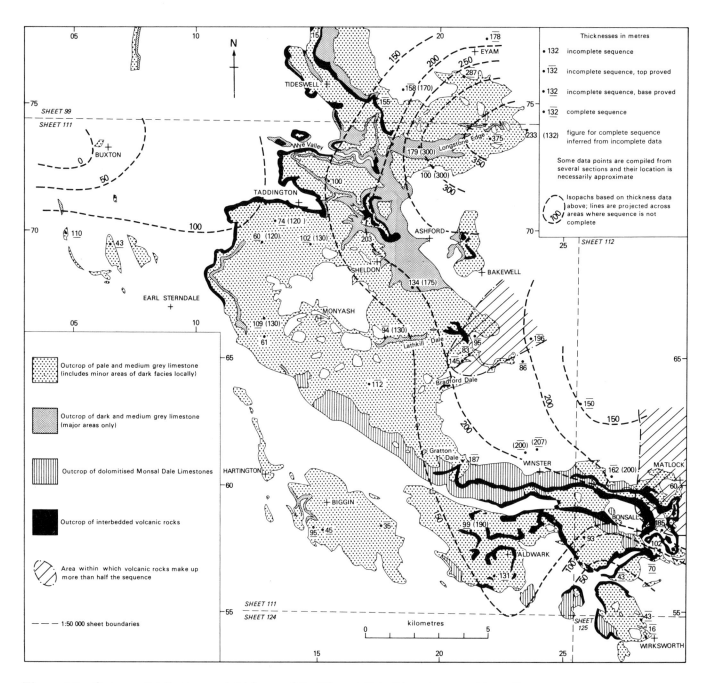

Figure 10 Outcrop distribution and thickness of the Monsal Dale Limestones and laterally equivalent volcanic rocks.

Local erosion surfaces within the formation are present in many areas (Butcher and Ford, 1973, pp. 188–189) but none of these is of regional extent, and there is no evidence to support the view (Shirley, 1959, p. 412) that a widespread unconformity exists. The variations in the sequence that were attributed to this cause by Shirley are more satisfactorily explained by lateral facies changes.

Lithology and facies variations

The formation consists of laterally variable pale, medium and dark grey limestones with many fossiliferous bands.

Chert is common but not universally present. Lateral passage between the various facies can be seen in many of the dale sections as, for example, at Headstone Head ([SK 17 SE/22]; Figure 12). In general, the pale and medium grey limestones are thickly bedded and mottled by bioturbation; most are calcarenites, ranging from well sorted biosparites and pelsparites to poorly sorted biomicrites. Emergent surfaces, marked by pedogenic tubes and crusts of brown laminated micrite, are common. The dark limestones are commonly thinly bedded with dark argillaceous partings; they are generally finer grained than the pale rocks, most being ill-sorted bioclastic calcisiltites with much micritic matrix, though calcarenites also occur. Mottling by bioturbation is common, but is of a more streaky nature and on a smaller scale than in the paler beds. Emergent surfaces are rare. Purely micritic rocks are not common, but a distinctive fine-grained lithology found in some dark sequences is finely laminated, with syndepositional breccias and convolution structures. The best known example is the Rosewood Marble (Adams and Cossey, 1978). Calcirudites more commonly occur in the paler lithologies than in the dark; they range from beds of brachiopods usually in growth position in a fine-grained matrix, to current-sorted deposits of coarse crinoid or shell debris.

Westward thinning away from the type area is accompanied by a decrease in the proportion of dark limestones (Figure 14), but in some other areas, dark limestones pass laterally into pale without a change in thickness. The thickest sequence, at Longstone Edge, does not contain a greater proportion of dark limestone than the thinner Monsal Dale sequence (*compare* Figures 14 and 13). On the maps the outcrops of the darkest beds are distinguished where possible from those of the more widespread medium and pale grey beds, but only one dark unit, the 'Station Quarry Beds' at the base of the Wye valley sequence (Cope, 1937), has been named. It is now regarded as a member of the formation.

Volcanic rocks

Interbedded volcanic rocks comprising basaltic lava, hyaloclastite and tuff (Chapter 4), are common among the Monsal Dale Limestones, forming useful marker bands in many places. Where they are subordinate to the limestones they are regarded as members of the formation, but where, as near Alport, their proportion in the sequence exceeds that of the limestones they are regarded as a separate lithostratigraphical unit, the Fallgate Volcanic Formation (Aitkenhead and Chisholm, 1982).

Figure 10 shows the outcrop distribution of the volcanic members and also the approximate extent of the areas where volcanic rocks make up more than half the sequence. The stratigraphical positions of all these members, except the Shothouse Spring Tuff which lies at the same horizon as the Lower Miller's Dale Lava, are shown in Figure 11.

Environment of deposition

The warm shelf sea in which the Bee Low Limestones were deposited continued in existence during Brigantian times but the marked facies variations in the Monsal Dale Limestones point to local changes of environment. The pale and medium grey limestones are normally poorly laminated and are mottled by bioturbation. There is evidence of subaerial emergence at times, indicating deposition in very shallow water, presumably above wave base; however, organic activity was in general so vigorous that traces of current action have not been preserved. Bioturbation is also common in the dark facies, but emergent surfaces are few, suggesting that deposition took place in somewhat deeper water. Graded beds, whether interpreted as turbidites originating from slump-sheets or as storm-generated layers, point to deposition below wave base. The depth was probably not great, however, for lateral passage between the facies takes place over quite short distances, and shelly faunas in growth position are locally found in dark limestones. Thicknesses of volcanic rocks suggest that water depths of about 73 m were attained in early Brigantian times around Youlgreave (p. 60).

At the shelf margins, growth of the apron-reef facies was halted in late Asbian or early Brigantian times, and boulder beds, breccias and conglomerates of probable P_{1a} age were formed, probably by wave action against the temporarily emergent reef (p. 24). Early Brigantian erosion within the shelf province was mainly by karstic processes, and evidence for a complex series of events in the Wye valley area has been published by Walkden (1977).

Faunal marker horizons

The biostratigraphy is discussed separately on pp. 55–58, but the following fossils and fossil bands of particular stratigraphical value within the formation are worthy of special mention. *Girvanella* and *Saccamminopsis* are more common and have a greater range than has previously been reported for the Wye valley area (George and others, 1976, p. 32); *Girvanella* tends to occur at three levels in the lower part of the formation (for example, Cox and Bridge, 1977, figure 6). *Saccamminopsis* ranges from just above the base to a level, in one borehole [SK 16 NE/10] near Sheldon, estimated to be about 13 to 17 m below the top of the formation; it is, however, particularly abundant near the base. Brachiopods are common at several levels in the formation but only the Lathkill Shell Bed (Figure 11) has been distinguished by name. Several coral bands are present in the Wye valley area, and some have their type localities there. These include the Upperdale Coral Band, the Hob's House Coral Band, the White Cliff Coral Band (*Lonsdaleia duplicata* Band) and the *Orionastraea* Band. The latter is less well developed than to the north (Stevenson and Gaunt, 1971, p. 31), and *Orionastraea* also occurs sporadically at a slightly lower level.   JIC, NA, IPS

DETAILS

BUXTON

In Buxton, Monsal Dale Limestones of pale facies are present in a restricted area near Hardwick Mount. The base of the formation is seen in the railway-cutting [SK 07 SE/59] on the south side of the town, where 12.12 m of pale grey bioclastic limestone rest on a potholed surface at the top of the Miller's Dale Limestones; a high horizon in the Monsal Dale Limestones is indicated by the presence of *Orionastraea indivisa* 2.13 m above the base and *O. placenta* 6.18 m higher. The main stratigraphical break here lies at the base of the Monsal Dale Limestones, which appear to occupy a hollow in the underlying Miller's Dale Limestones.

At Harpur Hill, a small faulted outlier of Monsal Dale Limestones of the pale facies overlying a lenticular development of Upper Miller's Dale Lava is present, though poorly exposed.   IPS

BUXTON TO EARL STERNDALE

Several outliers of the formation are present between Buxton and Chrome Hill. About 43 m of limestones at the base have been proved in the extensive outcrop in the Greensides Syncline [065 691]. A kilometre to the east lies an equally extensive outlier between Hind Low and Hillhead Farm, and a smaller one is present a little beyond. Both have been worked at Buxton Quarry [078 693; 084 689], where medium to dark grey limestones at the base of the formation lie with slight discordance on pale thickly bedded Bee Low Limestones [SK 06 NE/26]. Fossils collected nearby from the top of Hind Low [0811 6892] include *Striatifera striata* and *Lonsdaleia floriformis*, indicative of a Brigantian age.

Other smaller outliers lie along the margin of the Dinantian outcrop between Turncliff and Chrome Hill. The limestone lithology is variable and exposure poor, making correlation difficult. Erosional

bedding surfaces with abrupt upward changes in lithology are present at several places suggesting that this was a tectonically unstable marginal area of the shelf during Brigantian times. Steep dips (up to 63°) prevail along much of this tract, and up to 110 m of beds are estimated to be present where the outcrop is widest near Turncliff. Scattered exposures here [0473 6992 to 0474 6988], mapped as knoll-reef, contain a rich brachiopod fauna similar to that described from the P_{1b} reef at Castleton by Mitchell (*in* Stevenson and Gaunt, 1971, pp. 140–152). Another knoll-reef is present [0590 6828] north-west of Stoop and is unusual for the occurrence of white chert nodules in the pale fossiliferous limestone. However, the best

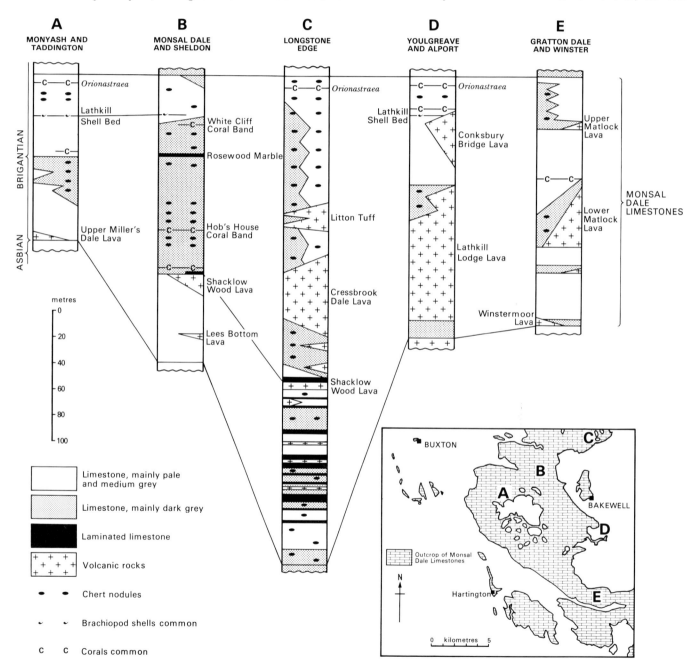

Figure 11 Generalised sections of the Monsal Dale Limestones and part of the Fallgate Volcanic Formation. (The volcanic sequence shown at the right hand side of column D is assigned to the Fallgate Volcanic Formation.)

exposure in the vicinity [0639 6812 to 0639 6814] consists of 10.1 m of well bedded dark grey limestone with dark chert nodules and common brachiopods including *Gigantoproductus edelburgensis* and with *Girvanella* encrusting shell fragments in the basal 2.44 m. NA

WYE VALLEY, WESTERN AREA

Around Miller's Dale some 10 m of dark limestones, the 'Station Quarry Beds' of Cope (1936, p. 180), lie beneath the Upper Miller's Dale Lava. They die out north and west of Station Quarry and also in the lower reaches of Blackwell Dale; to the east, with the disappearance of the Upper Miller's Dale Lava, they merge with overlying beds of similar lithology. Station Quarry at Miller's Dale [SK 17 SW/67] shows their full thickness in a 6.63 m section in grey and dark grey thin bedded limestones with some chert. A band with *Saccamminopsis* is present close to the base. Nearby [1377 7330], close to the former station, Cope (1933, p. 132) records a 1.83 m pothole at the base of the Monsal Dale Limestones, cut into the underlying Miller's Dale Limestones; the pothole was filled with pyritous mudstone and contained a 'large waterworn block of limestone' near its base (*see also* Walkden, 1977, pp. 350–351).

At Miller's Dale Lime Works [SK 17 SW/68], beds in this position comprise: grey limestone, part bioclastic 5.75 m; on 4.93 m of dark limestone, cherty at the top, and with *Saccamminopsis* in bands at the base and at 0.73 m and 3.73 m above it. These beds rest on the Miller's Dale Limestones; the top of the section lies close to the base of the Upper Miller's Dale Lava.

The Upper Miller's Dale Lava crops out between Hurdlow and Miller's Dale, with an isolated occurrence at Harpur Hill. Details are given separately on p. 62.

Around Miller's Dale the beds above the lava consist mainly of pale massive limestones. The disused Miller's Dale Lime Works [SK 17 SW/69] shows a 57.3 m section in predominantly pale massive limestones (the 'Priestcliffe Beds' of Cope, 1933, p. 134) with some crinoidal and shelly bands. A 4.5 m dark bed, part cherty and with a *Saccamminopsis* band, occurs at 34.75 m above the base; the top of this bed marks the base of the 'upper leaf' of pale limestone present in the area of interdigitation farther east (*see below*). The succeeding pale limestones contain shelly bands and some cherty layers.

Near Litton Mill, a section [SK 17 SE/18] along the disused railway line shows 19.32 m of dark thinly bedded limestone, cherty in places, resting with overlap on the flow-front of the Upper Miller's Dale Lava. The limestones contain thin bands with *Saccamminopsis* at two main levels, and the lower of these is cut out by a non-sequence in the immediate vicinity of the lava front. The lowest beds in contact with the lava contain intraclasts derived from it (Cope, 1933, p. 137; 1936, pp. 183–185; Walkden, 1977, pp. 354, 356). At the western end [1617 7289] of Litton Tunnel, Walkden (1977, p. 354) noted the presence of a 10 cm K-bentonite band lying 4 m above the base of the formation; the rock showed relict vitroclastic structure.

WYE VALLEY, EASTERN AREA

Here, with the general disappearance of the Upper Miller's Dale Lava, the Monsal Dale Limestones are considered as a whole.

To the east of Litton Mill, the formation shows an overall eastward passage from a predominantly pale sequence into dark thinly bedded cherty limestones (Figure 12). In the lower part of the succession, the passage into dark beds is complete to the east of a general line through Castcliff [185 739], Cressbrook Mill and the junction of Taddington and Monsal dales. However, near the top of the formation, a more persistent 'upper leaf' of pale limestone extends farther east, dying out to the east of a north-east to south-west line through Headstone Head. The dark strata are in many places

poorly exposed, while the pale tend to form features on the valley sides. There is also an eastward thickening of the sequence.

The base of the formation is seen [1661 7269] at the eastern end of Litton Tunnel, where dark cherty limestones rest on an erosion surface in the Miller's Dale Limestones. A pothole here, up to 1.83 m deep, was figured and described by Cope (1933, pp. 132–134); it contains pebbles of dark grey limestone in a crumbly calcareous and pyritous matrix overlain by a thin mudstone band. Walkden (1977, p. 350) describes the pothole as 7 m deep and 25 m wide. This, and the similar feature at Station Quarry, have been described variously as washouts (Cope, 1933, p. 132), channels (Butcher and Ford, 1973 — referring to the Litton Tunnel feature), and palaeokarstic solution hollows (Walkden, 1977, pp. 350–351): the last explanation is preferred.

At Cressbrook Mill, an isolated outcrop of Upper Miller's Dale Lava rests directly on the Miller's Dale Limestones, the 'Station Quarry Beds' being absent.

UPPERDALE TO FIN WOOD

Around Upperdale and Hay Dale the Monsal Dale Limestones comprise a sequence of partially exposed dark cherty limestones beneath the pale limestones of the 'upper leaf' (Figure 12). The Upperdale Coral Band is present [SK 17 SE/20] near the base of these beds as 1.37 m of grey cherty limestone with abundant *Lithostrotion junceum* and other corals. It is separated from the higher Hob's House Coral Band by 22.11 m of partly exposed dark limestones, with some pale intercalations, one of which includes a band of silicified shells including *Gigantoproductus crassiventer*. The Hob's House Coral Band comprises 1.63 m of pale cherty limestone with common *Dibunophyllum konincki*. The band is overlain by dark cherty limestones, poorly exposed except close to the base of the 'pale upper leaf'. An isolated occurrence of the band occurs at Crossdale Head [SK 17 SE/19] where it is 0.8 m thick.

At White Cliff [SK 17 SE/21] the 'pale upper leaf' comprises 28.4 m of the 32.97 m section; the pale beds are predominantly grey cherty limestones, with the White Cliff Coral Band, 2.44 m thick and yielding *Lonsdaleia duplicata*, *L. floriformis* and *Palaeosmilia regia*, at 22.16 m above the base of the pale beds.

The section [SK 17 SE/22] at Headstone Head shows a sequence of dark thin-bedded cherty limestones with a 0.84 m laminated bed with small-scale slump folds near the base. The top of the section lies at about the position of the base of the 'upper pale leaf' which has here died out (*see below*). The White Cliff Coral Band occurs within a thin grey limestone, part cherty, close to the top of the section; the coral fauna includes *Lonsdaleia duplicata*.

A comparison between the White Cliff and Headstone Head sections brings out the diachronous character of the 'upper pale leaf' by its relationship to the White Cliff Coral Band. The latter lies within the pale beds at White Cliff, then passes into predominantly dark strata in the vicinity of Headstone Head.

To the east of Headstone Head, a section [SK 17 SE/13] in the disused railway-cutting exposes the uppermost 14.37 m of Monsal Dale Limestones. These show unusually thick laminated limestones and beds of related lithology (one 6.09 m thick), some silicified, interbedded with dark thinly bedded limestones. The uppermost 0.15 to 0.18 m is a grey-brown limestone with common *Spirifer bisulcatus* and some crinoid debris (Butcher and Ford, 1973, pp. 184–186).

At Hob's House, the 51.73 m section [SK 17 SE/24], mainly in the dark facies, contains the Hob's House Coral Band, 0.4 m thick and with *Dibunophyllum spp.*, *Diphyphyllum sp.* and *Lithostrotion spp.*. A 0.68 m band of laminated limestone occurs at 33.8 m above the coral band, but is not the band of similar lithology that occurs near Headstone Head. The top of the section is estimated to lie some 38 m below the base of the 'upper pale leaf', while the base lies above the presumed position of the eastern extremity of the Lees

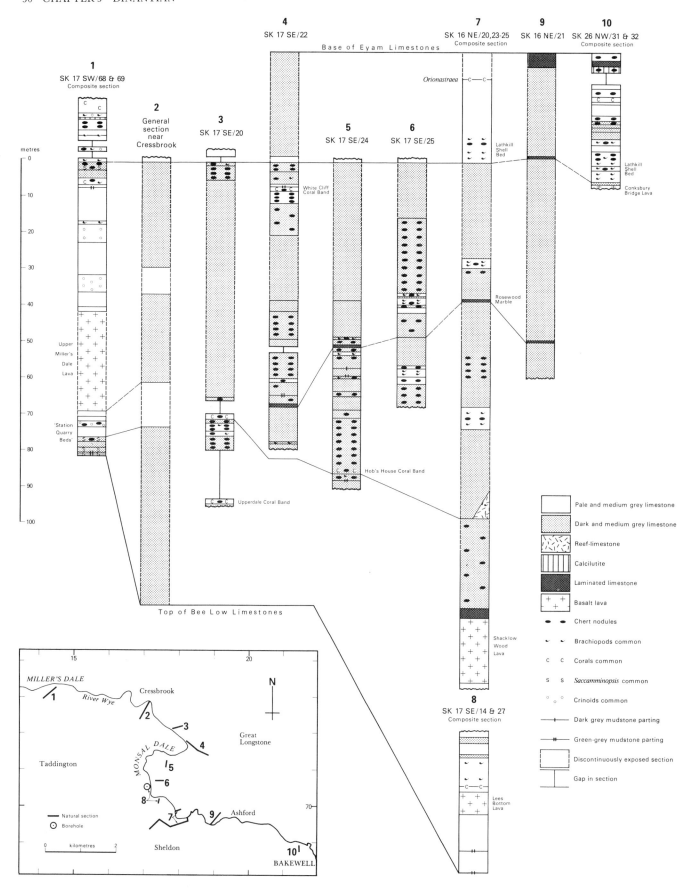

Figure 12 Sections of Monsal Dale Limestones in the Wye valley.

Bottom Lava in the landslip of Hob's House. The original description of Cope (1933, p. 140) was apparently based on material from within the landslip.

The White Cliff Coral Band is present nearby [1785 7115], close to the base of the 'upper pale leaf', as a 0.95 m bed of pale shelly limestone with *Palaeosmilia regia*.

LONGSTONE EDGE

The Longstone Edge Anticline exposes dark cherty limestones in the western part of the axial area around Watersaw Rake; these are flanked by pale limestones corresponding broadly to the 'upper pale leaf' of Monsal Dale and extending directly into it at Little Longstone. In the eastern part of the structure, the exposed Monsal Dale Limestones are entirely of pale facies.

In the axial area of the anticline, a section [2184 7366 to 2211 7365], now inaccessible, on the south side of Deep Rake, showed 29.76 m of mainly pale limestone with cherty and crinoidal bands. The *Orionastraea* Band, 0.61 m thick, was present 2.13 m from the top; it contained *Orionastraea*, seen *in situ* only. A band referred to the same level is exposed [2190 7346] near Bleaklow at the top of a 1.52 m section in pale cherty limestone; it contained *Amplexizaphrentis sp.*, *Diphyphyllum lateseptatum* and brachiopods. To the south, the Buskey Cottage Borehole [SK 27 SW/26] proved 4.76 m of grey and grey-brown limestone with some chert, below the base of the Longstone Mudstones at 34.64 m.

On the northern limb of the anticline, the *Orionastraea* Band is traceable above a thin band of dark limestone east of Sallet Hole Mine. A section [SK 27 SW/29] shows 54.38 m of pale limestone with some darker beds, cherty horizons in the top 40.41 m, and some shell bands. The *Orionastraea* Band is represented by an occurrence of *Palaeosmilia regia* within mainly unexposed beds 3.35 m from the top; it is well exposed nearby, north of the district boundary (Stevenson and Gaunt, 1971, p. 82, loc. 77).

The formation is known at depth in the Longstone Edge area from a number of boreholes summarised in Figure 13, which the following notes are designed to supplement. All section numbers refer to boreholes.

a Strata below Shacklow Wood Lava. These are known only in borehole [SK 27 SW/20]. The base of the Shacklow Wood Lava is at 233.58 m and the presumed base of the Brigantian and Monsal Dale Limestones at 358.98 m. Below a depth of 339.52 m is a subdivision of predominantly pale limestones, cherty in places and with darker intercalations, especially near the base. *Saccamminopsis* is abundant at 354.71 m. From 339.52 m to the base of the Shacklow Wood Lava the borehole proved a very variable sequence, comprising extensive laminated limestones and an associated calcilutite (p. 34), dark and pale limestones, some cherty, pale crinoidal limestones, thin tuffs and a thin vesicular lava. The macrofauna was non-diagnostic.

The upper part of the subdivision was also proved in [SK 27 SW/10] between the Shacklow Wood Lava (234.42 m) and the bottom of the hole at 272.64 m; here it contained substantial tuffaceous limestones.

b Strata between base of Shacklow Wood Lava and base of Litton Tuff. The thickness of the subdivision varies from 79.23 m [SK 27 SW/10] to 132.18 m [SK 27 SW/1], the latter applying to an area where the Cressbrook Dale Lava is particularly thick.

The Shacklow Wood Lava is persistent at the base (p. 61). The overlying limestones are mainly cherty dark and pale, with some pale crinoidal beds in the eastern boreholes; there is a persistent laminated limestone at the base. Tuffs are present near the base of the subdivision at the western end of the Longstone Edge Anticline and there are also higher and more extensive ones, best seen in [SK 27 SW/1].

The Cressbrook Dale Lava reaches a thickness of 80.88 m in

[SK 27 SW/5], where it comprises hyaloclastites (p. 61) throughout. The lava or its equivalents are absent in the westernmost borehole [SK 17 SE/2].

The uppermost subdivision comprises dark and pale limestones with varying amounts of chert and crinoidal limestones.

c Litton Tuff and higher strata. Nearly the full thickness of the subdivision, 158.67 m, was proved in [SK 27 SW/10]. The Litton Tuff is present at depth over most of the Longstone Edge area (p. 61).

The overlying strata are mainly dark limestones in the west; elsewhere they show alternations of pale and dark lithologies with a tendency for the former to predominate in the central area [SK 27 SW/4 to SK 27 SW/10]. Chert is present in both pale and dark beds in places and there are pale crinoidal limestones and common pale shelly horizons. At Sallet Hole Mine, coarsely crinoidal dark limestones overlie the Litton Tuff. In the Eastern Decline [2198 7367] these limestones, some 3 m above the tuff, contain rounded clasts of pale green igneous rock.

EYAM

The Eyam Borehole [SK 27 NW/15] (Dunham, 1973, pp. 84–85) proved a nearly complete section of the formation. It started at the top of the Black Bed of Middleton Dale (Stevenson and Gaunt, 1971, p. 78) and passed through a variable limestone sequence with the Litton Tuff, 1.08 m, at 47.8 m and the Cressbrook Dale Lava, 76.59 m at 136.39 m.   IPS

SHELDON AND ASHFORD

The complete succession, 203 m thick, was proved in a borehole [SK 16 NE/10] north of Sheldon. The same sequence, except for a few metres at the base, crops out in the sides of the Wye valley nearby [SK 16 NE/20, 23, 24, 25 and 17 SE/26, 27], though the exposure is discontinuous. The sequence is in three parts (Figure 14). The lowest 50 to 60 m are pale and medium grey limestones with two lenticular volcanic members, the Lees Bottom and Shacklow Wood lavas (p. 61). The overlying 120 to 130 m are mainly dark and medium grey cherty limestones; the top 30 m around Sheldon (the 'upper pale leaf' of Monsal Dale, p. 29) are paler, with the Lathkill Shell Bed at the base. To the east, the top unit passes into dark cherty limestone at Ashford, and the proportion of volcanic rocks in the underlying beds increases (Figures 12, 14). Westwards the entire sequence thins, and pale limestones increase at the expense of the dark. Both thin and thick sequences yield *Koninckopora*, an alga not known from strata younger than the oldest Brigantian (Chisholm and others, 1983), which is present in the lowest few metres in the two sequences; the Lathkill Shell Bed and *Orionastraea* Band can also be recognised near the top of both.

The lowest unit crops out around Lees Bottom [171 705]. The limestones below the Lees Bottom Lava are poorly exposed alongside the river [1718 7034], but were penetrated in the Lees Bottom Borehole [SK 17 SE/14] which, beneath 4.9 m of weathered lava, proved 16 m of partly tuffaceous medium and pale grey limestone with *Koninckopora* resting on pale beds assigned to the Bee Low Limestones. In the Sheldon Borehole the basal limestones contained *Koninckopora*, *Saccamminopsis* and *Lithostrotion* cf. *arachnoideum*. About 30 m of pale and medium grey limestone are present between the lavas, increasing to about 60 m in the south-west, where the lavas die out. Brigantian coral-brachiopod faunas have been collected near the base [SK 17 SE/27] and from a 0.9 m band about 34 m higher up [SK 17 SE/26]. A localised mass of reef-limestone, described by Worley (1976) in mine workings near Sheldon, probably lies among these limestones.

The middle unit of mainly dark thinly bedded and cherty limestone, 120 to 130 m thick, north of Sheldon, is discontinuously

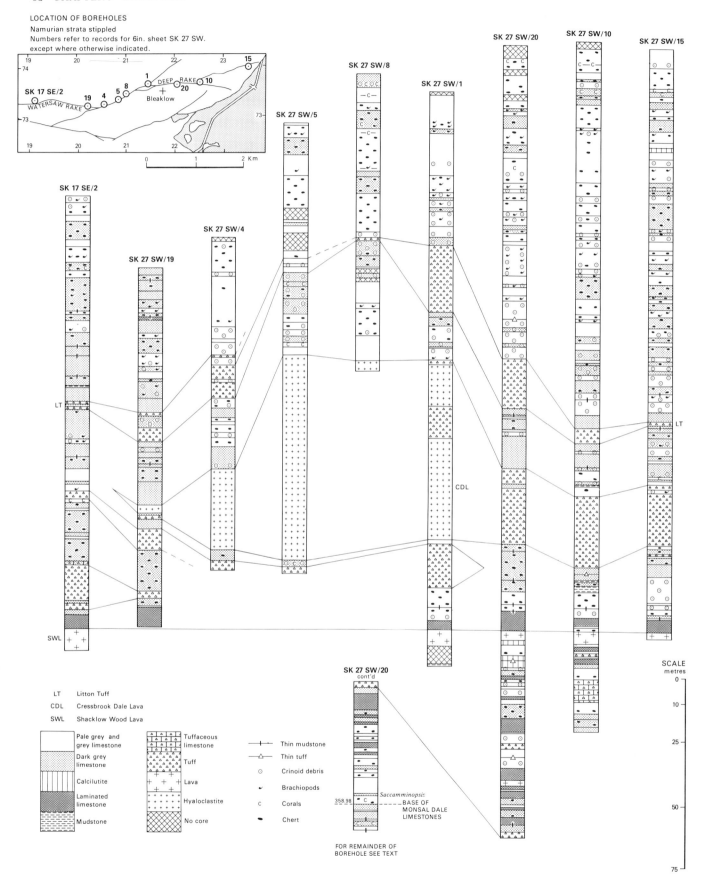

LOCATION OF BOREHOLES
Namurian strata stippled
Numbers refer to records for 6in. sheet SK 27 SW.
except where otherwise indicated.

LT   Litton Tuff
CDL   Cressbrook Dale Lava
SWL   Shacklow Wood Lava

Pale grey and grey limestone
Dark grey limestone
Calcilutite
Laminated limestone
Mudstone

Tuffaceous limestone
Tuff
Lava
Hyaloclastite
No core

Thin mudstone
Thin tuff
Crinoid debris
Brachiopods
Corals
Chert

Figure 13   Boreholes in the Monsal Dale Limestones of the Longstone Edge area.

exposed in Shacklow Wood [168 701 to 184 695]. To the south-west in Deep Dale [162 696], the proportion of pale limestone increases progressively and the sequence thins markedly (Figure 14). This lateral facies change is also visible in a tributary dale [163 704] west of Lees Bottom, where the sequence thins to about 90 m and there are three pale intercalations. On the eastern side of Monsal Dale, below Fin Cop [174 790], the sequence is mainly dark and cherty; the upper part of section [SK 17 SE/25] shows 53 m of this lithology

with two thin bands, partly paler in colour, containing silicified productoids.

At the base of the middle unit, 1.4 m of finely laminated limestone rest on the Shacklow Wood Lava in a roadside section [SK 16 NE/20]; the same bed was recorded by Butcher (1975, fig. 3) in Magpie Sough [1775 6920], but in other sections nearby (Figure 14) it was absent. About 27 m above the lava in Little Shacklow Wood, a prominent lens of pale reefy limestone

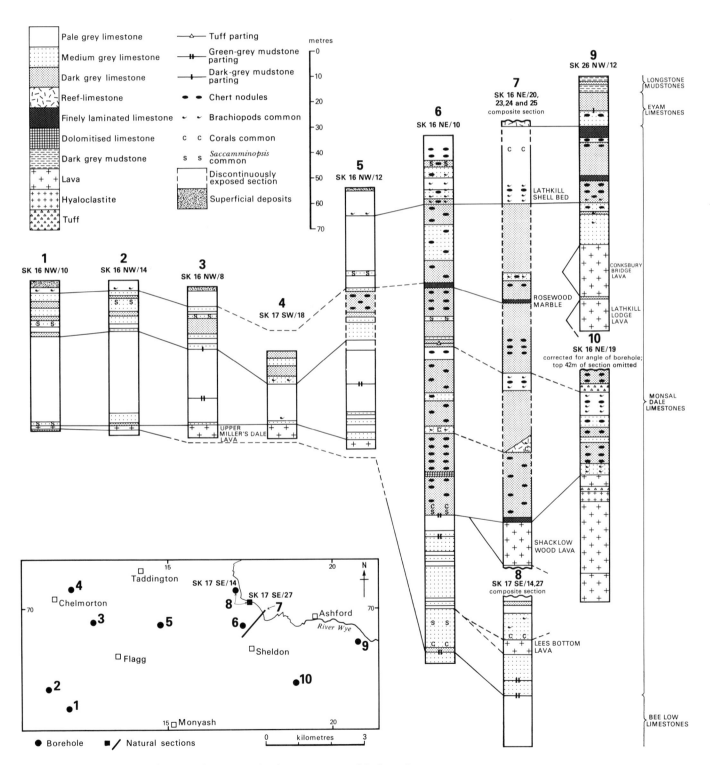

Figure 14   Sections of Dinantian strata in the area west of Bakewell.

[1814 6950] with a rich coral-brachiopod fauna, including *Dibunophyllum bipartitum*, *D. craigianum*, *D. konincki*, *Diphyphyllum lateseptatum*, *Lithostrotion junceum*, *Gigantoproductus edelburgensis*, *G.* cf. *striatosulcatus* and *Productus hispidus*, may correlate with the Hob's House Coral Band (*see above*). The Rosewood Marble, 0.45 m thick in outcrop workings [1794 6916] at the type locality in Little Shacklow Wood, is a finely laminated limestone once used as an ornamental stone (Ford, 1964, p. 184). In the Sheldon Borehole it was 1.69 m thick. At Ashford the same lithology recurs at several levels in the sequence, and the Rosewood Marble, as defined here, probably correlates with a laminated bed 0.3 m thick in Dutch Barn Quarry [1890 6955] near the base of the Rookery Plantation section [SK 16 NE/21]. The bed has been studied in detail by Adams and Cossey (1978), who conclude that it was deposited in no great depth of water, on a slope inclined towards the south-east. The convoluted structures in the bed are explained by synsedimentary slumping movements down the slope.

The highest unit ('upper pale leaf' of Monsal Dale) is generally about 30 m thick but varies greatly in lithology. At the base the Lathkill Shell Bed, consisting of about 6 m of limestone with abundantly shelly bands and some chert, is well exposed at intervals along the top of Great Shacklow Wood [174 696], and the overlying pale limestones are widely exposed on the adjacent plateau between Sheldon and Nether Wheal [150 692]. *Orionastraea* has been collected at several localities, the most notable being in a band about 7 m below the base of the Eyam Limestones on the east side of Deep Dale [1623 6935]; this probably represents the *Orionastraea* Band of the country to the north. South-east of Sheldon, the Lathkill Shell Bed was formerly exposed in the sides of open workings in Magshaw Rake [1820 6791], where 3 m of medium and dark grey cherty shelly limestone rested on dark limestone overlain by paler beds. North of here, the shell bed dies out and the overlying pale beds pass into a dark cherty facies (Figure 12) well exposed in natural sections around Ashford [SK 16 NE/21, 22]. Some of the dark beds have been worked, as Black Marble, for ornamental stone (Ford, 1964). Laminated cherty beds similar to the Rosewood Marble are also present, the highest being taken to mark the local top of the Monsal Dale Limestones. The best exposure [SK 16 NE/22] shows 4.3 m of dark cherty finely laminated limestone with some convoluted structures. The dark facies extends eastwards to the outskirts of Bakewell, where a typical section was proved in the Field House Borehole [SK 26 NW/12] overlying the Conksbury Bridge Lava. Beyond here, however, there is a reversion to paler beds (*see below* and Figure 10).   JIC, IPS

BAKEWELL

Around Bakewell the Monsal Dale Limestones crop out in an anticline extending northwards to the vicinity of Rowdale House. The beds in the northern part of this structure are pale, commonly bioclastic and cherty in places, but north-west of Bakewell pass by interdigitation into the dark beds of the Ashford area (*see above*); around the town there is an exceptional development of laminated limestones and calcilutites, some cherty, and all prone to replacement by chert.

The lowest strata seen include the Conksbury Bridge Lava (p. 61). A section [SK 26 NW/31] below Endcliff Wood, starting immediately above the lava, shows 27.4 m of mainly pale limestone with some chert; dark intercalations are present near the base. The section finishes close to the base of the Eyam Limestones with the *Orionastraea* Band, 1 m thick, at 4.5 m below the top; in addition to *Orionastraea placenta*, the band contains *Diphyphyllum sp.* and *L. junceum*.

To the south-west of Bakewell, the best section in the Monsal Dale Limestones is in a quarry [SK 26 NW/33] near Bank Top House. Here the lowest part of the Eyam Limestones rests on 20.53 m of pale limestones, some cherty, with conspicuous

calcilutite beds, some laminated, with chert and much silica replacement; potholed surfaces are present at several horizons in the section.   IPS

WEST OF MONYASH

South of Chelmorton, the prominent escarpment associated with the Upper Miller's Dale Lava continues southwards, with only small exposures, to the vicinity of Great Low [105 683], and finally fails at the assumed termination of the lava just north of Hurdlow Town [1118 6693].

The overlying predominantly pale crinoidal biosparite sequence has been proved in boreholes [SK 16 NW/8 – 14], four of which are represented in Figure 14. The beds are chert-free; crinoid debris, brachiopods and *Girvanella* are common at some levels, and *Koninckopora* is present in the basal few metres of the formation [SK 16 NW/10, 12 and 14] (p. 57). The lowest beds are best exposed near Chelmorton [1167 6757] and Greatlow [1051 6806], and somewhat higher strata are present in the 18 m section in a disused railway-cutting near Street House [1131 6731 to 1171 6726] (Cox and Bridge, 1977, p. 62).

Dark limestones are present in the top 7.2 m of the last section; the facies, which tends to be micritic and to contain scattered chert nodules in places, forms a unit traceable from above Chelmorton to near Cronkston Grange [120 659], with an isolated outcrop farther south near Cotesfield [137 644]. The unit is also delineated in the boreholes referred to above, except in SK 16 NW/9 and 13 which started below this level.

The remaining upper part of the formation is predominantly medium to pale grey. There are few boreholes through this part of the sequence, the best being SK 16 NW/12 near The Jarnett (Figure 14) in which the Lathkill Shell Bed (*see below*) occurs at a depth of 10.88 m. This marker band is probably represented by small exposures of pale limestone with white chert nodules and abundant silicified brachiopods, widely scattered across the outcrop [e.g. 1296 6554, and 1428 6993]. At four localities south and south-west of Monyash [such as 1448 6540, and 1496 6577], coral bands are present, representing possibly two levels in the uppermost 5 m of the formation; the lower is taken to represent the *Orionastraea* Band. The fauna at the first of these localities includes *Orionastraea placenta* and *Diphyphyllum lateseptatum*, and latter being the dominant coral at all four places. The fauna at the last locality includes many brachiopods of reef affinity, heralding the development of an overlying flat-reef at the base of the Eyam Limestones (p. 42).   NA

LATHKILL DALE AND BRADFORD DALE

The higher parts of the Monsal Dale Limestones crop out over a wide area between Monyash, Over Haddon, Alport and Youlgreave, and there are good sections in the dale sides. Borehole sections (Figure 15) indicate that all but the highest limestones pass eastwards into lavas and tuffs of the Fallgate Volcanic Formation (p. 60). There is little sign of this transition on the map, for only the feather-edges of the lavas come to crop in Lathkill Dale and Bradford Dale; most of the volcanic rocks are concealed by younger sediments.

The oldest strata in the vicinity are exposed in Lathkill Dale [173 656 to 198 661]. The most complete section [SK 16 NE/27] (Figure 15), at Haddon Grove, shows 23.6 m of dark well bedded cherty limestone with argillaceous partings, the Lower Lathkill Limestones of Shirley (1959, p. 413). A coral band near the middle of these beds contains *Aulophyllum pachyendothecum*, *Dibunophyllum bipartitum*, *D. konincki*, *Lithostrotion junceum*, *L. martini* and *L. pauciradiale*: brachiopods are also common in places. To the south, at Long Rake [1715 6398], 30 m of dark cherty limestone were

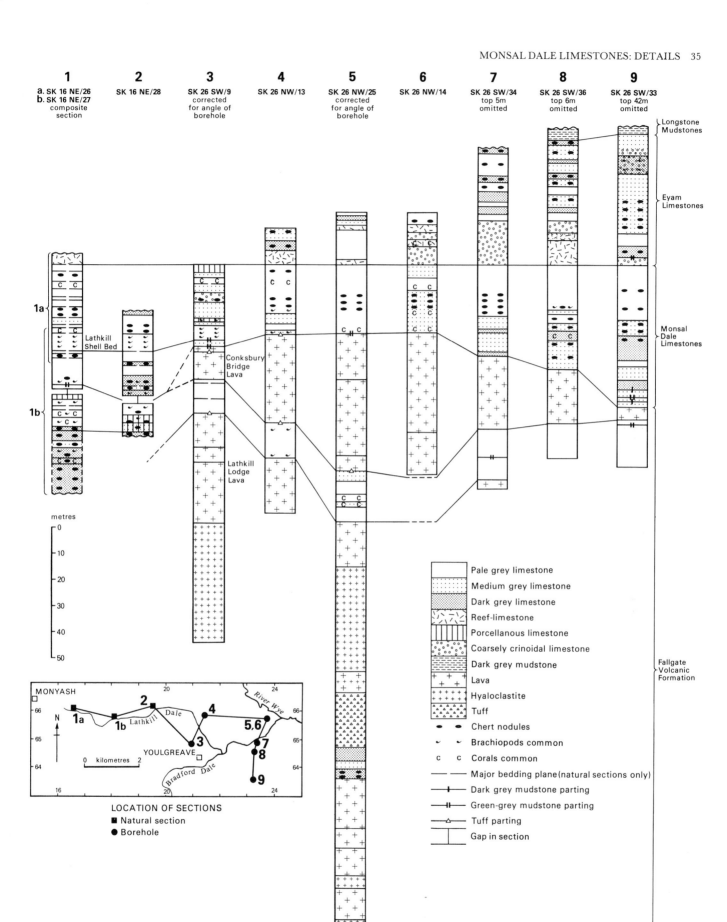

Figure 15   Sections of Dinantian strata in Lathkill Dale and adjacent areas.

proved at depth without reaching the base. South of here a borehole [SK 16 SE/5] proved a similar sequence, though to a lesser depth (Cox and Bridge, 1977, pp. 108–111).

In the Haddon Grove section, the overlying 43.2 m of mainly pale grey bioclastic limestone, with the Lathkill Shell Bed at the top, are well exposed. The lowest 12 m are particularly rich in fossils, with *Dibunophyllum bipartitum*, *Diphyphyllum lateseptatum*, *Lithostrotion maccoyanum*, *L. pauciradiale*, *Lonsdaleia floriformis*, *Palaeosmilia murchisoni*, *Syringopora* cf. *geniculata* and *Gigantoproductus* cf. *sarytscheffi*. West of Over Haddon, the top few metres of the dark limestones are exposed [SK 16 NE/28], overlain by 11 m of paler beds (Figure 15), but farther east the Lathkill Lodge and Conksbury Bridge lavas appear in this part of the sequence (*see above*). Limestones below the Lathkill Lodge Lava are exposed south of Over Haddon, where they are pale and cherty with some calcilutites; a shell bed, 2.6 m thick, is present on the north side of the valley [2026 6619]. Limestones between the lavas are about 30 m thick and are in general poorly exposed, though a section [2058 6609] near the base shows 13.3 m of pale crinoidal and foraminiferal limestone, shelly in part.

In the Haddon Grove section, a laminated pedogenic crust, 17 m above the top of the dark limestones, is overlain by a thin clay wayboard, and this is separated from the base of the Lathkill Shell Bed by 13.4 m of pale thickly bedded calcarenite, shelly at the base. Traced eastwards towards Over Haddon the lower part of the pale calcarenite passes into dark cherty limestone [SK 16 NE/28]: the underlying wayboard is not exposed here but there can be little doubt that it lies at the horizon of the Conksbury Bridge Lava (Figure 15).

The Lathkill Shell Bed is typically a pale grey bioclastic limestone with abundant brachiopods, mainly *Gigantoproductus crassiventer*. Many are packed together in a concave-up position. It is 5 m thick in the Haddon Grove section [SK 16 NE/27], and 7 m in the sections below Ricklow Quarry [SK 16 NE/26] and west of Over Haddon [SK 16 NE/28]. It forms a useful marker horizon (Shirley, 1959, p. 416) along the sides of Lathkill Dale and its tributary valleys, especially Cales Dale, and acts as a correlating link between the sections mentioned. It is present in a borehole [SK 26 SW/9] near Youlgreave, where it rests directly on the Conksbury Bridge Lava, but it dies out to the east of here (Figure 15). To the south it persists and can be traced along the sides of Bradford Dale northeast of Middleton [195 632]. Small exposures of shelly limestone between Middleton, Calling Low [180 650] and Cales Farm [169 644] probably belong in the main to this horizon.

The 26.3 m of strata between the top of the shell bed and that of the Monsal Dale Limestones in the section below Ricklow Quarry [SK 16 NE/26] are mainly pale grey calcarenites. Two coral bands are present, one at the base and another 16.8 m higher in the sequence. A prominent band, 4.4 m thick, characterised by large irregular nodules of pale chert, lies between the coral bands, its base being 6 m above the top of the shell bed. The lower coral band contains *Diphyphyllum lateseptatum* and *Palaeosmilia regia*; the upper one is richer, with *Dibunophyllum bipartitum*, *D. craigianum*, *D. konincki*, *Diphyphyllum lateseptatum*, *Koninckophyllum interruptum*, *Lithostrotion junceum* and *L. portlocki*. Shirley (1959, p. 416) recorded *Orionastraea placenta* from this section, but it is not known in which of the two coral bands it was found. Both bands, with the intervening cherty bed, can be recognised at intervals throughout Lathkill Dale and Cales Dale, as well as in boreholes farther east (Figure 15). Near Conksbury Bridge, a coral band 1.2 m thick is exposed [2094 6571] 10.8 m above the Conksbury Bridge Lava; it is probably the lower of the two bands. Higher beds are well seen [2127 6548 to 2130 6551] in a section 21.49 m thick, of pale limestone with some cherty layers, crinoidal towards the base. Shirley (1959, p. 416) records *Orionastraea* from these strata 'in the cliff in the wood on the left bank of the stream above Conksbury Bridge', but this occurrence has not been confirmed by the present work. A borehole near-

by [SK 26 NW/13] proved the upper coral band, with *Orionastraea*, at a level 19.9 m above the lava. Both coral bands were encountered in a borehole at Nutseats Quarry [SK 26 NW/14], the lower being thicker here than elsewhere, and in two leaves. The lower leaf lay directly on the lava and contained *Nemistium edmondsi*, a coral also found in the lower coral band in a borehole [SK 26 SW/36] near Bowers Hall at a level 10.2 m above the lava (Figure 15); this probably represents the *Lonsdaleia duplicata* Band better developed to the north of the district (Stevenson and Gaunt, 1971, pp. 30, 31). South of these boreholes the only section of note [SK 26 NW/35] is in Shining Bank Quarry (Plate 6), north-east of Alport, where 23 m of pale massive limestone, the lowest 5 m with common chert, underlie the Eyam Limestones.

In Bradford Dale a section [SK 16 SE/16] east of Middleton shows 29.6 m of mainly pale limestone between the top of the Lathkill Shell Bed and the base of the Eyam Limestones. A lower coral band, 9.4 m above the shell bed, is separated by 4.2 m of pale cherty limestone from an upper coral band in dark limestone. These bands are taken to be the two described from Lathkill Dale, though they are closer together here, and the lower band, not the upper, contains *Orionastraea*. Above, a band 0.9 m thick of thinly bedded cherty limestone is present 5.1 m below the base of the Eyam Limestones. This rather unusual lithology increases in importance towards the north-east, for at Moatlow Knob [SK 26 SW/88] the top 10 m of Monsal Dale Limestones contain 4.9 m of cherty thinly bedded porcellanous limestone, with coarse crinoid and shell debris, interbedded with more normal calcarenites containing corals. JIC, IPS

MILLCLOSE MINE

The workings of Millclose Mine extended through the full thickness of the formation beneath the Namurian cover at the eastern edge of the district, between Watts Shaft [2578 6183] and Pilhough [250 650]. The sequence was described by Traill (1939; 1940) and Shirley (1950), and summarised by Smith and others (1967, pp. 27–32). The base of the Matlock Group (now Monsal Dale Limestones) was drawn at the top of the Upper 129 Toadstone, but following a new assessment of the faunal evidence (Chisholm and others, 1983), the line is now drawn at the base of the Lower 129 Toadstone. JIC

PILSBURY, HARTINGTON AND BERESFORD DALE

Monsal Dale Limestones are present in small outliers to the north and south of Hartington (Figure 16), where they occupy erosional hollows cut into the Asbian reef belt, mainly at the junction of the shelf and apron-reef facies. Depositional dips concordant with the floors of the hollows are believed to be mainly responsible for the synclinal arrangement of strata in the outliers, though the possibility of some tectonic folding cannot be excluded. The steepest dips, up to 45°, are found at the western sides of the hollows, where the limestones are banked against the steep sides of eroded apron-reef masses. Shallower dips, nearly concordant with those in the Bee Low Limestones beneath, are found on the eastern sides but here too the basal Monsal Dale Limestones can locally drape over topographical features that result from the erosion of the underlying formation. The sequences in the hollows are laterally variable, but in general consist of interbedded dark, medium and pale grey limestones with abundant shells at several levels. The fauna includes undoubted Brigantian forms. An interpretation of the facies relationships put forward by Ludford and others (1973) was based on the belief that 'shelf' facies (now Bee Low Limestones) and 'marginal' facies (now Monsal Dale Limestones) were both of Asbian age: the existence of the disconformity between them was not recognised.

On the north-eastern flank [1207 6364] of the outlier at Pilsbury (Aitkenhead and Chisholm, 1979) small exposures of dark shelly

Plate 6 Shining Bank Quarry near Alport: thick boulder clay overlies a small thickness of Longstone Mudstones.

Dark thin-bedded Eyam Limestones form the upper 'lift' of the quarry and paler thicker-bedded Monsal Dale Limestones form the lower 'lift' (L 2280).

limestones at the base of the formation overlie pale Bee Low Limestones. A marker band up to 10 m thick, consisting of pale micritic limestone with abundant brachiopods such as *Gigantoproductus edelburgensis* and *Linoprotonia hemisphaerica*, is exposed in small crags overlooking Pilsbury [1200 6329]. This band can be traced three-quarters of the way around the outlier, and when taken as a datum serves to indicate the variation in thickness of the underlying sequence and the relief of the surface on which it rests. Thus in the northern and eastern flanks of the inlier the band closely overlies the Bee Low Limestones; 210 m south-east of Pilsbury it lies about 9 m above the top of an inlier of the apron-reef facies [1195 6325]; in the gully [1209 6325] a little to the east, the sequence is at least 22 m thick. A conglomerate bed up to 1.6 m thick in this gully (*see* section SK 16 SW/15 and Figure 16) contains rounded pebbles of micritic limestone probably derived from the apron-reef by wave action in early Brigantian times. Coral faunas collected from a band immediately overlying the conglomerate indicate a Brigantian (D_2) age as they include *Lonsdaleia?* and *Palaeosimilia regia*. The D_1 age

assigned to these beds by Ludford and others (1973) is thus unlikely to be correct.

Dark limestones forming a thin cover on the Bee Low Limestones occur in a small outlier to the south of Pilsbury and may correlate with similar limestones exposed in a section 1.8 m thick, at Lud Well [1236 6249].

The disconformity at the base of the Monsal Dale Limestones is well exposed [1280 6133] in an outlier south of Banktop. A variable sequence of medium and dark grey fossiliferous limestones with undulating bedding planes rests on a marked erosion surface cut into pale massive Bee Low Limestones. There is up to 11 m of relief on the erosion surface, which is almost vertical in places. Ludford and others (1973, fig. 2) did not recognise the existence of the disconformity, and were forced to explain the distribution of outcrops by faulting.

The sequence in the outlier consists of a laterally variable assemblage of pale, medium and dark grey limestones including lenticular shell beds up to 10 m thick. Faunas collected at the north

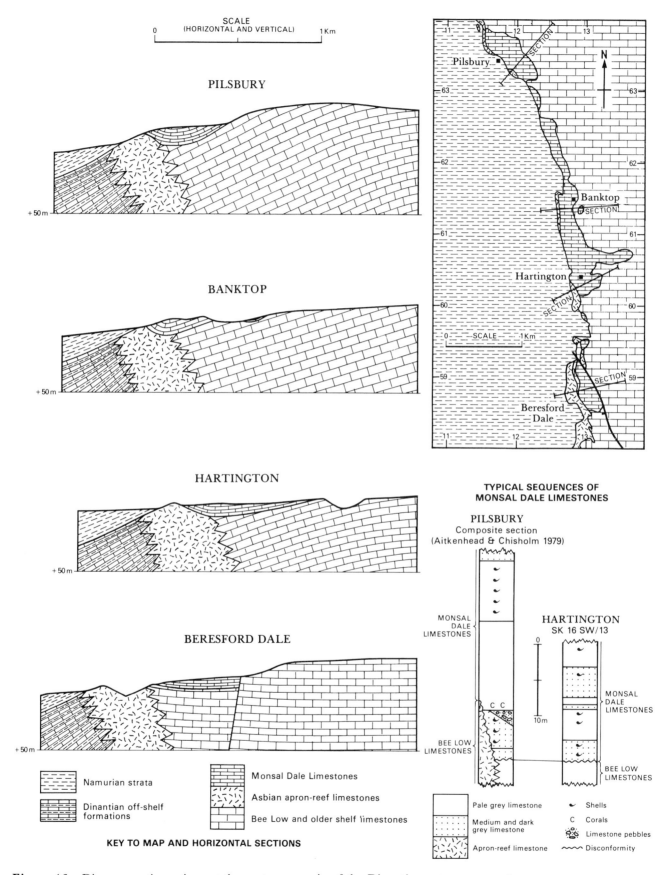

Figure 16   Diagrammatic sections at the western margin of the Dinantian outcrop around Hartington.

end of the outlier [SK 16 SW/14] are of Brigantian age, with *Gigantoproductus edelburgensis, G.* cf. *gigantoides, G.* sp. *maximus* group, and *Linoprotonia hemisphaerica.* At the west side [1260 6159], the Monsal Dale Limestones are banked at angles of up to 45° against reef-limestones from which a rich Asbian fauna has been collected. Part of the dip may be due to subsequent folding at the shelf margin.

Behind the school house at Hartington, a section [SK 16 SW/13] shows 16 m of interbedded pale, medium and dark grey Monsal Dale Limestones containing Brigantian faunas, resting sharply on pale massive Bee Low Limestones; the contact is concordant, but there is no evidence of the interbedding or lateral passage between the formations reported by Ludford and others (1973, p. 219). On the west side of the outlier [1283 6028], near Nettletor Farm, shelly beds at the base of the Monsal Dale Limestones dip at angles up to 30° towards the north-east; they rest on a steep slope at the north end of the Asbian apron-reef outcrop of Bullock Low [129 600].

JIC, NA

Three small outliers are present between Hartington and Beresford Dale; they appear to have a similar relationship to the marginal shelf and apron-reef facies of the Bee Low Limestones as the outliers north of Hartington. There are several exposures, [1300 5979, 1304 5953, and 1304 5857] of up to 19 m of mainly medium grey fine-grained shelly limestone in thin undulating beds. The abundant brachiopods are mainly gigantoproductoids, linoproductoids and *Antiquatonia,* together with a few forms such as *Productus hispidus, Pugilis pugilis* and *Striatifera striata,* which indicate an early Brigantian age.

NA

BIGGIN AREA

Up to 95 m of Monsal Dale Limestones are present in an outlier between Middlemoor Farm [179 569] and Heathcote [146 602]. In a borehole at Biggin [SK 15 NE/4], 45 m of pale and medium grey partly mottled calcarenite, with *Saccamminopsis* sp. in the lower part, rests with a sharp contact on pale Bee Low Limestones. The same beds are exposed in an old railway-cutting north-east of Coldeaton [SK 15 NE/11], where a brachiopod fauna including *Antiquatonia* cf. *insculpta, Gigantoproductus* sp., *Linoprotonia* cf. *hemisphaerica* and *Spirifer bisulcatus* occurs near the base of the formation. The corals *Diphyphyllum lateseptatum, Lithostrotion martini* and *L. pauciradiale* are present at a similar stratigraphical level in a borehole [SK 15 NE/2] (Bridge and Kneebone, 1983) north-west of Uppermoor Farm. Dark grey limestones, locally forming a mappable unit at the base of the formation, have been traced around the head of Biggin Dale where they are well exposed in several places [e.g. 1449 5830, and 1461 5940]. Faunas in a coarsely bioclastic bed, 0.60 m thick, at the first of these localities include *Pugilis pugilis* and *Dibunophyllum bipartitum,* indicating a Brigantian age.

JIC, NA

WINSTER AND ALDWARK

North of the Bonsall Fault, the limestones at outcrop are widely dolomitised, and are known best from boreholes drilled through the cover of Namurian mudstone north of Winster, where the dolomitisation is only partial (Figure 17, A, B, C). The sequence is about 200 m thick and differs little from that known farther east (Figure 17, D; Chisholm and others, 1983). South of the fault only the lowest 130 m or so are preserved; limestones and volcanic rocks, including the Winstermoor and Lower Matlock lavas (p. 60), are well exposed, and parts of the sequence are known also from boreholes.

The limestones below the Lower Matlock Lava average about 65 m in thickness, and are mainly pale and medium grey with darker bands at the base and near the top (Figure 17, A, B, D). At Bonsall, not far beyond the eastern boundary of the district, they pass by interdigitation into a similar thickness of volcanic rocks

(Chisholm and others, 1983) which form the lower part of the Lower Matlock Lava. Towards the south-eastern corner of the district the whole sequence thins, however, and around Upper Golconda Mine [249 551] a thin Lower Matlock Lava rests directly on the Bee Low Limestones.

The basal contact of the limestones is sharp, whether they rest on the Winstermoor Lava or on Bee Low Limestones. In a borehole [SK 25 NW/22] near Aldwark, dark limestones a few metres above the base of the formation contained a fossil assemblage including *Aulophyllum pachyendothecum, Diphyphyllum lateseptatum* and *Koninckopora*; Chisholm and others, 1983). The same beds in a borehole [SK 15 NE/3] at Pikehall yielded a fauna including *Davidsonina septosa* (gerontic form) and *Lonsdaleia floriformis.* At outcrop, the basal beds are seen in small exposures [2003 5723] north of Roystone Grange; they are shelly, with *Gigantoproductus* cf. *crassiventer, G.* sp. *edelburgensis* group, *Megachonetes siblyi* and *Palaeosmilia murchisoni.* In a quarry [1923 5862] at Gotham they contain *Lonsdaleia floriformis, Palaeosmilia murchisoni* and a rich assemblage of reef brachiopods including *Productus productus* (towards *hispidus*).

Slightly higher beds are exposed at Minninglow Quarry [2053 5763], where about 12 m of medium grey limestone, thinly bedded and cherty at the top, contain *Clisiophyllum keyserlingi, Lithostrotion martini* and *Gigantoproductus* cf. *crassiventer.* At Hoe Grange Quarry [SK 25 NW/33] 38 m of pale, medium and dark grey limestone have yielded *Palaeosmilia murchisoni, Gigantoproductus* sp. and *Megachonetes siblyi.* Higher beds are poorly exposed, but have been proved in a borehole [SK 25 NW/22] (Cox and Harrison, 1980, p. 93) south-west of New Barn.

North of the Bonsall Fault, the limestones above the Lower Matlock Lava are all dolomitised at outcrop, but boreholes (Figure 18) show that the sequence consists of between 80 and 125 m of mainly pale limestone, with darker beds above the Lower and Upper Matlock lavas. The latter is extensively developed in the adjacent district (Smith and others, 1967, p. 12). It dies out near Winster but the associated darker limestones persist beyond the margin of the flow (Figure 17, A, B, C). The top of the formation is marked in the boreholes near Winster by a sharp upward change from pale grey limestone to dark; in some of these boreholes (Figure 17, A) a thin band of greenish-grey mudstone is present at the contact, and in one hole [SK 26 SW/66] a thin coal was present above the mudstone.

South of the Bonsall Fault, limestones cap outliers of the Lower Matlock Lava around Aldwark, and are commonly exposed in small crags. A typical example [214 574] shows up to 5 m of medium grey bioclastic limestone with close-set undulating bedding planes. A similar craggy escarpment caps the Shothouse Spring Tuff outcrop north and east of Grangemill. Higher limestones are exposed at Ivonbrook Quarry [SK 25 NW/34], where about 45 m of bioclastic medium and dark grey limestone contain several clay 'wayboards'. The fauna includes *Chaetetes depressus, Diphyphyllum* cf. *fasciculatum, Lonsdaleia floriformis, Gigantoproductus dentifer, G.* sp. *edelburgensis* group and *Linoprotonia hemisphaerica.* A sequence of similar bioclastic limestones, resting at a depth of 57.30 m on the Shothouse Spring Tuff, was proved in a borehole [SK 25 NE/41] near the district boundary.

JIC

EYAM LIMESTONES

A period of emergence ended the deposition of the Monsal Dale Limestones, and the submergence that followed in late Brigantian times (P_2 Zone) led eventually to the formation of the extensive central Pennine deep-water basin of Namurian deposition. The earliest deposits were the Eyam Limestones and Longstone Mudstones, both of P_2 age. The contact between these formations is normally conformable, with

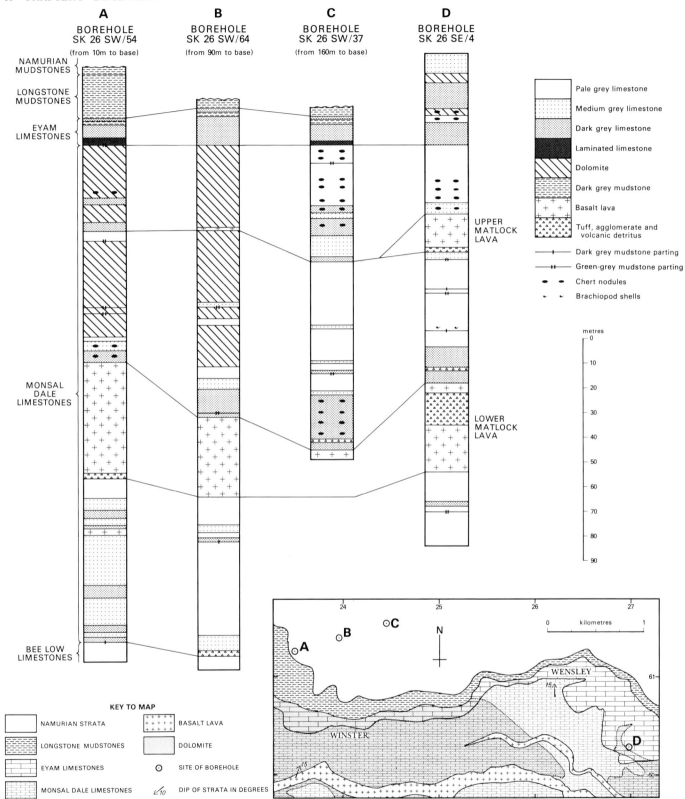

Figure 17 Sections of Dinantian strata near Winster.

Figure 18 Sections of Dinantian strata around Gratton Dale.

interbedding of lithologies in places. Erosion surfaces recorded locally within the sequence (Sibly, 1908, p. 63; Smith and others, 1967, p. 23) are due probably to scour by currents and are not considered to have any regional importance. However, an emergent surface within the Eyam Limestones (p. 44) is probably related to a minor oscillation of sea level superimposed on the main transgression. Though the Eyam Limestones, the Longstone Mudstones and the overlying Namurian mudstones lie in conformable sequence, each unit rests in places on the surface of the Monsal Dale Limestones. The relationship is most satisfactorily explained by overlap of the older beds by the younger against an uneven surface, the late Brigantian disconformity, though lateral passage between Eyam Limestones and Longstone Mudstones may also play a minor part. Knoll reefs, formed soon after the onset of the transgression, complicate the outcrop pattern for they, too, give rise to upstanding masses against which later mudstones overlap earlier beds. Knolls may thus, as at Hassop, be surrounded at the present level of erosion by Namurian mudstones. In the west of the area the same late Brigantian disconformity is believed to be responsible for the overstep of Namurian mudstones on to Monsal Dale and Bee Low Limestones of the apron-reef belt (p. 70). At Buxton, the Eyam Limestones rest in most places on Miller's Dale Limestones, though high Monsal Dale Limestones are present locally, apparently preserved in a hollow in the surface of the unconformity.

A formal definition of the Eyam Limestones, with type sections, has been given by Aitkenhead and Chisholm (1982). In the present area, the formation is up to 50 m thick. The limestones are of two main types, the bedded facies and the reef facies (Figure 19); both are of shallow-water origin. The bedded facies is the more widespread and consists of variable dark and medium grey limestones, commonly thinly bedded and cherty, with some pale beds. The reef facies generally forms small knolls ('buildups') in which a core of pale rubbly vughy fine-grained and poorly bedded limestone with a characteristic brachiopod fauna is surrounded by bedded flank deposits of coarsely crinoidal limestone. Most of the knoll-reefs are found at the base of the formation but there are some at higher levels, as near Monyash. A caliche crust has been found on two of the knolls, indicating emergence, but the regional extent of this phenomenon is not known. In the present district the distribution of the knoll-reefs appears to be random.

Emergence prior to the deposition of the Eyam Limestones is indicated by the occurrence near Winster (Figure 19) of a thin coal at the base of the formation, a feature recorded also in adjacent districts (Smith and others, 1967, p. 29; Stevenson and Gaunt, 1971, p. 98).

DETAILS

Buxton

On the northern edge of the district, dark Eyam Limestones are present at Fairfield, where they rest unconformably on Miller's Dale Limestones. At St Peter's Church [0663 7414], a knoll-reef at the base of the formation yields a P_2 brachiopod fauna similar to

that described (Stevenson and Gaunt, 1971, p. 105) from Brook House just to the north of the district. A section [0640 7382] by Fairfield Road shows a band of grey limestone 0.99 m thick, with clisiophyllids and gigantoproductoids on 2.54 m of dark cherty limestone.

In Buxton, an area of dark, mainly thin-bedded cherty limestones, shown on the 1978 edition of the 1:50 000 map as Monsal Dale Limestones, is now referred to the Eyam Limestones as it overlies beds close to the top of the Monsal Dale Limestones A section [SK 07 SE/16] in Holker Road shows 8.08 m of dark cherty limestones and pale limestones with a coral fauna at the top and a shell bed, 0.71 m thick below, with a fauna including the P_2 form *Gigantoproductus elongatus*. To the west of Buxton, the formation is present in the anticline west of Burbage.

GREAT LONGSTONE TO BAKEWELL

In this area the Eyam Limestones range up to about 20 m in thickness but are locally absent, probably as a result of overlap by the overlying mudstone formations.

In an isolated outcrop at the eastern end of the Longstone Edge Anticline a typical section in an openworking [2361 7405] on Red Rake shows 3 m of dark thinly bedded limestone. These beds are overlapped to the south at Hassop Common. At Hassop several knoll-reefs occur as inliers surrounded by mudstones of probable Namurian age. The knolls show quaquaversal dips and are commonly crinoidal as well as shelly. One [2240 7276] is peculiar in containing a substantial development of limestone with corals including *Clisiophyllum keyserlingi* and *Dibunophyllum bipartitum*. To the north of Hassop the belt of knolls extends on to the outcrop of the Monsal Dale Limestones as a number of outliers: a tendency for reef-limestones to form at the base of the Eyam Limestones has been noted in the ground to the north (Stevenson and Gaunt, 1971, p. 32). The dark bedded facies reappears near Rowland, and is seen in small exposures [such as 4.22 m at 2046 7191]; however, in the Buskey Cottage Borehole [SK 27 SW/26] these limestones are absent (p. 45). At Little Longstone the limestones thin, apparently by lateral passage into the Longstone Mudstones. In a disused railway-cutting [SK 17 SE/23] the total thickness is exposed as 7.09 m of dark thinly bedded limestone with a mudstone band, 0.5 m thick, containing bivalves and trilobite fragments.   IPS

At Ashford about 20 m of the bedded facies of Eyam Limestones are present [SK 16 NE/22] above the laminated cherty bed (p. 34) which has been taken to mark the top of the Monsal Dale Limestones in this area. The lowest 2.5 m are exposed south-west of Ashford, resting on the laminated bed, the section [1916 6938] showing dark grey fine-grained limestone, in beds up to 0.5 m thick, separated by dark argillaceous partings. The top is exposed [1994 6935] south-east of Ashford; 2.1 m of dark limestone with bands of dark grey mudstone up to 0.7 m thick are overlain by Longstone Mudstones. A borehole [SK 26 NW/12] between Ashford and Bakewell (Figure 19) proved 13.41 m of Eyam Limestones. Graded beds made up the lowest 1.91 m, the remainder being dark and medium grey fine-grained limestones.   JIC

Around the Bakewell Anticline the formation is thin and is locally overlapped or replaced by Longstone Mudstones. In the north-east, a knoll-reef occupies nearly the full thickness; it has been worked in Cracknowl Quarry, where a section at the south end [2144 7035] shows 6.1 m of reef-limestone, massive and shelly in the lower half, better bedded and with some chert above. The reef-limestones are overlain by 0.99 m of dark bioclastic cherty limestone. A smaller knoll-reef forms an outlier [215 696] near Eweclose. Farther south, and in Bakewell itself, the Eyam Limestones are absent and Longstone Mudstones rest directly on the Monsal Dale Limestones. They are, however, present in an outlier [215 688] above the town.

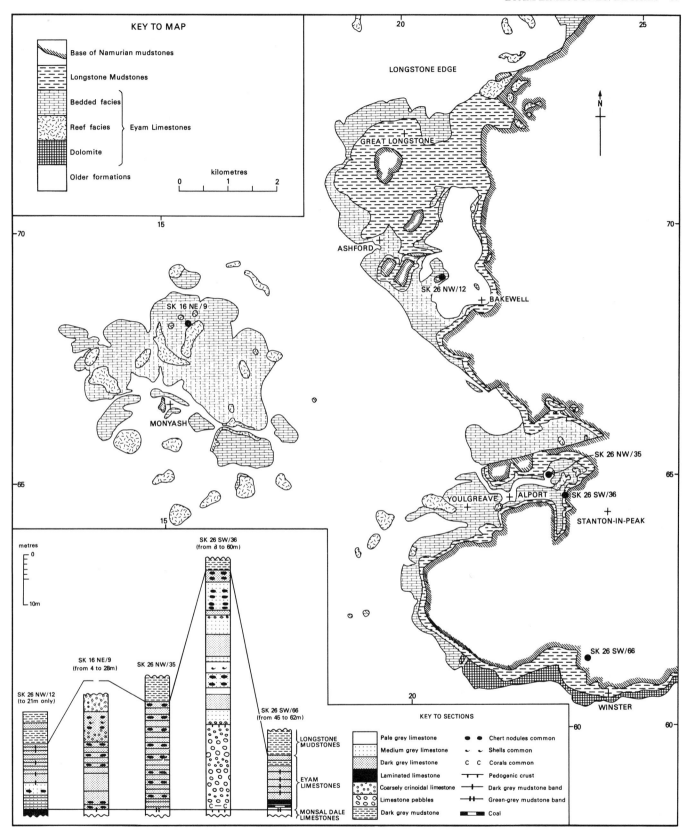

Figure 19   Outcrop distribution of the Eyam Limestones and Longstone Mudstones, with sections.

They reappear south-west of Bakewell and are best seen in a quarry [SK 26 NW/33], where 2.97 m of dark limestone with mudstone partings overlie Monsal Dale Limestones. IPS

MONYASH

Outliers of Eyam Limestones occupy the central part of the Monyash Syncline. The sequence is up to 30 m thick, and bedded and reef facies are both present.

Most of the knoll-reefs lie at the base of the sequence, the best exposed examples being in Lathkill Dale, around Ricklow Quarry [1646 6615]. At the west side of this, a core of pale rubbly-weathering limestone, with vague bedding dipping at angles up to 30°, is underlain by coarsely crinoidal reef-flank facies. To the east the core facies appears to pass laterally into flank facies, which has a caliche crust developed on it (Adams, 1980). The flank facies extends beyond the quarry to link up with neighbouring knolls. Flat-bedded limestones overlie the reef-flank deposits in the quarry and wedge out against the knoll core, but details of this contact are not exposed. The bedded limestones consist of pale, medium and dark grey calcarenites with some chert, the top 2 m with abundant productoids. A smaller reef is exposed in the dale side about 400 m east of the quarry: the flank deposits wedge out against the reef core, which lies in a slight hollow eroded into the top of the Monsal Dale Limestones and has yielded a rich brachiopod fauna. Bedded limestones, with productoids at the top, lap against the knoll. Similar knoll-reefs are scattered across the surrounding plateau, many of them as outliers separated from the main outcrop.

North of Monyash, the reef facies is found higher in the sequence. Roadside exposures [1479 6786 and 1487 6769] of the bedded facies, comprising dark grey calcisiltites with chert nodules and lenses and irregular bedding planes with dark shaly partings, pass up into a more varied section, 13 m thick, in a disused quarry [1491 6774] (figured in part by Cox and Bridge, 1977, plate 3). Here beds of reef core, reef flank and bedded facies are seen in a complex interfingering relationship.

A borehole [SK 16 NE/9] (Figure 19), north of the above quarry, proved 26.5 m of medium and dark grey bedded limestone, with 0.10 m of mudstone at the base, resting on Monsal Dale Limestones. The top 12 m of the Eyam Limestones contain bands of shells, coarse crinoid debris and small limestone pebbles, probably derived from a reef. JIC, NA

HADDON FIELDS, ALPORT AND YOULGREAVE

The formation is absent, probably by overlap, north of Haddon Fields but elsewhere the bedded facies is up to 47 m thick. Where the reef facies is present, normally at the base of the sequence, the maximum known thickness is 25 m.

Typical sections are shown in Figure 19. There are exposures of both facies at intervals along Bradford Dale between Youlgreave and Alport, and in Lathkill Dale between Conksbury Bridge [212 656] and Bowers Hall [235 649].

The bedded facies is best seen in Shining Bank Quarry, north-east of Alport (Plate 6). The section [SK 26 NW/35] (Figure 19) shows 18.5 to 21 m of dark thinly bedded cherty limestone on 0.03 to 0.04 m of dark calcareous mudstone on 0.40 m of grey clay, above the Monsal Dale Limestones. The Longstone Mudstones lie conformably above. A section [SK 26 SW/92] comprising 15.7 m of similar limestone can be seen in the river bank below Bradford, but the base and top are not exposed. The lowest beds are also seen on the east side of Lathkill Dale in a section 10.85 m thick [2155 6512] of dark cherty limestone with a band, 0.6 m thick, containing *Lithostrotion junceum* in the upper part. In Stanton Mill Quarry [SK 26 SW/91] 7.7 m of well bedded medium and dark grey cherty

limestone rest on 8.7 m of paler massive coarsely crinoidal limestone of the reef-flank facies. A disconformity at the base of the overlying mudstones, illustrated by Sibly (1908, p. 62), is probably due to local scour; it is unlikely to have any regional or tectonic significance.

A knoll of reef-limestone is well exposed, near Alport, at Rhienstor Rock [2187 6441], a crag about 24 m high consisting of pale massive shelly limestone of typical reef-core facies. Another mass exposed in crags [2324 6488] by the River Lathkill, near Shining Bank Quarry, is at least 25 m thick and contains both core and flank facies. The latter, as seen in a borehole [SK 26 SW/36] (Figure 19), is made up of coarse crinoid debris and poorly rounded pebbles of reef-core limestone up to 6 cm across. The reef is capped by an emergent surface, shown by a rind, 5 mm thick, of brown laminated micrite with associated rootlet tubes.

Farther north, around Haddon Hall [235 663], knoll-reefs of very variable size have been mapped, some projecting up through the mudstone cover. The Hall is built on one of these, a typical section [2346 6636] showing 1.9 m of crinoidal limestone with a little chert. Nutseats Quarry shows a section 14.6 m thick [2373 6578] in dark to pale grey crinoidal and shelly limestones, grey cherty limestones and dark grey cherty limestones. The inclusion of bands of reef-flank facies among these bedded limestones may indicate the former presence of a knoll-reef, now quarried away. A borehole [SK 26 NW/14] (Figure 15) in the floor of the quarry proved a further 5 m of the bedded facies overlying 15 m of reef-core and reef-flank facies. JIC, IPS

ELTON AND WINSTER

Along the south and west sides of the Stanton Syncline the formation varies little, consisting generally of 10 – 15 m of dark grey thinly bedded sparsely cherty limestone with dark grey mudstone bands up to about 20 cm thick. Typical sections are shown graphically in Figures 17, 18 and 19. Most of the outcrop is dolomitised but there are unaltered areas west of Gratton Dale.

The best sections are in boreholes north of Winster (Figures 17, 19), where the lowest 1.2 m of limestone are finely laminated with convoluted structures, the same lithology as the Rosewood Marble (p. 34). The laminated beds rest either on a greenish-grey mudstone wayboard above the Monsal Dale Limestones or directly on these limestones, except in borehole SK 26 SW/66 (Figure 19), where a seam of bright coal 2 cm thick is present at the top of the wayboard. A similar sequence was recorded in the Millclose Mine workings, which straddle the district boundary north-east of Winster (Smith and others, 1967, pp. 29–30). Farther east the sequence at outcrop thickens and the lithology becomes more varied (Smith and others, 1967, p. 14). An example is borehole SK 26 SE/4 (Figure 17).

West of Winster, discontinuous exposures at Dale End [SK 26 SW/89] (Figure 18) show the lowest 3.5 m to be dark with argillaceous bands, overlain by 8.5 m of medium and pale grey limestone. There is no clay wayboard at the base, the boundary being drawn at the bottom of the argillaceous beds. JIC

LONGSTONE MUDSTONES

The formation, up to 20 m thick, consists of dark grey mudstone, calcareous in places, with marine faunas. The type area is around Great Longstone and Little Longstone (Aitkenhead and Chisholm, 1982, p. 6); the relationship to the Eyam Limestones is discussed on p. 42. The top is defined by the appearance of the basal Namurian goniatite band containing *Cravenoceras leion*.

DETAILS

The formation reaches a thickness of some 20 m in the area south of Longstone Edge but is poorly exposed. The lowest part is seen in the disused railway-cutting [SK 17 SE/23] at Little Longstone, where 4.3 m of dark mudstone with bands of dark fine-grained limestone rest on the Eyam Limestones; the mudstones yield *Neoglyphioceras sp.* The top is defined by the presence of the basal Namurian *C. leion* Band in the railway-cutting near Thornbridge Hall (p. 70), though the underlying mudstones are not exposed here. South-east of Great Longstone, the Buskey Cottage Borehole [SK 27 SW/26] proved Longstone Mudstones between the base of the boulder clay at 17.23 m and the top of the Monsal Dale Limestones at 34.64 m, the Eyam Limestones being absent. Near Churchdale Hall an exposure [2063 7013] shows 0.3 m of dark calcareous mudstone with *Lyrogoniatites sp.* and *Sudeticeras sp.*

Near Ashford, the base is seen in an old quarry [SK 16 NE/22], where 1.0 m of mudstone with dark grey limestone bands rests on Eyam Limestones, which here are dark grey with bands of mudstone. A trench section nearby [2000 6939] showed the full thickness of 15.0 m. Calcareous bands near the base contained *Posidonia membranacea* and *Sudeticeras sp.*, and the top was defined by a calcareous band with *C. leion*. A small outlier west of Bakewell [206 690] has been mapped on the evidence of a borehole [SK 26 NW/12] (Figure 19), which proved 6.76 m of dark grey fossiliferous mudstone, with a 1.3 m band of dark limestone near the middle, resting on Eyam Limestones.

To the east of the Bakewell Anticline and south to Long Rake, the Longstone Mudstones form a narrow outcrop whose presence is inferred from exposures of the *C. leion* Band (p. 70). On the south side of Long Rake the openworking at Raper Mine [2169 6523], now filled, proved a full thickness of 14.26 m of mudstone with a limestone 0.4 m thick containing *Sudeticeras sp.* at 4.66 m above the base. In Shining Bank Quarry (Figure 19, SK 26 NW/35Œ 5 m of dark calcareous mudstone with thin limestone bands and a fauna including *Neoglyphioceras sp.* and *Sudeticeras sp.* overlie the Eyam Limestones (Plate 6). The top is not exposed.

Between Haddon Fields and Stanton the full thickness has been proved in boreholes. One of these [SK 26 NW/21] showed 8.48 m of dark fissile mudstone, fossiliferous in part, between the top of the Eyam Limestones and the *C. leion* Band. Another [SK 26 SW/46] proved 7.75 m of partly calcareous mudstone with *Posidonia corrugata* and *P. membranacea* at the base, and *Sudeticeras?* at the top, overlain by calcareous mudstone with *C.* aff. *leion*. Boreholes farther south gave a thickness of 9.07 m [SK 26 SW/33] north of Eagle Tor, and of 19.78 m [SK 26 SW/57] near Winster. The outcrop around the Stanton Syncline has been drawn on the basis of these figures.

JIC, IPS

STRATIGRAPHY OF OFF-SHELF PROVINCE

MILLDALE LIMESTONES

The Milldale Limestones, re-defined as a formation by Aitkenhead and Chisholm (1982, p. 10), include the oldest rocks cropping out in the district and are found in two anticlinal inliers around the Dove and Manifold valleys (Figure 3). Strata now assigned to the Milldale Limestones were penetrated by the Gun Hill Borehole [SJ 96 SE/18] (Hudson *in* Hudson and Cotton, 1945), in the axial part of the Gun Hill Anticline 13 km north-west of the Manifold valley inlier, from a depth of 619.35 m to the bottom at 1410.90 m. Around Alstonefield, the upper part of the formation is inferred to pass laterally into the Woo Dale Limestones, which here include the Iron Tors Limestones.

A thickness of about 305 m of Milldale Limestones is estimated to be present at outcrop in the Dove valley and about 218 m in the Manifold valley; the Gun Hill Borehole proved 791.55 m. The base of the formation is known only from the Caldon Low Borehole in the Ashbourne district (Institute of Geological Sciences, 1978, p. 11; Aitkenhead and Chisholm, 1982).

The present work has largely confirmed the broad stratigraphical conclusions of Parkinson (1950) and Ludford (1951), but there are marked differences in detail and emphasis which may be seen by comparing the relevant parts of their composite 1964 map (Parkinson and Ludford, 1964) with the current BGS 1:50 000 sheet.

Lithology, facies variation and environment of deposition

The Milldale Limestones consist essentially of knoll-reefs or buildups (p. 7) and inter-reef (off-reef) sediments, that pass laterally and vertically into each other.

The knoll-reefs crop out extensively around the Manifold valley, and to a lesser extent around the Dove valley, where the main outcrops lie south of the district (Aitkenhead and Chisholm, 1982, fig. 5). The characteristic rock type is a poorly bedded or massive, medium grey, very fine-grained or porcellaneous limestone with rather irregular masses of sparry calcite formerly described as 'reef-tufa' (Parkinson, 1957, p. 516). These are mostly from about 0.5 to 15 cm across, and either have an elongate form, giving a banded appearance to the rock, or occur in small irregular masses crudely aligned to the inferred dip direction. Small flat-bottomed forms known as *Stromatactis* are similar to those illustrated by Lees (1964, fig. 22). Since the work of Bathurst (1959) it has become widely accepted that most of these structures were originally cavities, and Bathurst (1982) has since concluded that these were probably formed by erosion beneath successive submarine lithified crusts which were subsequently disrupted by bioturbation, compaction and shearing.

Petrographical studies by Ludford (1970) showed that the knoll-reef limestone is composed largely of calcilutite and calcisiltite with some coarser bioclasts including crinoid debris, foraminifera tests and ostracod shells plus reef 'tufa'. Morgan (1980), in a study of 46 thin sections mainly from knoll-reefs ('mounds') around the Manifold valley, found that overall they comprise 20.5 – 27.3% micrite, 19.3 – 22.7% microspar (slightly coarser recrystallised micrite), 12.1 – 12.7% clotted micrite in the form of small peloids, 11.6 – 23.2% bioclasts, and 14.4 – 36.6% sparry calcite, including radiaxial fibrous spar (Bathurst, 1959) filling cavities and growing on and in fossils.

Thach (1964) recognised growth forms similar to those described by Lees (1964) from the Waulsortian 'reefs' of Eire, and concluded that the larger masses had either grown by lateral and vertical accretion from one simple knoll or, as in the case of Wetton Hill, resulted from several simple knolls growing so large that they merged together to form a large composite mass. The late Tournaisian to early Viséan Waulsortian reefs at the type locality in Belgium appear to be of similar facies to those of near equivalent ages in Eire and England, including the present district. Detailed work by Lees and others (1977) in Belgium and the comparative study by Morgan (1980) have tended to confirm this similarity, but a great deal of detailed petrographical work still needs to be done on each knoll-reef before the term 'Waulsortian' can be confidently applied. Moreover, as these authors point out, there are still major unsolved problems regarding the origin of these 'reefs', particularly the origin and mode of accumulation of the lime mud.

The only fossil organisms found in any quantity which may have had a sediment-trapping effect are fenestellid bryozoans and crinoids. However, neither of these seems to occur in sufficiently dense concentrations to be effective. In the absence of a clearly preserved organic framework to support the mounds of lime mud, it has long been assumed that algae were responsible for trapping and holding the sediment (Prentice, 1951, p. 179). Delicate lamination of possible algal origin has been noted in many exposures, but neither Lees in Belgium nor Morgan in the present area have been able to find algal microstructures to confirm the field evidence. This absence has led these authors to suggest that the mounds developed in relatively deep water in the aphotic zone. Morgan has cited the presence of clotted micrite and cavities lined with radiaxial fibrous calcite as evidence of early submarine lithification (*see also* Bathurst, 1982). Further evidence of this early submarine lithification is provided by Morgan's semi-quantitative palaeoecological study of comprehensive collections of faunas from 24 localities around the Manifold valley, where it was concluded that 66% of the mainly brachiopod-bivalve genera were adapted for life on firm substrates, 19% for soft substrates and 16% had indeterminate preferences. It should be noted that seven of these localities are on the Wetton Hill east knoll-reef which is thought to be at least partly of Asbian age (p. 49). The concentrations of shells, particularly brachiopods, at certain places or pockets on these reefs is one of their most noticeable features, but the reason for such concentration is not clear.

The inter-reef sediments occur around and between the reef masses. Dark grey limestones form mappable units in small areas in the Dove valley and they predominate in the Manifold valley where they correspond in part to the 'Cementstone Series' of Prentice (1951). These limestones are predominantly thinly bedded calcisiltites or calcilutites with only minor bioclastic laminae containing crinoid debris. Macrofossils are generally rare but are present in a few beds. Shaly intercalations and irregular chert nodules are common, and partially silicified and dolomitised limestones also occur sporadically. Sandy limestones are recorded by Hudson (*in* Hudson and Cotton, 1945, p. 328) in the lowest 75.9 m of the Gun Hill Borehole. Quartz grains, together with fresh feldspar, limonite and chlorite of possible volcanic origin were also noted by Prentice (1951, p. 175) in the lowest beds ('Cementstone Series') between Thors Cave and Waterhouses in the district to the south. The age and correlation of these beds and those in the borehole are uncertain, but both may represent terrigenous influxes at an early stage in Dinantian sedimentation before the sea had transgressed to its full extent.

The inter-reef facies probably represents deposition from mainly weak bottom currents in relatively deep water with only occasional periods when stronger currents brought in enough nutrients to sustain benthonic faunas.

Well bedded medium grey peloidal bioclastic limestones, with conspicuous crinoidal laminae and lenses, comprise the bulk of the inter-reef facies in the Dove valley inlier. They were probably deposited more rapidly at somewhat shallower depths than were the dark lithofacies. Crinoid thickets grew periodically and were flattened, winnowed, and dispersed by intermittent bottom currents.

As Lees and others (1977) have pointed out in their study of the Belgian Waulsortian facies, the reefs themselves appear to have had little or no direct influence on their immediate surroundings. There is no clear evidence in the inter-reef facies of the Manifold valley, for instance, of coarse material derived from the reef slopes. Exposed unfaulted contacts between the knoll-reef and inter-reef facies are uncommon; they appear to be gradational over a few metres at the lowest parts of the reef margin, whereas at somewhat higher levels around Wettonmill the well bedded inter-reef limestones overlap against the reef surface. At higher levels, in the Dove valley, the transition appears to be gradational. The nature of the contact depends on such factors as the rate of reef growth relative to that of accumulation of the inter-reef sediments, the degree of early lithification, and the ability of bottom currents to deposit sediment or to winnow and erode the sediment surface from time to time.

DETAILS

MANIFOLD VALLEY

Inter-reef facies The outcrop of the dark lithofacies in the Manifold valley and its tributary south and east of Wettonmill [0996 5638 to 0985 5503] contains probably the oldest limestones exposed in the district (*but see below*). The beds lie mostly on the western limb of a north–south-trending anticline squeezed between two elongate knoll-reef masses. They clearly underlie the reef-limestones exposed immediately to the west, but the relationship with those to the east is obscured by faulting. The lithology comprises thinly interbedded dark grey calcisiltites and grey crinoidal calcarenites with some thin shaly partings. Irregular chert nodules and silicified bioclasts are fairly common, especially in the dark beds. These limestones are best seen in a section 21 m thick in the valley east of Wettonmill [0983 5607] beneath a small exposure of knoll-reef. Farther south they are seen in the bed and left bank of the River Manifold [SK 05 NE/9] where a section 28.40 m thick includes calcarenites in the top 4.40 m, made up largely of finely comminuted crinoid debris and with foraminifera including cf. *Eblanaia sp.* and *Paraendothyra sp.*, indicative of a Chadian or Tournaisian age. The crinoidal beds pass rapidly and irregularly into crinoidal biomicrites with fenestellids, and then into massive micrite of the Thors Cave knoll-reef containing dispersed crinoids and brachiopods. An exposure on the river bed [0963 5500] just south of the district in the same reef mass shows many fenestellids encrusted with fibrous calcite. The lower part of this reef, is probably similar to the Waulsortian facies of Belgium (Lee and others, 1977). An unusual fauna of silicified brachiopods from roadside exposures [098 554] and loose blocks between the above localities has been recorded by Brunton and Champion (1974). The content and preservation of this fauna is comparable with that from the Tournaisian part of the sequence at Brownend Quarry near Waterhouses, in the Ashbourne district (Morris, 1970).

Strata lying at a higher stratigraphical level, lateral to the Wetton Hill west knoll-reef mass, are well exposed at only one locality [1009 5522], a roadside cut west of Wetton. They consist of 11 m of thinly interbedded dark grey fine-grained and grey crinoidal limestones with chert nodules and a few orthotetoid and megachonetid brachiopods, together with foraminifera indicative of a Chadian age (*see also* Parkinson and Ludford, 1964, p. 174).

A junction between inter-reef beds and a knoll-reef is exposed behind the house and outbuildings at Wettonmill [0958 5612]. The former comprises 6 m of dark grey uniformly fine-grained limestone with chert nodules, in parallel sharp-based beds up to 40 cm with some weak internal lamination and dark shaly intercalations up to 5 cm. The beds are near horizontal but are turned up slightly towards the sharply defined surface of the massive grey micritic reef-limestone which is inclined at about 50°. A zone, about 30 cm wide, of disturbed shale with broken chert nodules, separates the reef from the undisturbed inter-reef beds and is indicative of minor movement between the two. However, onlap of the latter on to the former in relatively deep water is the most likely relationship. Both facies have yielded the conodont *Scaliognathus anchoralis*, indicative of a late Tournaisian (Ivorian) age. Similar limestones with shaly intercalations are exposed in the bed of the River Manifold and at the sides of tracks north-west and north of Wettonmill.

In the inlier at Ecton, the best exposure occurs in a roadside adit [0956 5804]; here the section comprises about 15.0 m of dark grey thinly bedded rather shaly limestone with one massive bed, 0.75 m thick, lying about 1.0 m above the base and forming the roof of the adit. The massive bed contains a sparse non-diagnostic brachiopod fauna but has yielded the Tournaisian (Ivorian) conodont *Polygnathus communis carina*. The section lies only about 15 m below coarser, thicker bedded limestones at the base of the overlying Ecton Limestones. Nearby, in Clayton Level [adit at 0961 5808], a 47 mm band of tuffaceous gritty clay [E42631] (*see* Appendix 3) lies in a position estimated to be slightly below the base of the above section. Berridge and Siddiqui (written communication) conclude that this is a reworked basic tuff consisting mainly of kaolinite both in the clasts and clay matrix, the whole being transected by a stockwork of calcite veinlets. Relict volcaniclastic structures such as pseudomorphs after shards, amygdales and possibly feldspar microlites and phenocrysts are visible in some grains.

Knoll-reef facies Amalgamated knoll-reefs form two irregular elongate masses around the core of the Ecton Anticline; a western outcrop extending 2.4 km from near Top of Ecton [101 573] to the southern boundary of the district [097 550], and an eastern one extending 1.4 km SSW from near Manor House [104 565]. There are extensive exposures of the poorly bedded grey micritic limestone on the hillsides [such as 102 570, 097 562], and steep cliffs of this rock are present at a lower level above the river [095 561, 097 559, 096 557 and 096 554]. The convex outer margins of the indented western boundary probably reflect the form of individual knolls which have coalesced to form the larger mass. In a few places, such as Sugarloaf [098 578], this is confirmed by quaquaversal dips. The eastern boundary appears to coincide closely with the base of this reef mass, the dips being mainly to the west or north-west; however, the only clear exposure of the junction is in the bed of the River Manifold in the extreme south [0985 5504]. The eastern outcrop, which includes Wetton Hill, is bounded on its west side by a probable reverse fault along half of its length, implying that this knoll-reef mass lies at a somewhat higher stratigraphical level than the western outcrop. This is supported by the presence of the goniatite *Dzhaprakoceras sp.*, indicative of a Chadian – Holkerian age, at two exposures [1024 5648 and 1029 5551]. Both these exposures also yielded ten to fourteen brachiopod genera and the second included in addition the corals *Clisiophyllum cf. multiseptatum* and *Cravenia rhytoides*, diagnostic of the Arundian Stage. In general, the dip directions from the scattered exposures on Wetton Hill [104 562] are similar to those of present slopes, and are assumed to be quaquaversal. However, near the foot of the slope south of Manor House [1048 5657], the reef biomicrites are partially covered with a thin drape of poorly cemented coarsely crinoidal limestone. According to

Lees (oral communication, 1978), the orientations of *Stromatactis* cavities indicate that the reef was eroded somewhat prior to the deposition of the crinoidal bed and that therefore, the latter is not a flank facies of the reef but a later deposit (p. 51).

DOVE VALLEY

Inter-reef facies In this area, in contrast to the Manifold valley, the predominant inter-reef lithology of the Milldale Limestones comprises medium grey fine-grained calcarenites with thin lenses and irregular laminae of crinoid debris. Beds are usually thin and irregular and scattered chert nodules are sporadically present. Such rocks occur in many scattered exposures on the steep sides of the Dove valley below Coldeaton Bridge, and are best exposed in a 32.5 m roadside section immediately west of Lode Mill [1450 5509], where they pass up into the knoll-reef facies. On the opposite side of the valley, the 50 m crag of Shining Tor [146 550] is formed of similar limestones but with many chert nodules in the lower 25 m. The apparent lack of bedding in the upper part of the section is probably due to weathering rather than a real absence of bedding planes.

The dark lithofacies comprises only a subordinate part of the sequence, forming restricted outcrops at Coldeaton Bridge [146 561] below Iron Tors, at Fishpond Plantation [146 554] (Plate 7), and near Lode House [142 551]. The limestones in the first of these areas form an inlier within the paler inter-reef sequence and are inferred on structural grounds to be the oldest in this part of the dale. Crags south-east of the bridge [1461 5608] provide a section 30 m thick consisting of thinly bedded dark grey calcisiltites and fine-grained calcarenites with irregular chert lenses and nodules, dark shaly intercalations, and scattered brachiopods. These beds pass up into paler, more bioclastic, peloidal calcarenites with a more plentiful fauna including *Levitusia humerosa*. This brachiopod is indicative of the Chadian Stage showing that these beds are somewhat older than a knoll-reef 300 m to the east, which has yielded Arundian macrofossils (*see below*).

Two other dark limestone sequences form laterally restricted outcrops within the paler lithofacies. The first is well seen by the main Dovedale path at Fishpond Plantation [1455 5536] where some 9 m of irregularly bedded cherty limestones are exposed; the second includes an exposure, 12 m thick, underlain by about 9 m of grey finely peloidal limestones 450 m SSE of Lode House [1420 5507].

The most easterly exposure of the inter-reef facies is in the Tissington Trail cutting [154 551] west of Alsop en le Dale, where thin beds of grey-brown dolomitised limestone occur with irregular chert nodules and much veining, minor faulting and cavitation. These beds lie in a narrow outcrop between two knoll-reefs and, probably as a consequence, are tightly folded.

Knoll-reef facies The main Dovedale reef mass lies just south of the boundary of the district, and the knoll-reef facies in the present area occurs mainly in scattered oval-shaped outcrops. One of these, east of Alstonefield, was penetrated by a borehole [SK 15 NW/10] (Bridge and Kneebone, 1983) drilled to a depth of 100 m without reaching the bottom of the reef. A detailed study of the cores by Morgan (1980) showed that the predominant rock type is medium grey micrite with some darker grey and paler pinkish mottling in places. Peloidal micrites and biomicrites are also common at some levels. A darker variety of biomicrite, characterised by calcite spar-encrusted fenestellid bryozoans and resembling the Waulsortian 'Blue Vein Facies' of Lees and others (1977), was noted in places, with a few thin bands at other levels. Dips measured from the alignment of crudely stratified bioclastic debris and sparry calcite masses including stromatactoid cavities, vary from 0° to 30°, the differing values probably representing differences in the orientation of growth surfaces of successive individual mounds, stacked to form a composite knoll-reef in the manner suggested by Lees (1964, fig. 3A).

Plate 7 Crag in the Dove valley near Lode House; dark lithofacies of the Milldale Limestones comprising well bedded dark grey calcisiltite with chert lenses and nodules and with shaly intercalations (L 1150).

The best exposed knoll-reef, and the only one that has yielded diagnostic fossils, is in the dale east of Coldeaton Bridge [1499 5610], where grey poorly bedded micritic limestone, with sparry calcite cavity fillings, forms crags 32.5 m high. Brachiopods and some corals are sporadically present, apparently in pockets, and include the Arundian coral *Clisiophyllum multiseptatum*. The reef is immediately overlain by grey to pale grey-brown peloidal limestones with many brachiopod shells and shell debris.

Gun Hill Borehole

Strata below the base, of the black conglomerate limestone, which is 0.61 m thick at 619.35 m (Hudson *in* Hudson and Cotton, 1945), are now assigned to the Milldale Limestones. The presence of sandy limestones in the lowest 75.9 m is confirmed by a thin section (Bg 1053) of well laminated partially dolomitised calcarenite from a depth of 1301.5 m, which contains about 5–10% of quartz grains.

Igneous rock described as 'green tuff and ?lava' by Hudson is present in the borehole between 709.0 and 770.5 m. A thin section (E 46369), from 731.52 m, confirms that the altered rock represents a vesicular basaltic lava.

HOPEDALE LIMESTONES

The name was introduced during the present survey for limestones around Hopedale [122 550] at the southern boundary of the district (Aitkenhead and Chisholm, 1982, p. 10). The formation includes some outcrops assigned by Prentice (1951, plate XV) to the 'Waterhouses Limestone' and 'Upper Reef Limestone'.

The Hopedale Limestones lie in a broad synclinorium between the Manifold and Dove valleys, and are inferred to pass laterally into the Bee Low Limestones to the north of Alstonefield. The lower part of the formation is not exposed and is known in only the district to the south. The total thickness is estimated to be about 60 m in Hopedale, and 64.46 m was penetrated by a borehole [SK 15 NW/9] near Gateham Farm without proving the base.

The Hopedale Limestones are highly variable both vertically and laterally and, like the underlying formation, comprise a knoll-reef and an inter-reef facies. The latter includes thick coarsely crinoidal and pebbly beds, and thinner-bedded grey to dark grey finely bioclastic or peloidal calcarenites with chert nodules. Some of the beds are graded and probably represent occasional storm-surge deposits or other turbidite influxes off the nearby knoll-reef slopes. Both coarsely crinoidal and fine-grained calcarenites are found interbedded in the few sections at or near the reef margins, suggesting that any crinoidal flank facies developed only intermittently. Knoll-reefs are a major part of the formation in the district, forming the large and presumably composite knolls of Wetton Hill east and Narrowdale Hill, and the smaller knolls of Gateham Grange and Steep Low. The latter has an elongate outcrop which appears to lie in the core of a steep-sided pericline. The reefs consist mainly of grey to pale grey-brown biomicrite or micrite with sporadic irregular masses of sparry calcite ('reef tufa') partially or wholly filling cavities and enveloping fossils. The fauna is widely distributed but rich in places, brachiopods being specially abundant. However, there has been little palaeoecological or sedimentological investigation. An exception was the study by Morgan (1980) of a few exposures on Wetton Hill east, where the knoll was found to be of similar facies to the older knoll-reefs nearby but with less sparry calcite 'cement' and less diversity of fauna. The latter contradicts the general impression gained from previous faunal collections, which tend to show greater diversity in the younger reefs. Although delicate lamination has been noted in many exposures, its association with fossil algae that might have trapped sediment or formed a framework during reef growth is yet to be proved. The presence of reef-derived clasts in off-reef beds (*see below*) may indicate contemporaneous wave-brecciation and a more shallow-water origin for these reefs than is postulated for the older reefs.

The Hopedale Limestones were deposited in a belt lying between the Bee Low and Monsal Dale limestones of the Asbian–Brigantian shelf province and the deeper-water off-shelf deposits represented by the Ecton Limestones. Significantly, the shelf edge to the east is not clearly defined by a marginal apron-reef belt, and the thinness of the succession suggests that deposition took place in an area relatively starved of sediment, perhaps due to vigorous currents. The reef masses in the underlying Milldale Limestones may have continued to influence sedimentation by providing a foundation for this relatively upstanding area.

DETAILS

Of the two major facies which comprise the Hopedale Limestones, the inter-reef beds are relatively poorly exposed, even the best section – that at the disused quarry [1078 5570], north of Wetton – shows only 10.9 m of strata. These consist of medium to pale grey peloidal bioclastic limestone in well defined beds from 0.03 to 2.58 m thick, some showing fining-upwards grading. The bases of beds are sharp and, in some cases, erosive. The highest bed, 2.40 to 2.58 m thick, contains rounded limestone pebbles up to 20 cm across. The beds have yielded a rich brachiopod and coral fauna, the latter including cerioid forms of *Lithostrotion* and *Emmonsia parasitica*, indicating an Asbian to possible early Brigantian age. This exposure lies within 100 m of the outcrop of a knoll-reef, forming Wetton Hill east, and consists partly of bioclasts and pebbles

probably derived from the flanks of the reef. The only other extensive exposure occurs in old quarries [1189 5552 to 1192 5550] west of Hope Marsh: 8.6 m of dark grey fossiliferous limestone, thinly bedded with chert lenses in the top 1.8 m, and thicker bedded and coarsely bioclastic below. The fauna includes the coral *Diphyphyllum lateseptatum* and numerous well preserved brachiopods, including *Productus hispidus* and *Pugilis pugilis*, indicative of the Brigantian Stage. A Brigantian fauna has also been collected from an exposure, 3.5 m thick, of medium to dark grey coarsely crinoidal limestone near Under Wetton [1167 5601].

The best record is provided by a borehole [SK 15 NW/9] near Gateham Farm (Bridge and Kneebone, 1983). This proved about 50 m of the formation (allowing for steep dips present in the lower part), from its upper junction with the Mixon Limestone-Shales at a depth of 35.60 m to the bottom of the borehole at 100.06 m, and well illustrates the variable lithologies. Coarsely crinoidal limestones predominate from 35.60 to 55.66 m with pebbles of micritic reef-limestone common in some beds. Below this level the grain size becomes finer downwards, though some coarse crinoidal bands are present in places, and chert nodules become increasingly common. The lowest 1.85 m is again very coarsely crinoidal and contains some large brachiopod shells and many small clasts of micritic 'reef' limestone. Several varicoloured mudstone bands, 0.02 to 0.5 m thick and similar to the clay wayboards in the Bee Low and Monsal Dale limestones (p.   ), occur between 60.81 m and 69.20 m. Patchy dolomitisation and silicification is also evident throughout much of the sequence.

There are several large and numerous small exposures of the knoll-reef facies on the slopes of Wetton Hill east, the little knoll north of Gateham Grange, the hill south-west of Narrowdale, and Steep Low. All of these show unbedded or irregular discontinuous beds of grey to pale grey-brown biomicrite with scattered sparry masses and an abundant fauna in places. Age-diagnostic forms are rare. Brachiopods are particularly common, and exposures at the Gateham Grange knoll, for instance, have yielded twenty different species together with five species of bivalve, two of simple corals, one compound coral and fenestellid bryozoa. An exceptional concentration of the colonial coral *Diphyphyllum* together with some *Lithostrotion* occurs in a band revealed in scattered exposures on the south-west slope [118 570] of Narrowdale Hill. The age-diagnostic evidence from previous faunal collections from Narrowdale and Gateham hills is summarised by Parkinson (1950, p. 276). The upper B$_2$ or late Asbian age indicated by these tends to be confirmed by the goniatites *Bollandoceras* cf. *micronotum* and an indeterminate Beyrichoceratid collected from crags on Wetton Hill east [1090 5686] and Narrowdale Hill [1175 5724 to 1175 5717] during the present work.

ECTON LIMESTONES

Hudson (*in* Hudson and Cotton, 1945, p. 324) first introduced the name Ecton Limestones for the beds exposed in the main outcrop of the Manifold valley around Ecton [096 583]. The unit was formally defined as a formation by Aitkenhead and Chisholm (1982, p. 11). It includes some outcrops mapped as Apestor and Warslow Limestones, Waterhouses Limestone, and Brownlow Mudstones by Prentice (1951, plate 15).

Less extensive outcrops occur as anticlinal inliers surrounded by the Mixon Limestone-Shales around Butterton and north of Mixon, where they were described first by Challinor (1928) under the heading 'Mountain Limestone', and later by Hudson (*in* Hudson and Cotton, 1945) who called them the Mixon Limestones: the Mixon outcrop has been described by Morris (1969). The Gun Hill Borehole [SJ 96 SE/18] also proved 182.5 m of beds assigned to the formation. The total thickness around Ecton is estimated to be about 225 m.

The Ecton Limestones are predominantly medium grey or brownish grey to dark grey, thinly bedded, and bioclastic limestones. Locally, however, especially south of Swainsley Farm, the middle part of the succession is very fine-grained or micritic with some lamination and flaggy bedding. Part of these beds was included in the Brownlow Mudstones of Prentice (1951).

The bioclastic beds are sharp-based, and many are massive. Some parallel lamination is present in the thinner beds and in the upper part of some of the thicker ones. Interbedded shaly partings, chert lenses and nodules are common. Thick beds of conglomeratic limestone are found at several levels, commonly with subangular clasts of fine calcarenite dispersed in a coarsely bioclastic matrix. A huge block of reef-limestone is present in one bed (p. 51), and fragmentary brachiopod shells, corals and crinoids generally form conspicuous bioclasts. Evidence of shallow-water deposition is lacking, and most of these bioclastic and conglomeratic beds were probably deposited by turbidity currents, presumably flowing from a rising shelf area to the east. In places, lenticular beds with irregular erosive bases may represent turbidite feeder channels.

The dark very fine-grained beds contain *Chondrites*-like burrows in places and largely lack bioclastic debris. They probably formed in areas and at times of minimum turbidite activity.

DETAILS

Manifold valley and Manor House

Parts of the formation are well exposed in the type area around Ecton (Figure 20), but the monotonous nature of the lithology and the scarcity of persistent marker bands make correlation difficult.

The type section [SK 05 NE/13], alongside the former track of the Manifold Valley Light Railway, covers the basal 108 m of the sequence, including the junction with the underlying Milldale Limestones, which is taken at the base of a 0.50 m grey-brown partially silicified coarsely bioclastic bed exposed on the adjacent hillside [0951 5819]. The lowest 74 m are mainly brownish-grey, thinly to thickly bedded, fine to medium grained, bioclastic and peloidal calcarenites with some chert nodules and mudstone partings. The bases of the beds are sharp; some are erosive. Some beds are also graded, the thicker ones showing a lower massive division and an upper parallel-laminated division. Graded conglomeratic limestone beds 2.06 and 2.0 m thick, with brachiopod and coral fragments, occur about 31 and 71 m above the base. The coral assemblage from the lower bed includes *Clisiophyllum* cf. *rigidum*, *Koninckophyllum* cf. *divisium* and *Palaeosmilia murchisoni*, suggesting an Asbian age for the formation as indicated on the 1:50 000 map. Since this was published, however, foraminiferal assemblages have been examined from several levels; they include *Eoparasraffella sp.*, *E. restricta*, *E. simplex*, *Eotextularia diversa* together with *Koninckopora inflata*, and suggest that the section is of Arundian age.

These lowest 74 m of beds are overlain by about 18 m of dark grey fine-grained limestone, in laminae and thin beds up to 0.15 m thick, overlain by about 16 m of grey to dark grey, thickly bedded, fine to coarse bioclastic limestone with some graded bedding and scattered chert nodules. The sharp upward change from bioclastic

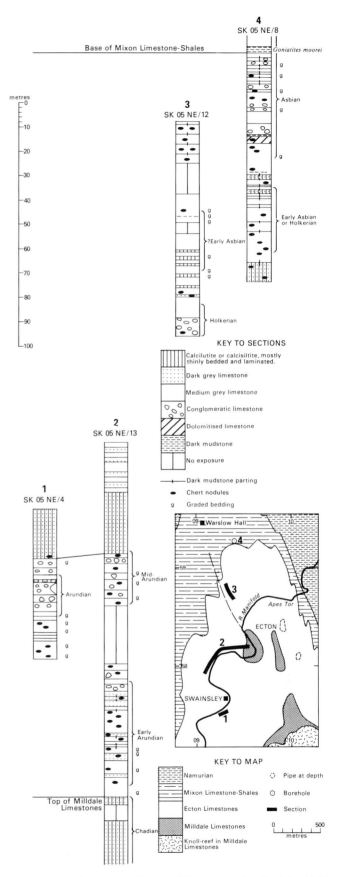

Figure 20 Sections of Ecton Limestones in the Manifold valley.

and conglomeratic limestones to these dark thinly bedded and laminated beds is best seen in a section by the roadside south-east of Swainsley Farm [SK 05 NE/14]. Here 31.18 m of the bioclastic facies include a conglomeratic bed, 4.3 m thick, containing a grey limestone block, 3.8 m thick, with patches of ?algal biolithite (Folk, 1959). According to Dr D. J. C. Mundy (personal communication) this block contained attached productoids, ?pseudomonotids and abundant large myocopid ostracods (including *Entomoconchus*), the latter filling primary growth-cavities. This association is typical of algal frameworks developed in Asbian reefs. No age-diagnostic macrofossils were recovered from this sequence, but microfossils including the foraminifera *Eblanaia sp., Eoparastaffella restricta, E. simplex, Eotextularia diversa* and *Rectodiscus sp.* sand conodonts including *Gnathodus* cf. *commutatus,* indicate an Arundian age for these beds. There is an abrupt upward change to 13.6 m of dark laminated and thin-bedded micritic limestones; these are unfossiliferous except for the basal 1.6 m, which yielded brachiopods and foraminifera of no stratigraphical value. Similar limestones have been traced only as far as 1.2 km to the south [0925 5636] of this section. Sporadically exposed and strongly folded examples occur in the River Manifold to the south-west [0915 5750 to 0899 5738] and pass up into thicker beds of mainly dark fine-grained limestone with dark burrows and shaly intercalations but with a few crinoidal bioclastic beds. A good section of about 24 m of such strata is exposed by the old railway track [0897 5739] south of Swainsley Tunnel.

Good exposures extending to higher levels in the Ecton Limestones also occur in the old Dale Quarry east of Warslow [093 587]. A section 43 m thick [SK 05 NE/12] is estimated to have its base roughly 140 m above that of the formation. It consists mainly of grey to dark grey bioclastic calcarenite, with some interbedded grey calcisiltite or calcilutite; shaly intercalations are present from 14.61 to 22.80 m above the base. A poorly delineated conglomeratic bed at the base contains rounded pebbles of mainly fine-grained calcarenite dispersed in a coarsely bioclastic matrix. Crinoid and brachiopod shell fragments are the commonest bioclasts, and coral fragments occur in a few beds. Some beds also contain thin chert lenses or irregular bands. Most beds are from 0.05 to 0.60 m thick, with a maximum of 2.6 m, while some are laterally discontinuous and with irregular erosive bases, locally cutting down through one or more of the underlying beds. At 12.06 m above the base, five thin beds amalgamate to form one thick bed. Foraminifera present include *Archaediscus* aff. *gigas, A. krestivnikovi, A. stilus eurus,* cf. *Eblanaia sp., Endothyra* ex. gr. *excellens, Eoparastaffella simplex, Koskinotextularia cribriformis,* and *Pojarkovella nibelis,* indicating a late Holkerian to early Asbian age.

To the west of Apes Tor [0983 5868 to 1005 5865] these beds are seen in a number of old roadside quarries, where strong folding (p. 113 and Plate 8) makes it difficult to estimate the total thickness (probably in the order of 60 to 65 m); the lowest bed lies at about the same stratigraphical level as the base of the Dale Quarries section. Similar lithologies are present, with darker thinly bedded and finer grained limestones in the middle of the section (Prentice, 1951, fig. 5).

A borehole [SK 05 NE/8] (Figure 20; Bridge and Kneebone, 1983), in the axial part of the Ecton Anticline north-east of Warslow, proved the top 64 m of the formation (thickness adjusted for dip) from a depth of 29.83 m to the bottom at 99.81 m. Between 62.86 and 72.82 m there was a distinctive lithology consisting mainly of dark grey thinly bedded very fine-grained calcarenites with shaly intercalations, chert nodules and common bioturbation. This sequence probably correlates with similar beds 14.61 to 22.80 m above the base of the Dale Quarries section; also with the 11 m (approximately) of fine-grained limestones in the middle of the section at Apes Tor Quarries (*see above*).

The top 18.4 m of the formation and its junction with the Mixon Limestone-Shales has also been seen [SK 15 NW/16] in a short trial level on the east side of Ecton Hill.

To the south [1048 5657], near Manor House and not far from the eastward termination of the Ecton Limestones, lies the only outcrop in the district where the Ecton Limestones are in unfaulted contact with an older knoll-reef of probable Chadian age. The basal coarsely crinoidal deposit forms a drape over the eroded reef surface (p. 47) and is overlain in discontinuous exposures by thinly interbedded graded bioclastic calcarenites and dark grey fine-grained or argillaceous limestones with dispersed bioclasts. The basal crinoidal bed appears to be devoid of foraminifera, though these fossils are present in the overlying beds 9 to 11 m above the base. The assemblage includes *Archaediscus* at the 'angulatus' stage and *Pseudolituotuba extensa,* indicative of the Asbian Stage.

BUTTERTON AREA

Although there are many exposures in Warslow Brook, the stream south of Kirksteads, and Hoo Brook, the intensity of folding is such that the compilation of composite sections has not proved feasible.

In the small inliers around Butterton, the best exposure is in the bed of Hoo Brook, just downstream from the ford at this village [0862 5619]. It consists of dark grey thickly bedded fine-grained limestone with a little chert.

MIXON AREA

Exposures in the eastern and most extensive of the two inliers, north of Mixon, are mainly confined to localities on or near the banks of the River Hamps [0461 5832; 0465 5815]. The former shows about 10 m of interbedded dark grey argillaceous and grey bioclastic limestone with a few shaly partings and thin chert lenses. Some beds show slight graded bedding. At the latter locality a conglomeratic bed, 1.3 m thick, contains sub-rounded pebbles, up to 0.12 m in diameter, of finely granular limestone, together with crinoid, brachiopod and coral fragments, set in a partially dolomitised matrix and overlying a dark grey argillaceous limestone bed 2.0 m thick. The only exposure of note [0391 5817] in the smaller western inlier west of Manor Farm contains a similar conglomeratic bed in a section 3.30 m thick. Coral-brachiopod faunas collected during this and previous surveys (Morris, 1969), although not diagnostic, are consistent with an inferred Asbian age for these strata. It is uncertain whether the conglomeratic beds lie at the same horizon, but the field relationships suggest that the last-mentioned bed correlates with similar lithologies at the top of a sequence of Ecton Limestones proved in a borehole [SK 05 NW/11] south-west of New Mixon Hay from a depth of 122.9 m to the bottom at 147.7 m.

GUN HILL BOREHOLE

Outside the areas described above, the Ecton Limestones are known only from the Gun Hill Borehole [SJ 96 SE/18] (Hudson *in* Hudson and Cotton, 1945), where the top and base of the formation are now taken at depths of 467.9 m (1535 ft) and 650.4 m (2134 ft) respectively. Recently determined foraminiferal assemblages from 512.37 m and 616.92 to 619.35 m indicate an early Asbian or Holkerian age.

MIXON LIMESTONE-SHALES

This formation was named by Hudson (*in* Hudson and Cotton, 1945, p. 321) and redefined by Aitkenhead and Chisholm (1982, p. 12). It has extensive outcrops in the inlier around Mixon, and around Butterton (Figure 3). The outcrop continues eastward in a narrow belt round the northern culmination of the Ecton Anticline, before dying out

Plate 8 Roadside quarries near Apes Tor, Ecton; Ecton Limestones folded into a subsidiary anticline and syncline within the Ecton Anticline (L 1180).

near Gateham Farm. The junction with the Namurian is conformable in most of this area (Coffey, 1969), though farther east, near Gateham Grange, the formation is overlapped by Namurian mudstones which are in contact with knoll-reefs in the Hopedale Limestones.

The Mixon Limestone-Shales include one named member, the Onecote Sandstones. These calcareous sandstones occur in the upper part of the formation, and were first named by Hudson who included them in the 'Morredge Grits' (Challinor, 1929). This term has now been abandoned because, as Holdsworth and others (1970) pointed out, it obscures a fundamental distinction between the calcareous sandstone and siltstone facies (Onecote Sandstones), and the non-calcareous protoquartzites (Minn Sandstones).

The formation is highly diachronous. In the extreme eastern outcrop it is probably mainly of P_{2b} age and is overlapped by Namurian mudstones. Around Butterton and Warslow it ranges from near the Asbian – Brigantian boundary to the top of the Brigantian, while around Mixon it appears to extend well down into the Asbian.

Outside the outcrop, the formation is known in the Gun Hill Borehole [SJ 96 SE/18] (Hudson *in* Hudson and Cotton, 1945), where strata between the depths of 258.2 and 467.9 m are now assigned to the formation. Sandy beds in the upper part of this sequence, between the depths of 259.1 and 339.2 m, represent the Onecote Sandstones. The formation also shows many similarities to the Widmerpool Formation, its correlative in the Widmerpool Gulf to the south of the present district (Aitkenhead, 1977).

The total thickness of the formation is estimated to be about 183 m in the Mixon inlier, about 193 m around Butterton, about 134 m in the eastern limb of the Ecton Anticline, and only about 24 m near Gateham Grange. The Onecote Sandstones are estimated to reach a maximum thickness of about 119 m north and west of Butterton. The succession consists for the most part of dark mudstone, limestone and sandstone, the argillaceous component probably forming slightly over half the total. Most of the formtion is probably of turbidite facies, although an alternative interpretation of deposition or reworking of beds by storm-wave processes is also possible. The sequence is characterised throughout by mainly thin sharp-based graded beds, each unit having a bioclastic calcarenitic (or in some cases conglomeratic) basal division, that passes abruptly to a darker argillaceous upper division, usually showing fine parallel lamination or long wave-length low amplitude ripple lamination. The lamination is also in places highly irregular, convoluted, or even chaotic. The basal bioclastic division is absent in some beds or represented only by a few thin laminae. The bioclastic limestone has a mainly fragmented, and presumably derived, crinoid-brachiopod-coral fauna, and a few beds contain limestone pebbles and flakes of mudstone derived from previously deposited sediments within the basin. Dark grey to black fissile 'pelagic' mudstones form thin intercalations between the turbidite beds, and these in places contain a poorly preserved goniatite-bivalve fauna. Chert nodules and lenses are present in some of the turbidites, and rare cone-in-cone calcite bands occur at the base of some beds. Finely disseminated pyrite or, less commonly, other sulphides such as chalcopyrite and sphalerite are sporadically present in all lithologies but are particularly conspicuous in the intercalated dark mudstone.

The Onecote Sandstones appear to form an elongate depositional lobe thinning gradually westwards (119 m around Butterton and 80 m at Gun Hill), though the thinning is abrupt near Warslow and at Ford to the south of the district. They consist of mainly thin sharp-based beds of turbidite character, generally of even fine to medium-grained texture with poor graded bedding. Many beds contain an irregularly ripple-laminated upper division. The sandstone is commonly pale grey-brown and calcareous. The few sole structures measured suggest palaeocurrents from a general easterly source. Thin section examination shows that these rocks are calcareous quartz arenites, with monocrystalline quartz forming the dominant grain type and with minor proportions of polycrystalline quartz, chert and calcite in the form of shell fragments and ooliths or peloids; feldspar is absent. The matrix is either silty clay or sparry calcite.

DETAILS

Mixon inlier

Detailed descriptions of this area have been given by Challinor (1928), Morris (1969) and Coffey (1969). Additional details have been provided by a borehole [SK 05 NW/11] south-east of New Mixon Hay, which proved the lower part of the formation from a depth of 2.9 m to the junction with the Ecton Limestones at 122.9 m. The succession here consists of sharp-based turbidite units grading upwards from medium to dark grey bioclastic calcarenite to dark grey argillaceous calcisiltite or calcareous mudstone in varying proportions. Most of the beds are less than 0.10 m thick, and only a few exceed 0.35 m. Coarse crinoidal and shelly debris and small argillaceous limestone clasts and mudstone flakes are present in a few beds. The turbidite units are interbedded with subordinate dark grey to black fissile mudstone, which contains a sparse fauna including goniatites and bivalves with some brachiopods, drifted plant fragments, and sporadic phosphatic nodules or coprolites. The mudstone is commonly pyritic and non-calcareous. A typical sequence of beds in the borehole is illustrated graphically in Figure 21. The fauna, indicative of the B_2 Zone, includes Posidonia aff. becheri, cf. Beyrichoceras (rectangularum group), together with Goniatites sp.. Foraminifera from a depth of 56.13 m include Archaediscus gigas, cf. Neoarchaediscus sp. and Palaeotextularia ex gr. longiseptata, together with Koninckopora inflata diagnostic of the late Asbian stage, and the entire borehole sequence is probably Asbian in age. This accords with the age given by Coffey (1969) for exposures in the River Hamps, near Lower Green Farm [045 586].

The River Hamps and its tributary, West Brook, provide exposures in the inlier, though only a few of these have yielded age-diagnostic fossils. In a small waterfall [0471 5613] north of Onecote Grange, a bed, 1.0 m thick, of coarsely bioclastic limestone yields a late Asbian or early Brigantian fauna with many brachiopods, including Striatifera striata, and one goniatite, Bollandoceras sp.. A goniatite collected by Challinor (1928, p. 99) from a now filled-in quarry [0309 5775] south-east of Old Mixon Hay, and named by Bisat (in Hester, 1932, p. 42) as Goniatites dinckleyensis (see Hudson, in Hudson and Cotton, p. 322), is indicative of the P_{1a} Zone. A later fauna of P_{1c} age, including G. sphaericostriatus, was also recorded by Challinor (1928, p. 102) from the River Hamps [0414 5961] above Brindley Croft. A P_{2a} fauna, including G. granosus, was found farther upstream [0404 5967] by Morris (1969) and Coffey (1969), and confirmed during the present work in an exposure of 1.2 m of dark fissile fossiliferous mudstone overlying 0.4 m of dark fine-grained limestone. G. granosus was also found by Bisat (in Hester, 1932) and Morris (1969) beneath the Onecote Sandstones at several localities west of Onecote Grange. The youngest goniatite fauna here is of P_{2c} age according to Coffey (1969). The P_{2c} age of the Onecote Sandstones is also indicated by the presence of Sudeticeras cf. hoeferae, discovered by the former Anglo-Iranian Oil Company SSE of Old Mixon Hay, and inferred to lie at or near the top of the member. Many other goniatite finds are recorded in the works cited, but most are poorly located or from loose blocks on the river beds. However, they serve to indicate that the Mixon Limestone-Shales is probably a full unbroken sequence in which all the Brigantian and late Asbian goniatite biozones are represented.

The Onecote Sandstones crop out around the Mixon inlier and in an extensive area south-east of Newhouse Farm. The best exposures are in a tributary of Warslow Brook [0472 6003] and on the left bank of the main stream [0488 5992 to 0500 5987], where up to 9 m of graded sandy and argillaceous limestone beds with thin intercalations of fossiliferous dark mudstone are exposed. Similar sequences, 5.8 and 6.0 m thick, occur in gullies [0532 5951; 0539 5920] north-west of Upper Elkstone. Near Feltysitch [0401 5972], a unit, 1.83 m thick, at the base of the member is made up of a massive brown fine to medium-grained sandstone and an underlying cross-laminated fine-grained calcareous sandstone. In West Brook [0334 5640], a 14.0 m exposure of similar beds occurs near White Lee Farm; they are also seen in the tributary [0377 5599 to 0396 5656] south-east of Newhouse Farm.

The strata assigned to the formation in the Gun Hill Borehole are probably of similar (turbidite) facies to the exposed sequence. However, they also include rocks described by Hudson (in Hudson and Cotton, 1945, p. 323) as 'tuffs and tuffaceous limestones' between depths of 420.6 and 424.9 m, and a 'dark-green calcareous tuff' 7.9 m thick, at 432.8 m. Evans and others (1968, fig. 3) postulate that these rocks correlate with tuffaceous strata at Astbury in the Macclesfield district. Hudson suggests that the Gun Hill sequence ranges in age from upper B_2 to P_2, and determinations of late Asbian foraminifera in samples from depths of 453.24 and 454.15 m tend to confirm this.

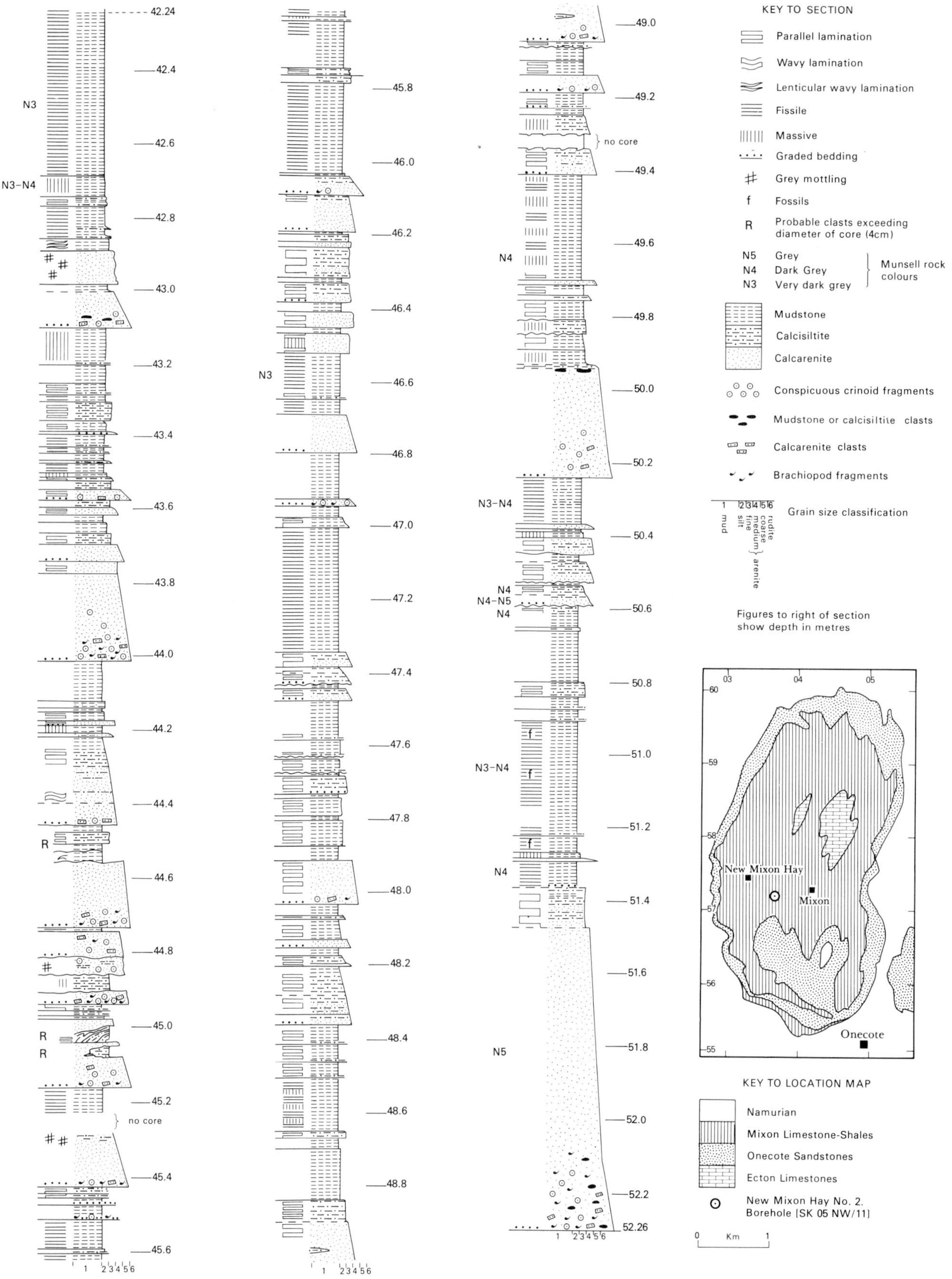
42.24
42.4
N3
42.6
N3-N4
42.8
43.0
43.2
43.4
43.6
43.8
44.0
44.2
44.4
R
44.6
44.8
45.0
R
R
45.2
no core
45.4
45.6
1 2 3 4 5 6
45.8
46.0
46.2
46.4
N3
46.6
46.8
47.0
47.2
47.4
47.6
47.8
48.0
48.2
48.4
48.6
48.8
1 2 3 4 5 6
49.0
49.2
no core
49.4
49.6
N4
49.8
50.0
50.2
N3-N4
50.4
N4
N4-N5
N4
50.6
50.8
51.0
N3-N4
51.2
N4
51.4
51.6
51.8
N5
52.0
52.2
52.26
1 2 3 4 5 6
KEY TO SECTION
Parallel lamination
Wavy lamination
Lenticular wavy lamination
Fissile
Massive
Graded bedding
Grey mottling
f Fossils
R Probable clasts exceeding diameter of core (4cm)
N5 Grey
N4 Dark Grey
N3 Very dark grey
Munsell rock colours
Mudstone
Calcisiltite
Calcarenite
Conspicuous crinoid fragments
Mudstone or calcisiltite clasts
Calcarenite clasts
Brachiopod fragments
Grain size classification
1 2 3 4 5 6
mud
silt
fine
medium
coarse
rudite
arenite
Figures to right of section show depth in metres
03
04
05
60
59
58
57
56
55
New Mixon Hay
Mixon
Onecote
KEY TO LOCATION MAP
Namurian
Mixon Limestone-Shales
Onecote Sandstones
Ecton Limestones
New Mixon Hay No. 2. Borehole [SK 05 NW/11]
0 Km 1

Figure 21 Logs showing a typical sequence of turbidites from the Mixon Limestone-Shales in New Mixon Hay No. 2 Borehole [SK 05 NW/11].

BUTTERTON–WARSLOW AREA

The formation is exposed in several streams in this area. However, intense folding and poor fossil preservation preclude the establishment of a composite section. The base is recognised only in a borehole [SK 05 NE/8] south-east of Warslow Hall, where it is taken beneath a mainly mudstone sequence, 5.51 m thick, at 29.83 m. The mudstone contains *G. moorei* at depths of 29.41 and 29.55 m indicating an upper B_2 age. A foraminifera assemblage from 71.15 to 90.80 m indicates an early Asbian or Holkerian age. Elsewhere, in stream sections, the boundary is arbitrarily taken where mudstone becomes a significant part of the sequence [such as 0876 5680; 0868 5559].

The boundary usually appears to be near the horizon of a conglomerate bed containing the P_{1a} Zone indicator *Goniatites crenistria*. This distinctive bed, first noted by Hudson (*in* Hudson and Cotton, 1945, p. 322), is exposed on either side of an anticline east of The Twist [0693 5607; 0701 5609]. It consists of 1.2 m of very coarsely bioclastic conglomeratic limestone with a dark grey fine-grained matrix, and has yielded the goniatite *Beyrichoceras sp.* as well as *G. crenistria*. A large brachiopod fauna has also been collected from this bed, including forms such as *Antiquatonia wettonensis*, *Buxtonia sp.*, *Echinoconchus subelegans*, *Gigantoproductus sp.*, *Pugilis pugilis* and *Striatifera striata* characteristic of the beds overlying the apron-reef limestones along the shelf margin to the east. The conglomerate bed may have been laid down as either a turbidite or a debris flow deposit derived from the shelf margin or knoll-reef area to the east during some high energy event such as a storm, tidal wave or earthquake. Probably the same bed is exposed [0880 5678] south of Kirksteads and in the Warslow Brook [0820 5772] north of Heathy Roods. Both contain *G. crenistria* according to Prentice (1951, locs. 83 and 62). During the present work, 0.4 m of dark grey platy mudstone overlying the bed in Warslow Brook yielded only *Goniatites sp.* [striatoid group]. *G. crenistria* was also recorded at or near these localities and also in Warslow Brook [0834 5781] by Hudson. Conglomeratic bioclastic graded beds are present at six levels in Borehole SK 05 NE/8. One such bed, 2.11 m thick at a depth of 11.08 m, is probably the same as that exposed nearby in a disused quarry [0940 5923], where it contains a non-diagnostic coral–brachiopod fauna together with a conodont assemblage of Brigantian age.

A P_{1c} goniatite fauna (Bisat, 1957) younger than that noted above, and including *G. warslowensis* and *Pronorites ludfordi*, was collected by D. Parkinson and A. Ludford from an exposure at the footbridge [0853 5777], north of Clayton House. In Hoo Brook [0840 5581; 0866 5556], interbedded limestones and fossiliferous mudstones have yielded *Dunbarella persimilis* and *Posidonia* aff. *becheri* at the first locality, and these species plus *G. sp.* [bisati/kajlovecense group] and cf. *Nomismoceras sp.* at the second; the faunas are indicative of a P_{1c} or P_{1d} age.

Mudstone with thin beds of argillaceous limestone is sporadically exposed in the stream near Warslow Hall. Coffey (1969) found goniatites at two localities: one [0916 5941] with *G. granosus*, indicative of the P_{2a} Zone, the other [0943 5951] with *Neoglyphioceras sp.* possibly indicative of the P_{2c} Zone. *Sudeticeras sp.* was collected at or near these localities during the present survey, confirming the P_2 age.

The highest faunal horizon beneath the Onecote Sandstones occurs in Warslow Brook [0801 5762] NNW of Heathy Roods, where some 6.0 m of sandstone and sandy limestone with shaly partings are underlain by ?.5 m of argillaceous limestone and dark fissile mudstone containing *G. granosus*, indicative of a P_{2a} age.

The Onecote Sandstones are best seen in a tributary of Warslow Brook [0660 5741 to 0684 5755] near The Hill, where about 9.0 m of sandy limestone and sandstone, in sharp-based massive beds up to 1.2 m thick, are exposed. This sequence passes up into thinly interbedded argillaceous limestone and mudstone, which has yielded

Sudeticeras sp. [splendens group] at a level 5.5 m above the top of the Onecote Sandstones [0697 5757]; this must be very near the basal Namurian *Cravenoceras leion* horizon [0640 5730] recorded by Holdsworth (Aitkenhead and Holdsworth, 1974) south-west of The Hills. There is also good exposure in the stream [0544 5606] northeast of Butterton Moor End, and about 2.8 m of calcareous sandstone or sandy limestone with interbedded silty mudstone are exposed in quarries on Butterton Moor [0598 5603] and Grindon Moor [0601 5596].

EAST OF THE ECTON ANTICLINE

Two of the scattered exposures of thinly interbedded dark grey argillaceous limestone and mudstone, to the west [1085 5695] and south of Gateham Grange, contain sparse P_2 faunas, including *Neoglyphioceras sp.* and *Sudeticeras sp..* However, the best section is provided by a borehole [SK 15 NW/9] near Gateham Farm (Bridge and Kneebone, 1983). The basal junction with the Hopedale Limestones is taken at a depth of 35.60 m, at the base of the lowest thick mudstone band in the succession. Above this, the sequence is predominantly mudstone with thin sharp-based laminated argillaceous limestone beds, some with basal bioclastic divisions. A few pale grey-brown silty laminae, occurring between depths of 8.12 and 13.29 m, may be the sole representatives of the Onecote Sandstones. The mudstones are fossiliferous in places. Goniatites collected include *Sudeticeras sp.* [stolbergi group] at 13.70 m, *S. sp.* [splendens group] at 17.99 m, *S. sp.* [delepinei group] at 19.32 m, *Neoglyphioceras sp.* at 19.83 m and *S. sp.* at 27.26 m, indicating that the beds here are largely of P_{2b} age.   NA

BIOSTRATIGRAPHY

The outcrop of the main Dinantian formations is continuous with that in the Chapel en le Frith district to the north, and the Chesterfield and Derby districts to the south-east. Since the faunas are broadly similar, it is not considered necessary to repeat the detailed palaeontological accounts and lists of fossils published in the sheet-memoirs for those districts (Mitchell *in* Stevenson and Gaunt, 1971, pp. 128–154 and 384–409; Ramsbottom *in* Smith and others, 1967, pp. 48–52; Mitchell *in* Frost and Smart, 1979, pp. 131c and d).

In the present district the distribution of fossils in the Dinantian limestones has been studied to provide biostratigraphical data to support the classification and correlation of the lithostratigraphical units that have been mapped. A considerable number of macrofossils has been collected and identified, and this work has been supported by conodont studies and, towards the end of the survey, by examination of thin sections for foraminifera and algae. All the material is stored in the Biostratigraphy Research Group collections in the Keyworth office of BGS. Full lists of identifications are recorded on data cards and fossil matrix tables which can be consulted by reference to the Group Manager.

The more stratigraphically significant fossils are recorded in the detailed account of each formation (pp. 8–55), and drawn together into a table of distribution (Table 1). In the following outline of the biostratigraphy, a list of the characteristic fossils is given for each formation with comments on the age of the beds.

Table 1   Stratigraphical distribution of the significant corals and brachiopods from the Woo Dale Limestones, Bee Low Limestones, Monsal Dale Limestones and Eyam Limestones of the shelf province of the Dinantian

Genera and species	Woo Dale Limestones	Bee Low Limestones	Monsal Dale Limestones	Eyam Limestones
Composita ficoidea	X			
Davidsonina carbonaria	X			
Daviesiella derbiensis	X			
Linoprotonia corrugatohemispherica	X			
Axophyllum vaughani	X	X		
Daviesiella spp.	X	X		
Megachonetes pailionaceus	X	X		
Lithostrotion martini	X	X	X	
Lithostrotion portlocki	X	X	X	
Linoprotonia hemisphaerica	X	X	X	
Dibunophyllum bourtonense		X		
Haplolasma cf. *densa* (Hudson & Cotton, 1945, p.306)		X		
Lithostrotion aranea		X		
Lithostrotion sociale		X		
Davidsonina septosa		X		
Delepinea cf. *comoides*		X		
Gigantoproductus cf. *maximus*		X		
Gigantoproductus semiglobosus		X		
Gigantoproductus sp. nov. (of Mitchell, 1971, p.131, pl.14, fig.2)		X		
Lithostrotion pauciradiale		X	X	
Palaeosmilia murchisoni		X	X	
Siphonophyllia benburbensis		X	X	
Gigantoproductus dentifer		X	X	
Striatifera striata		X	X	
Dibunophyllum bipartitum		X	X	X
Lithostrotion junceum		X	X	X
Aulophyllum pachyendothecum			X	
Dibunophyllum craigianum			X	
Dibunophyllum konincki			X	
Dibunophyllum lateseptatum			X	
Koninckophyllum interruptum			X	
Koninckophyllum magnificum			X	
Lithostrotion maccoyanum			X	
Lonsdaleia dulpicata			X	
Nemistium edmondsi			X	
Orionastraea placenta			X	
Palaeosmilia regia			X	
Pseudozaphrentoides juddi			X	
Davidsonina septosa (gerontic form)			X	
Gigantoproductus crassiventer			X	
Gigantoproductus edelburgensis			X	
Gigantoproductus giganteus			X	
Gigantoproductus cf. *gigantoides*			X	
Gigantoproductus striatosulcatus			X	
Semiplanus sp. (latissimoids)			X	
Clisiophyllum keyserlingi			X	X
Lonsdaleia floriformis			X	X
Productus hispidus			X	X
Pugilis pugilis			X	X

Only a limited number of Eyam Limestones faunas have been identified, so many species show an apparently atypical range (*see* Mitchell *in* Stevenson and Gaunt, 1971, table 2).

Full lists of fossils are held by the Biostratigraphy Research Group of the Survey.

The use of cf. in this table indicates similarity to the species named.

SHELF PROVINCE

Woo Dale Limestones

The coral faunas from the exposed parts of the Woo Dale Limestones do not include any species diagnostic of age; *Axophyllum vaughani* and *Lithostrotion martini* are the most typical forms. The brachiopods, however, include the Holkerian taxa *Davidsonina carbonaria*, *Daviesiella derbiensis* and *Linoprotonia corrugatohemispherica*. Specimens of thick-shelled *Daviesiella* are common but are usually difficult to identify specifically due to difficulties in exposing the internal features. The presence of *Dibunophyllum bourtonense* at the top of the Woo Dale Limestones in the Dam Dale – Hay Dale section to the north of the present district (Mitchell *in* Stevenson and Gaunt, 1971, p. 130) indicates that the topmost part of the formation falls just within the Asbian; however, this fossil is not recorded in these beds farther south.

An assemblage of foraminifera from the type locality of the Iron Tors Limestones Member includes *Dainella sp.*, *Eoparastaffella sp.* and *Spinobrunsiina sp.*. These, together with the absence of any archaediscids and the alga *Koninckopora sp.*, suggest a Chadian age for the unit; no macrofossils have been found.

Higher strata spanning the highest 19.5 m of the Woo Dale Limestones and the lowest 1.8 m of the Bee Low Limestones in Biggin Dale [SK 15 NW/13] contain *Pojarkovella nibelis*, *Florennella sp.*, *Archaediscus grandiculus* and cf. *Holkeria sp.*, suggesting a Holkerian or very early Asbian age.

The Woo Dale Limestones in the Lees Barn Borehole [SK 15 NE/5] contain a typical Holkerian assemblage between 63.00 and 98.85 m, where *Florennella rectiformis*, *F. moderata*, *Bessiella sp.*, *Pojarkovella nibelis*, *Holkeria topleyensis*, *Brunsia spirillinoides*, *Glomospiranella sp.* and *Eostaffella sp.* are common consistuents of the fauna. The beds between 39.10 and 63.00 m are less rich in diagnostic foraminifera.

Bee Low Limestones

Macrofossils, mostly corals and brachiopods, are confined mainly to a few discrete bands. Characteristic Asbian assemblages include *Axophyllum vaughani*, *Dibunophyllum bourtonense*, *Clisiophyllum rigidum*, *Koninckophyllum vanghani*, *Lithostrotion arachnoideum*, *L. aranea*, *L. junceum*, *L. martini*, *L. pauciradiale*, *L. portlocki*, *Palaeosmilia murchisoni*, *Gigantoproductus maximus*, *G. semiglobosus* and *Linoprotonia hemisphaerica*. There are few changes in the composition of the assemblage through the formation but, particularly towards the top, a number of fossiliferous bands have yielded *Davidsonina septosa*. The apron-reef limestones (p. 54) contain both rich brachiopod faunas similar to those described from Treak Cliff (Mitchell *in* Stevenson and Gaunt, 1971, pp. 141–152, table 5), and goniatites that confirm the age as late Asbian (Upper *Beyrichoceras* (B$_2$) Zone). A notable addition to the reef-brachiopod faunas is the productoid *Cinctifera medusa*, not previously recorded in the United Kingdom but common in the fore-reef limestones at Chrome Hill (D. J. C. Mundy, personal communication).

The Bee Low Limestones in the Lees Barn Borehole con-tain a diagnostic Asbian foraminiferal assemblage which includes *Neoarchaediscus sp.*, *Nudarchaediscus* aff. *concinnus*, *Palaeotextularia* ex gr. *longiseptata* and large 'angulatus' stage *Archaediscus spp.* A typical Asbian foraminiferal assemblage comprising *Bibradya inflata*, *Cribrostomum sp.*, *Lituotubella magna*, and *Endothyranopsis crassa* was also recovered from a bed 41.85 m above the base of the formation in Biggin Dale [SK 15 NW/13].

Strata equivalent to the Bee Low Limestones and uppermost Woo Dale Limestones were examined between 358.98 and 611.48 m in a borehole [SK 27 SW/20] at Longstone Edge. The presence of *Gigasbia gigas*, *Koskinobigenerina sp.*, *Palaeotextularia* ex gr. *longiseptata*, *Cribrostomum sp.*, *Nudarchaediscus concinnus*, *Endothyra* aff. *spira* and *Koskinobigenerina sp.* between 358.98 and 407.70 m indicates a late Asbian age. Below 407.70 m the foraminiferal assemblage is very poor, but *Pojarkovella nibelis* was located at 416.13 to 425.04 m and this taxon occurs only in Holkerian or earliest Asbian strata. No further diagnostic foraminifera were observed below this level but a few stunted, poorly developed taxa were present. By comparison with the fauna in the rest of the shelf province, it is probable, though not certain, that these lower beds are of Holkerian age.

Monsal Dale Limestones

The Monsal Dale Limestones have a more varied and abundant fauna than the Bee Low Limestones. The assemblage is of Brigantian age and contains *Dibunophyllum bipartitum*, *D. craigianum*, *D. konincki*, *Diphyphyllum lateseptatum*, *Lithostrotion junceum*, *L. pauciradiale*, *L. portlocki*, *Londsdaleia duplicata*, *L. floriformis*, *Orionastraea placenta*, *Palaeosmilia regia*, *Gigantoproductus spp.* (with fluted trials), *Productus hispidus*, *Pugilis pugilis* and *Striatifera striata*.

Saccamminopsis and *Girvanella* occur through much of the Monsal Dale Limestones but concentrations of the former are present near the base, and the latter occurs locally at three levels in the lower third of the formation (Cox and Bridge, 1977, fig. 6).

Several fossil bands are present, especially in the upper part of the formation. They include the Lathkill Shell Bed with abundant *Gigantoproductus crassiventer*, the Upperdale Coral Band bearing mainly *L. junceum*, and the Hob's House Coral Band characterised by large clisiophylloids including *Dibunophyllum bipartitum*, *D. craigianum* and *D. konincki*. Towards the top of the formation the White Cliff Coral Band represents the *Lonsdaleia duplicata* Band of the ground to the north and *Orionastraea* tends to occur in two bands, a lower one with *O. indivisa* and an upper one with *O. placenta* Mitchell (*in* Stevenson and Gaunt, 1971, pp. 138–139).

Brigantian fossils, including foraminifera, make their earliest appearance at the base of the formation although a few typical Asbian forms, notably *Koninckopora inflata* and a single gerontic specimen of *Davidsonina septosa*, have been recorded in the lowest few metres (Chisholm and others, 1983).

Eyam Limestones

Many of the coral and brachiopod species present in the Monsal Dale Limestones range up into the Eyam

Limestones. Knoll- and flat-reef limestones are also present, with rich brachiopod faunas that differ from those in the Bee Low Limestones in that the number of species present is more restricted.

Longstone Mudstones

The goniatite-bivalve faunas present in the Longstone Mudstones include the goniatites *Neoglyphioceras sp.*, *Lyrogoniatites sp.* and *Sudeticeras sp.*, diagnostic of a late Brigantian (P_2 Zone) age.

OFF-SHELF PROVINCE

A variety of rock types occur in the off-shelf province (p. 7): many are poorly fossiliferous and difficult to correlate. In part, the evidence for the age of the formations was gained from the study of the more extensive outcrops of the beds in the adjacent Ashbourne district.

Milldale Limestones

Fossils are rare in the inter-reef facies of the Milldale Limestones. However, Tournaisian conodont faunas including *Scaliognathus anchoralis* and *Polygnathus communis carina* have been recovered from the lower part of the sequence, and the diagnostic Chadian brachiopod *Levitusia humerosa* occurs in the upper part. The reef-limestones locally yield rich brachiopod faunas with *Acanthoplecta mesoloba*, *Eomarginifera derbiensis* [small form], *Pugnax spp.*, and *Spirifer coplowensis* which, as an assemblage, is distinctive from the Asbian reef faunas. The goniatite *Dzhaprakoceras sp.*, which indicates a Chadian–Holkerian age, has also been collected from reef facies of the Milldale Limestones, and *Fascipericyclus sp.*, indicating an early–mid Chadian age, from the dark facies of the inter-reef limestones. The higher part of the formation contains the Arundian corals *Clisiophyllum multiseptatum* and *Cravenia rhytoides*.

Foraminifera from the Milldale Limestones in the Gun Hill Borehole include *Uriella sp.*, *Septaglomospira comblaini* and *Palaeospiroplectammina parva* at 1300.9 m; *Eoparastaffella simplex*, *Eblanaia michoti*, *Spinoendothyra sp.*, *Paraendothyra cummingsi* and *Rectodiscus sp.* at 943.3 m; and *Pojarkovella nibelis* at 674.2 m. These three assemblages indicate probable Tournaisian, Arundian, and Holkerian ages respectively. Exposures in the Manifold valley near Wetton have yielded *Eblanaia sp.* and *Paraendothyra sp.* of either Chadian or late Tournaisian age, and *Eblanaia michoti* and *Koninckopora sp.* indicating a Chadian age.

Hopedale Limestones

In the lower part of the Hopedale Limestones the coral faunas are similar to those of the Bee Low Limestones. Knoll-reefs are also present, and contain the brachiopod faunas typical of the late Asbian (Upper B_2 Zone), the age being confirmed by the presence of the goniatite *Bollandoceras* cf. *micronotum* (p. 49). The upper part of the formation contains a Brigantian fauna with *Dibunophyllum bipartitum*, *Diphyphyllum lateseptatum*, *Productus hispidus* and *Pugilis pugilis*.

Ecton Limestones

Corals are rare in the Ecton Limestones except for small fragments, but the presence of *Clisiophyllum* cf. *rigidum* and *Palaeosmilia murchisoni* suggests an Asbian age. However, the foraminiferal assemblages (*see below*) indicate an Arundian to early Asbian age for the formation. Ecton Limestones which flank both the Milldale Limestones and Hopedale Limestones knoll-reefs contain a coral fauna with *Michelinia egertoni*, *M.* [*Emmonsia*] *parasitica* and *M.* cf. *tenuisepta*, which suggests a late Asbian or early Brigantian age for these beds; the former stage is indicated by foraminifera (p. 51).

Beds in the lower part of the Ecton Limestones in the type area (p. 50) yield a diagnostic Arundian foraminiferal assemblage including *Eotextularia diversa*, *Glomodiscus miloni*, *Eoparastaffella simplex*, *E. restricta*, *Rectodiscus sp.*, and *Archaediscus* at the 'involutus' stage. Foraminifera become much less common in the overlying beds which yield assemblages including *Pseudolituotuba wilsoni*, *P. gravata*, *Koskinotextularia cribriformis*, *Archaediscus reditus*, *Pojarkovella nibelis*, *Archaediscus stilus eurus* and *A. krestovnikovi* of either Holkerian or very early Asbian age. This restricted and commonly stunted fauna is typical of Holkerian strata of off-shelf (basin) facies both here and elsewhere in Great Britain. The relatively abundant and diverse Holkerian foraminifera seen in the adjacent shelf province are absent here.

Higher beds of the Ecton Limestones in the Warslow Hall (29.83–62.90 m) and New Mixon Hay (56.13 m) boreholes [SK 05 NE/8, SK 05 NW/11] yield an Asbian assemblage of foraminifera incorporating *Archaediscus karreri*, *Palaeotextularia* ex gr. *longiseptata*, *Neoarchaediscus sp.*, *Koskinobigenerina sp.*, *Rectodiscus sp.*, *Nudarchaediscus sp.* and *Archaediscus gigas*. Ecton Limestones proved in the Gun Hill Borehole from 616.92 to 619.35 m contain an assemblage including *Plectogyranopsis convexa*, *Palaeotextularia* ex gr. *consobrina*, *Archaediscus stilus*, *A. krestovnikovi* and *Koninckopora spp.* which may be Asbian or Holkerian in age. Another sample from higher in the sequence (512.37 m) contains a more restricted, less diverse, assemblage of foraminifera together with the alga *Koninckopora spp.* This may be basal Asbian but is likely to be Holkerian in age.

Mixon Limestone-Shales

The basal Mixon Limestone-Shales in the Gun Hill Borehole (453.24–454.15 m) contain typical Asbian foraminifera including *Cribrostomum lecomptei*, *Archaediscus grandiculus*, *Archaediscus crux*, *Ammarchaediscus sp.* and the alga *Koninckopora inflata*.

The strata from 0 to 16.80 m in the Warslow Hall Borehole [SK 05 NE/8] contain *Bradyina sp.*, *Mikailovella gracilis* and *Howchinia bradyana*, all of which are common in the Brigantian and rare in the Asbian. Hence these beds are possibly of Brigantian age.

The intercalated mudstones of this formation contain goniatite faunas which range in age from the B_2 Zone to the P_{2c} Zone (*see also* pp. 53–55).　　MM, ARES

CHAPTER 4

Extrusive igneous rocks

Beneath the eastern part of the district basaltic lavas and tuffs, known collectively as the Fallgate Volcanic Formation, make up the bulk of the Brigantian and Asbian sequence (Figures 10, 11; Aitkenhead and Chisholm, 1982, pp. 6–7). Elsewhere, however, the volcanic rocks are subordinate to limestones, the individual volcanic units being regarded as members of the limestone formations. The names of the members (such as Lathkill Lodge Lava) should not be taken to imply that each necessarily represents a single flow; the majority contain several flows, and some 'lavas', such as the Lower Matlock Lava, also contain interbedded tuffs. Volcanic rocks are less common outside the shelf province, but have been recorded in the Milldale Limestones in the Gun Hill Borehole (p. 48) and the Manifold valley (p. 49).

Thin greenish-grey clay 'wayboards' in the shelf limestones (p. 6) are also of volcanic origin; they represent periodic accumulations of fine airborne volcanic dust that generally rest on emergent limestone surfaces (Walkden, 1972; 1974). Similar clays were deposited in deeper water during Namurian times (pp. 72 and 80).

The lavas are olivine-basalts, usually vesicular, except in the central parts of some flows; these rocks are apparently of subaerial origin. Submarine extrusion of basalt has produced hyaloclastites by the fragmentation of the outer skins of pillows; the term hyaloclastite is used here in the original and restricted sense of Rittmann (1962, p. 72). Hyaloclastites are most common in the thicker volcanic sequences, where they are known from boreholes; none has been seen at outcrop in the district. The close association of subaerially erupted lavas with submarine flows is a notable feature of the Derbyshire volcanicity and parallels the alternation of submergence and emergence recorded in the contemporaneous shelf-limestone sequence. The water depth attained during submergence can be estimated if it is assumed that the rate of volcanic rock build-up is rapid in relation to the subsidence rate of the basin; the water depth at the onset of an eruptive episode probably equals the thickness of subaqueous volcanic rocks (bedded tuffs and hyaloclastites) that lie between subaerially erupted flows. On this basis, water depths reached about 73 m in early Brigantian times around Youlgreave (Figure 10, boreholes SK 26 SW/9 and SK 26 NW/25).

Tuffs are less extensive at outcrop than lavas, the most important being the Shothouse Spring Tuff which is associated with the Grangemill necks (p. 98). The presence of certain tuffs in association with palaeokarsts (Stevenson and Gaunt, 1971, p. 24) suggests that they had an air-fall origin, and this may also be true of some of the clay wayboards (Walkden, 1972). Some of the coarser tuffs may include material resulting from the re-working of hyaloclastites (Plates 9.4, 9.5, 9.6), though it is not practicable to classify these deposits separately.

The geographical extent of the various units (*see* summary by Walters and Ineson, 1981) suggests four separate centres of volcanic activity within the shelf province. In the south, a centre near Bonsall gave rise to the Matlock lavas. Farther north a centre near Alport was apparently responsible for the Lees Bottom, Shacklow Wood, Lathkill Lodge and Conksbury Bridge lavas, as well as a group of flows at the north end of the Millclose Mine workings. A third centre, which produced hyaloclastites, tuffs and lavas, was located near Longstone Edge. A fourth centre near Tunstead is indicated by the distribution of the Miller's Dale lavas. The Alport and Bonsall centres were more active than the other two, as is shown by the location of areas where volcanic rocks make up more than half the Brigantian sequence (Figure 10).

These centres probably did not comprise single volcanic vents or fissures, and are better envisaged as local concentrations of eruptive sources. Whether vents or fissures predominated has not been established, although fissures were favoured by Stevenson and Gaunt (1971, pp. 20, 299) in view of the lack of any relationship between lavas and necks. Eruptions from the various centres were not contemporaneous and so it is impossible to correlate flows from one centre to another; each set of lava names is restricted to a single geographical area.

A well marked region of magnetic anomalies in the north and east of the district (Figure 46) corresponds broadly with the area where the Dinantian volcanic rocks are concentrated. The Bonsall centre coincides with a strong individual high within this general region of magnetic anomaly but the other centres of volcanism do not. Thus the thick volcanic sequence proved near Alport may lie well away from the nearest centre of eruption, if this is identified by the strong linear magnetic high (Figure 46) that runs from near Bakewell to beyond the district boundary near Ashover; however, in the present state of knowledge it seems better to name the centres of activity by reference to known concentrations of volcanic rocks, rather than from magnetic anomalies whose precise origin is uncertain.

Clay wayboards

The most widespread volcanic deposits in the shelf limestones are the thin beds of clay known in the Peak District as 'wayboards' (Ford, 1977, p. 40). They are best known in the Bee Low Limestones, where 30–40 are scattered throughout the formation (p. 15). Maximum thickness is 1.25 m, but probably about two thirds of them are less than 3 cm thick. Usually they are seen only in quarries or in borehole cores; in natural sections they are hidden beneath soil and vegetation or squeezed out by superincumbent pressure. The clay is generally grey-green when fresh, and yellow-brown when weathered or altered by groundwater. Dispersed crystals of pyrite are common. Walkden (1972) found that the clays are potassium-bentonites largely comprising mixed-layer illite–smectite; kaolinite is also present. From this mineralogy and the occurrence of similar clays in Asbian shelf limestones elsewhere in England and Wales,

Walkden concluded that the clays are the alteration products of widely dispersed volcanic dust, but whether this was derived from local volcanic centres or from more distant sources is not known. Wayboard clays commonly rest on palaeokarstic limestone surfaces; in such cases they have probably undergone subaerial weathering and may represent fossil soils (Walkden, 1974, p. 1239).

DETAILS

BONSALL CENTRE

The distribution of the Lower and Upper Matlock lavas (Traill, 1940, fig. 54; Walters and Ineson, 1981, figs. 3, 4) suggests the existence of a volcanic centre not far beyond the eastern limit of the district, near Bonsall, where there is also a concentration of volcanic necks. Hyaloclastites and tuffs are associated with the lavas. A strong magnetic anomaly (Figure 46) is located in the same general area, but may in part be due to the presence of dolerite sills younger than the extrusive rocks.

The **Winstermoor Lava** was recorded as 'toadstone' in mine shafts (Green and others, 1887, p. 145), but was not distinguished from the one now called the Lower Matlock Lava. It was first recognised as a distinct unit by Shirley (1950, 1959). There are no exposures but the outcrop can be traced by lava debris in the soil. West of Sacheveral Farm [228 593] it dies out, but to the east it can be traced across the axis of the Matlock Anticline until cut off by the Bonsall Fault. From scattered exposures it appears that the lava is underlain by pale limestone and overlain by darker beds, and on this evidence it is placed at the base of the Monsal Dale Limestones. Its distribution is patchy, for a borehole [SK 25 NW/21] near Elton, and not far north of the outcrop, penetrated the base of the Monsal Dale Limestones without encountering any lava. Likewise, in the area north of Winster, 2.3 m of pale grey red-stained clay of probable volcanic origin was proved at the appropriate level in one of the boreholes, but no volcanic material was detected in another hole close by (Figure 17, A, B). South of the main outcrop, a borehole near Aldwark [SK 25 NW/22] proved 4.4 m of green and brown clay with altered basalt fragments, and a lens of lava has been mapped, on surface indications, near Hoe Grange [215 559]. The present patchy distribution may be the result of penecontemporaneous erosion of a more extensive sheet, perhaps a leaf of the Lower Matlock Lava which, farther east (Chisholm and others, 1983), is known to extend down to the base of the Monsal Dale Limestones.

South of Winster, the outcrop of the **Lower Matlock Lava** can be traced eastwards into the type area (Smith and others, 1967, pp. 11–12), but within the present district exposures are rare. In boreholes north of Winster (Figure 17; A, B, C) the unit is about 50 m in thickness. Farther south, around Aldwark, the lava crops out in a number of outliers, where it forms the relatively steep slopes beneath a thin capping of harder limestone; exposures are few. The lava thins towards the west and south, and cannot be traced beyond Minninglow Hill [209 573], Curzon Lodge [234 560] and Golconda Mine [249 551]. North-east of Aldwark, however, the lava is replaced laterally by the Shothouse Spring Tuff. Arnold-Bemrose (1907, p. 263) described this as 'greenish volcanic tuff, consisting of fine and coarse layers of lapilli in a calcareous cement'. It is at present poorly exposed, but can be mapped by the presence of tuff fragments in the soil for about 1 km south-west and south-east of Shothouse Spring [242 589]. The outcrop is thus localised and lies close to the tuff-filled Grangemill necks (p. 98), which Arnold-Bemrose (1907, pp. 262, 263) and Walters and Ineson (1981, p. 90) have regarded as the source. A recent borehole [SK 25 NE/41], east of Grangemill and just east of the district boundary, supports this view in showing that the tuffs interdigitate with limestone as they are traced away from the necks.

Throughout the Winster–Aldwark area the lava rests on 60–70 m of Monsal Dale Limestones, but towards the east, near Bonsall, these limestones interdigitate with lava and tuff (Chisholm and others, 1983, fig. 2) which formed the lower part of the Lower Matlock Lava in that area: the thickness of this member accordingly increases there to about 100 m. South of Bonsall, the entire sequence thins markedly, however, and around Golconda Mine, at the south-eastern extremity of the district, a thin Lower Matlock Lava rests directly on Bee Low Limestones.

The **Upper Matlock Lava** is extensively developed in the adjacent district (Smith and others, 1967, p. 12). It dies out at outcrop [252 603] east of Winster, but the associated darker limestones persist beyond the margin of the flow and can be recognised in boreholes north of the village (Figure 17).

In the Millclose Mine workings the limestone sequence contained three lavas that were probably derived from the Bonsall centre. They are the **Lower 129 Toadstone** (Traill, 1940, p. 207), which thinned towards the north, and two flows that died out in the same general direction, the **Upper 129 Toadstone** and the **Upper Toadstone** (Traill, 1940, pp. 205, 196). The 129 toadstones may be equivalent to the Lower Matlock Lava (Chisholm and others, 1983, fig. 4); the Upper Toadstone is certainly the Upper Matlock Lava (Traill, 1940, p. 194).

ALPORT CENTRE

An elongate magnetic anomaly (Figure 46) that extends ESE from Bakewell to Ashover (well beyond the margin of the district) is tentatively interpreted as a large volcanic centre. A thick volcanic sequence (Ramsbottom and others, 1962, p. 138), now taken as the type section of the **Fallgate Volcanic Formation** (Aitkenhead and Chisholm, 1982, pp. 6, 7), has been proved near the eastern end of the anomaly. Provings of thick volcanic rocks assigned to the Fallgate Volcanic Formation within the present district are confined to the area around Alport [boreholes SK 26 SW/9 and SK 26 NW/25]: the Alport centre is named after these occurrences, even though they lie some distance away from the axis of the magnetic anomaly, the supposed source of eruption. There is no sign of the volcanic centre at the surface, for the thick volcanic sequence is concealed by limestones (Figure 15): farther west, where the lavas come to crop, they thin rapidly and die out among the limestones within a short distance. The name Fallgate Volcanic Formation is applied only where the proportion of volcanic rocks in the sequence exceeds that of limestone.

A borehole east of Haddon Fields [SK 26 NW/25] proved 228 m of lava and tuff with thin limestones, without reaching the bottom of the volcanic formation. The lowest 58 m were mainly amygdaloidal and compact basalt with a few bands of hyaloclastite. Above lay 13 m of medium and dark grey bioclastic limestone containing the Brigantian coral *Diphyphyllum lateseptatum*; on this evidence the base of the Brigantian stage has been drawn provisionally at the bottom of the limestone. Above the limestone lay a further 85.5 m of volcanic rocks, comprising in upward sequence: 20 m of well bedded tuff, 8 m of partly brecciated basalt, 40 m of hyaloclastite and 17.5 m of amygdaloidal basalt. This sequence can be interpreted as the filling by volcanic material of a body of water of progressively shallower depth, ending with subaerial lava flows. For the purpose of nomenclature, the extrusive rocks above the fossiliferous limestone are regarded as a single member, the *Lathkill Lodge Lava*. Its top parts are known in several boreholes (Figure 15) and the type locality is an outcrop in Lathkill Dale, where a section [2041 6617] shows 3.8 m of weathered vesicular basalt. The flow dies out among the limestones not far to the west, but details of the contacts are not visible. A lava, the top of which is poorly exposed in the floor of Bradford Dale [201 639], west of Youlgreave, is believed to be the Lathkill Lodge Lava, on the basis of its distance below the Lathkill Shell Bed. JIC

In the borehole east of Haddon Fields, referred to above, the Lathkill Lodge Lava was overlain by 19.1 m of limestone, and this by the **Conksbury Bridge Lava**, 51.71 m thick. The latter is here made up of three flows of compact and amygdaloidal basalt (Figure 15). Around the type locality the thickness is about 30 m, though the exposure is poor and the best section [2098 6600] shows only 3.8 m of vesicular lava. The member dies out westwards among the limestones near Over Haddon (p. 36). In an isolated outcrop north-west of Bakewell it is about 25 m, and an exposure [2129 6913] shows 4.3 m of vesicular lava. Farther north [209 697], an ill-exposed circular outcrop of igneous rock is interpreted as an inlier of the lava; this outcrop was formerly regarded as a neck (Arnold-Bemrose, 1907, pp. 261 – 272).

To the south of Alport, several lavas were encountered in the workings of Millclose Mine (Traill, 1939; 1940; Shirley, 1950). The four that thin or die out towards the south (the **144 Pump Station, 144 Pilhough, 103** and **Alport toadstones**) were probably derived from the Alport centre. The highest of them, the Alport Toadstone, is correlated with the Conksbury Bridge Lava and the others may represent leaves of the Lathkill Lodge Lava.

The lavas west of Ashford are also believed to have emanated fom the Alport centre, towards which they thicken (Figure 14). The **Lees Bottom Lava** is seen [1699 7050] near the A6 road in exposures of vesicular basalt totalling 8.5 m. It was also proved in the Lees Bottom Borehole [SK 17 SE/14]. The outcrop can be traced around the valley sides, but there is a gap at the south-west side of Lees Bottom where the lava dies out, and it was absent also from a borehole north of Sheldon [SK 16 NE/10]. The unit was correlated with the Lower Miller's Dale Lava by Butcher and Ford (1973, p. 180), but this is not acceptable, for around Lees Bottom it is overlain by 30 m of limestone with Brigantian coral-brachiopod faunas (p. 31).

The succeeding **Shacklow Wood Lava** is up to 20 m thick at outcrop; 17 m are exposed, part compact and doleritic, in a cliff by the A6 [1794 6975]. The lava is impersistent, dying out on the south side of the valley [1743 7010] and reappearing for a short distance above Lees Bottom, where an isolated lens [1686 7051] shows 0.55 m of vesicular lava. In Borehole SK 16 NE/10 it is represented by 0.55 m of greenish volcanic detritus at the junction of dark and pale limestones (Figure 14). South of the outcrop, the lava has been proved at depth in Magpie Sough as far south as Sheldon, but it dies out between here and Magpie Mine [1725 6816] where it is represented by the '480 ft wayboard' (Butcher, 1975, p. 69). It reappears to the east at True Blue Mine [1777 6799] (Carruthers and Strahan, 1923, p. 63), and thickens to at least 50 m in a borehole [SK 16 NE/19] near Bolehill Farm (Figure 14). The lowest of three volcanic horizons recorded at Dirtlow Mine [187 686] (Green and others, 1887, p. 140) is probably the Shacklow Wood Lava (Walters and Ineson, 1981, p. 101). Farther north, a thin but persistent lava encountered in boreholes at Longstone Edge (Figure 13) is also correlated with the Shacklow Wood Lava. It lies well above the base of the Monsal Dale Limestones, not at the base of that formation, as was supposed by Walters and Ineson (1981, fig. 18).   JIC, IPS

LONGSTONE CENTRE

Boreholes on Longstone Edge (Figure 13) have proved the existence of thick, but localised, volcanic rocks interbedded with the Monsal Dale Limestones. Most are presumed to be related to a volcanic centre near or to the north of Longstone Edge: these rocks thin out towards the west and south, and very few come to crop in Cressbrook Dale or the Wye valley. A minority may have been derived from other centres. The relationship between limestone, tuff, lava and hyaloclastite is complex, and not all the volcanic units have been named (Figure 13). No magnetic anomaly is present at

Longstone Edge (Figure 46), though there is one a few kilometres to the north which may mark the volcanic centre.

The earliest volcanic rocks are thin tuffs and a lava, all unnamed, in the lowest 125 m of the Monsal Dale Limestones. An overlying thin lava, present in several boreholes, is correlated with the Shacklow Wood Lava (*see above*). It is usually vesicular, but in one borehole [SK 27 SW/10] had a hyaloclastite, 0.76 m thick, at the base. Above this level the limestones contain lenticular volcanic rocks of very variable thickness; the lower tuff units are not named (Figure 13) but the main mass, which consists of hyaloclastite and tuff, is here correlated with the **Cressbrook Dale Lava**. The latter reaches a maximum thickness of 80.88 m in Borehole SK 27 SW/5, where it consists entirely of hyaloclastite, and it thins and disappears westwards; to the east it also thins but passes laterally into well bedded tuff (Figure 13) containing some limestone lapilli. To the south-west it also dies out, for it is not present among the limestones of the Wye valley between Cressbrook and Ashford. To the north, just beyond the district boundary, the lava has been encountered in several boreholes (Stevenson and Gaunt, 1971, figs. 8, 12; Dunham, 1973), one of the thickest records (76.59 m) being in the Eyam Borehole [SK 27 NW/15]. It crops out in a small area around the type locality (Stevenson and Gaunt, 1971, pp. 67 – 69). Normal lava is present in these sections, though with some hyaloclastite ('basalt breccia'; Walters and Ineson, 1981, fig. 18) in the Eyam and Wardlow Mires No. 1 boreholes. The existence of hyaloclastites among the tuffs beneath Longstone Edge was not known to Walters and Ineson (1981), however, and they consequently believed that the Cressbrook Dale Lava died out farther north.

In the Longstone Edge boreholes (Figure 13) the **Litton Tuff** lies 19 to 47 m above the Cressbrook Dale Lava and consists of one to three leaves of grey and green banded tuff with some limestone lapilli. The greatest recorded thickness [SK 27 SW/1] is 26.21 m; here it is well sorted with coarser bands of fine lapilli, elsewhere it is finer-grained and splits into two leaves. As seen in the Sallet Hole crosscut [2188 7370] the top 1.5 m are fine-grained and well banded, though the tuff is coarser-grained and unsorted below. In the Eastern Decline [2197 7366], the presence of rounded clasts of pale green igneous rock some 3 m above the tuff is perhaps an indication of a topographic feature in the tuff which underwent erosion.

To the west of Longstone Edge, the outcrop of the tuff can be traced by surface features southwards from the type locality (Stevenson and Gaunt, 1971, p. 75) into the present district, until it dies out on Hay Cop [178 739]. There are no good exposures. South of here, no outcrop has been mapped but a tuffaceous bed referred to this horizon occurred in a borehole near Great Longstone [SK 17 SE/12].   IPS

TUNSTEAD CENTRE

The known distribution of the Miller's Dale lavas suggests that they were derived from a centre just north of the district, near Tunstead [110 750], although no feeders have been identified (Stevenson and Gaunt, 1971, pp. 24 – 25, 27). A strong magnetic anomaly in this area (Figure 46) is probably associated with the Calton Hill Intrusion (p. 96); this is younger than the lavas, however, and of different composition, so that its geographical association with the centre may be coincidental.

The **Ravensdale Tuff**, up to 20 m thick, lies some 80 m from the top of the Bee Low Limestones in the lower reaches of Cressbrook Dale. An exposure in the stream [1724 7393] shows 2.44 m of banded tuff with both cognate and accidental lapilli. It may have resulted from the same volcanic event that gave rise to the little known igneous rock of the Litton Dale Borehole in the adjacent district (Stevenson and Gaunt, 1971, p. 50; Walters and Ineson, 1981, p. 105). Its distribution is very localised.

The **Lower Miller's Dale Lava** (Stevenson and Gaunt, 1971, p. 23) crops out extensively within the northern part of the district (Figure 10) at a stratigraphical horizon some 40 m below the top of the Bee Low Limestones. The outcrop extends from the eastern slopes of Grin Low in the west [057 717] to Tideswell Dale in the east [155 737], and from near Chelmorton [110 704] and Brierlow Bar [086 697] in the south to well beyond the district boundary in the north. At Buxton Quarry [0846 6902], 0.75 km south of Brierlow Bar, a lens, too small to be shown on the 1:50 000 and 1:25 000 maps, with maximum thickness of about 4.3 m and largely altered to clay, lies some 28.5 m below the top of the Bee Low Limestones and is regarded as a tongue of the lava. A borehole just north of Taddington [SK 17 SW/55], extends the limit shown by Walters and Ineson (1981, fig. 12) in that area (Harrison, 1981, fig. 5).

The mapped outcrops south of the River Wye tend to be scattered and lens-like, suggesting that the lava in this area was emplaced in a number of discrete tongue-like flows. An alternative but less likely explanation is that the outcrop represents the remains of a once continuous sheet left after dissection during a period of subaerial erosion. To the north, the outcrop is more continuous, indicating more sheet-like flows, a form which led Stevenson and Gaunt (1971, p. 20) to suggest that eruption was by extrusion along fissures. Walters and Ineson (1981) suggest that the vents in Monk's Dale in the adjacent district to the north were the main extrusive centres. However, as these vents contain tuff and agglomerate rather than the basalt or dolerite that would be expected to be associated with extensive basaltic lava flows, the first mode of eruption is preferred here.

On the east and south-east side of Buxton the lava, up to 15 m thick, forms a nearly continuous outcrop. In Ashwood Dale, a section in the railway cutting shows 7.62 m of vesicular lava [SK 07 SE/57]. A short distance to the north-west, however, the lava dies out locally. To the south-west, an exposure in a road-cutting [0640 7229] shows 3.35 m of pale green clay on 2.59 m of vesicular lava. The main outcrop terminates north of Harpur Hill, though an isolated lens occurs [062 709] to the south-west of this place. The lava reappears north of Hillhead Quarry, near which it is seen [0710 7023] in a railway-cutting. The lava forms isolated outcrops east of Calton Hill and in Blackwell Dale, a section [1290 7223] in the latter showing 3.3 m of mainly non-vesicular rock. At Wormhill, exposures continuous with the more extensive outcrops in the Chapel en le Frith district extend southwards to Miller's Dale, the lava dying out in the vicinity of Chee Tor. To the east, a small inlier is present in the dale immediately west of Ravenstor.

In Tideswell Dale the lava is intruded by dolerite (p. 97), which, in a quarry on the eastern side of the dale [1547 7378], overlies 0.61 m of purple prismatic clay-rock.

The **Upper Miller's Dale Lava** occurs in a nearly continuous outcrop between Hurdlow Town and Miller's Dale, whilst a thin and isolated outcrop is present at Harpur Hill. The lithology varies little, most outcrops showing amygdaloidal lava though there is minor variation in the proportion of calcite and chlorite in the amygdales.

At Calton Hill Quarry [1205 7115] (p. 96) some 6 m of vesicular lava are intruded at the base by dolerite. South-west of Miller's Dale, an exposure [1341 7299] shows 13 m of vesicular lava. To the east of Miller's Dale, the lava forms the outlier of Knot Low where 6.6 m of it are seen in a quarry [1339 7355]; the lowest 5 m are less vesicular and may represent the inner part of a flow. The junction with the underlying 'Station Quarry Beds' is visible in Station Quarry [1329 7347].

Near Litton Mill, the lava dies out eastwards along a line extending NNE–SSW for at least 1 km. It is best seen in Litton railway-cutting [SK 17 SE/18], which shows some 5.18 m of lava, the upper 1.52 m of which are poorly exposed. The lava is brown-weathered

Plate 9 Photomicrographs of clastic basalt textures from a borehole [SK 25 NW/25] at Haddon Fields.

Illustrations to contrast hyaloclastites *sensu stricto*, 1 to 3 on the left, with graded volcaniclastic rocks 4 to 6. The latter are ostensibly tuffs but may likewise be of hyaloclastite origin. Phyllosilicate and zeolite identifications in all samples were made or confirmed by X-ray diffraction analysis by R. W. O'B. Knox (see Internal Report PE/LD/82-17).

1 E 52568. Unnamed flow below the Lathkill Lodge Lava; depth 275.95 m. Ordinary light. Scale bar 5 mm.
Hyaloclastic with devitrified amygdaloidal microporphyritic basalt fragments in a matrix of polycrystalline phyllosilicate. The matrix consists of iron-poor chlorite with subordinate smectite. The clasts contain the same minerals in inverse proportion, together with plagioclase.

2 E 52559. Lathkill Lodge Lava; depth 170.0 m. Ordinary light. Scale bar 5 mm.
Hyaloclastite of coarser fabric but very similar composition to that of 1.

3 E 52559. An enlargement of part of 2. Plane-polarised light. Scale bar 1 mm.
The amygdales show the centrepetal succession of smectite zones (grey) and almost colourless chlorite. All former surface planes are lined by leucoxene (black) and a smectite palisade coats the outer surface of the clast. Partly fresh labradorite laths retain their zoned structure. The groundmass (upper left) shows the multispherulitic intergrowth of chlorite and smectite.

4 E 52563. Lathkill Lodge Lava; depth 205.75 m. Ordinary light. Scale bar 5 mm.
Laminated medium and fine-grained volcaniclastic rock, possibly a graded hyaloclastite. The clasts consist mainly of a smectite different from that of 1 to 3. The matrix is analcime intergrown with spherulitic smectite and cross-cutting veinlets are also filled with analcime. A solitary spheroidal bioclast of calcite appears near the centre of the top right quadrant of the field.

5 E 52564. Lathkill Lodge Lava; depth 218.85 m. Ordinary light. Scale bar 5 mm.
Part of the same lithological unit as 4. Though much coarser, the fabric is similar to that of the coarse laminae in 4; note the slight bedding orientation of clasts that is absent in 1, simple hyaloclastite of comparable coarsenesss. The mineralogy is similar to that of 4 but in the groundmass quartz, calcite and ?talc are additionally present.

6 E 52564. An enlargement of part of 5. Plane-polarised light. Scale bar 1 mm.
Note the contrast in size and abundance of the amygdales here with those of 3. The matrix consists of incrusting radiating intergrowths of analcime and smectite cemented by poikilotopic calcite and minor leucoxene. The clasts consist of smectite and leucoxene with amygdales filled by a colourless moderately birefringent phyllosilicate (?mixed layer illite-smectite) and analcime.

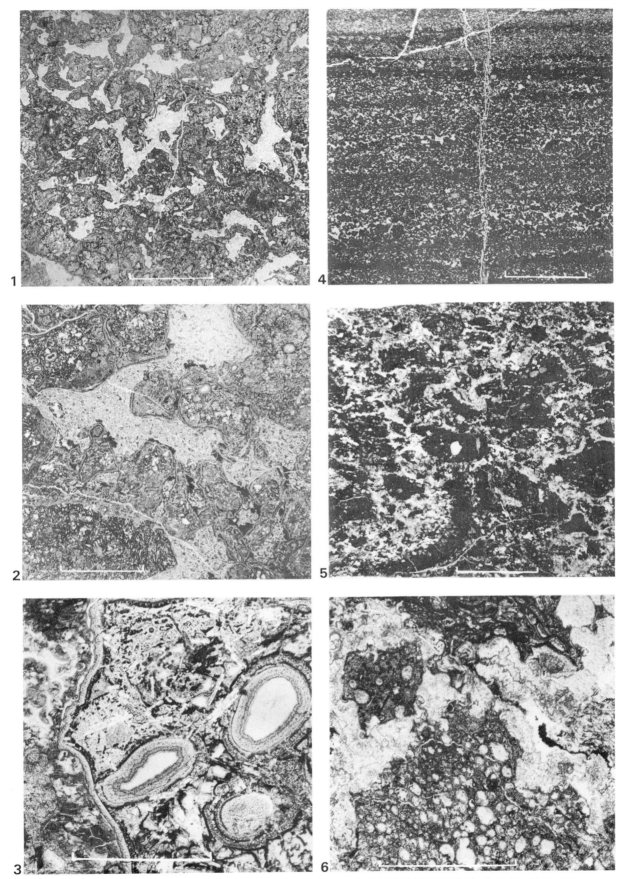

Plate 9   Photomicrographs of clastic basalt textures from a borehole [SK 25 NW/25] at Haddon Fields. (For detailed explanation see p.62)

and shows a rough flow-banding dipping eastwards at 40°; in the lava, rounded masses of harder material, up to 0.61 m in diameter, are interpreted as degraded lava blocks incorporated in a flow front. Dark thinly bedded limestone, cherty in places, rests with overlap on the lava.

An isolated area of lava, referred to this horizon, occurs near Cressbrook Mill, where an exposure [1737 7241] shows 3.66 m of the vesicular rock. The base is seen at the confluence of Cressbrook Dale with the Wye valley [1735 7264]. The correlation of this lava with that of the main outcrop has been discussed in detail by Walkden (1977, p. 358). At the western end [1617 7289] of Litton tunnel, this author (1977, p. 354) noted the presence of a K-bentonite band 10 cm thick, lying 4 m above the base of the Monsal Dale Limestones; the rock showed relict vitroclastic structure.

A borehole [SK 17 SE/13] north-east of Taddington proved 2.98 m of green-grey pyritic mudstone, which has been referred to the Upper Miller's Dale Lava horizon (Gatliff, 1982, p. 48); its base lay at a depth of 75.80 m.

IPS, NA

Petrography

(a) Coarse-grained lavas

The coarsest rocks of this suite strictly should be called dolerites. The Lower Miller's Dale Lava at Buxton [E 54993], for example, has plagioclase laths 0.45 mm to 1.35 mm long but a range of between 0.3 mm and 0.6 mm is more typical. Where fresh, the lavas consist of a typical basaltic fabric with interlocking calcic plagioclase laths, interstitial augite and accessory opaque oxide, one or more generations of olivine pseudomorphs and a little devitrified intersertal glass. A late-stage hydrothermal phase of crystallisation is generally expressed as interstitial asbestiform aggregates of mafic phyllosilicates, and in some cases has produced a stockwork of amoeboid amygdales. Some of the coarse-grained basalts also contain distinct, generally coarser and more spheroidal, true amygdales, defining a somewhat earlier phase of gas exsolution from the magma.

The plagioclase generally shows a random distribution of laths, and in no case is the mineral porphyritic. A few samples, however, show some slightly larger crystals, such as E 47809 from the Lower Matlock Lava, E 53780 and E 53781 from the Shacklow Wood Lava, and E 48017 from the Lathkill Lodge Lava; where fresh, such crystals tend to show traces of oscillatory zoning. Fresh feldspars otherwise display pronounced simple normal zoning from cores of sodic bytownite (An_{74}) to margins of sodic labradorite (around An_{54}) or even andesine. In less fresh basalts the plagioclase is albitised (An_{40} to about An_{20}).

Augite commonly occurs in a finely granular form with crystals of about 0.08 mm diameter, but it is sub-ophitic in samples E 52566 and E 52567 from the Fallgate Volcanic Formation. In samples E 48015 (Shacklow Wood Lava) and E 48019 (Lathkill Lodge Lava) the augite is fully ophitic, with crystals reaching 1.2 mm in diameter. Alteration is general and total in about half the samples examined; the products are mainly 'chlorite' or 'bowlingite' with minor carbonate and opaques. Recent studies by Walters and Ineson (1983) suggest that Fe-rich smectites are the dominant alteration products of basalts from Derbyshire. It is extremely difficult to identify these products on optical grounds alone, and XRD and electron microprobe analyses would be necessary for precise identification. Present XRD studies of basalts indicate that both smectite and chlorite occur (Plate 9).

Granular to subhedral olivine is totally pseudomorphed in all specimens examined. It occurs both in megacrysts of up to several millimetres across and a finer-grained type fully integrated in the groundmass. The megacrysts are in places relatively abundant (E 48018, Lathkill Lodge Lava), but the distribution is variable and some samples are almost megacryst-free (E 54947, Upper Miller's

Dale Lava). Replacement, especially in the case of the least altered basalts, is commonly by serpentiform aggregates of bowlingite. In places it is accompanied by serpentine, opaque oxide (including hematite), quartz, probable kaolinite and, more rarely, zeolite or fibrous amphibole. In some of the more thoroughly altered coarse basalts the place of bowlingite is taken by 'serpentine', smectite, 'chlorite' or other clay material and the occurrence of quartz and clay mineral aggregates is more common.

The interstitial to amygdular hydrothermal aggregates generally consist of similar minerals to those of olivine pseudomorphs in the same rock, except that iron oxides are rare or absent. In more thoroughly altered types the replacement products of augite are similar, and the former olivine is distinguished only where there is an extra strong opaque oxide association. The hydrothermal interstices are defined by their characteristic asbestiform or spherulitic crystal growth structure; walls of former voids are lined by a palisade of phyllosilicate crystallites with a linear or planar fabric aligned normal to the void wall. In lithologies with comparatively large former voids – including amygdales – an early layer of 'bowlingite' is in places succeeded by ?smectite in parallel growth (Plate 9.3) followed by calcite or microcrystalline clay. A minority of samples show the restriction of 'bowlingite' to olivine pseudomorphs and the presence of ?smectite in the former voids (E 48018).

Titaniferous iron oxide is an abundant accessory mineral. It shows a characteristic growth habit giving long slender cross-sections with ragged edges in detail. Some of the crystals are pierced by plagioclase, simulating 'ophitic' intergrowths.

In addition to the interstitial hydrothermal segregations most, if not all, of the coarse basalts contain traces of former interstitial tachylitic glass, pale olive-brown in colour and commonly almost isotropic, but typically containing opaque dust, colourless prismatic crystallites and needles of ?apatite.

(b) Fine-grained lavas

The fine-grained basalts are of two types. The commoner consists of a variolitic groundmass, rich in opaque oxide, set with microphenocrysts of olivine and plagioclase or their pseudomorphs. These rocks are commonly highly amygdaloidal, in contrast to the majority of the coarser basalts. The mineralogy is similar to that of the coarse basalts but alteration tends to be greater, with leucoxenisation of opaque oxides in addition to other changes. The relative dimensions and proportion of groundmass, megacrysts and amygdales are variable. The matrix normally exceeds megacrysts in amount.

The converse is true of the upper part of the Upper Miller's Dale Lava at Knot Low (E 54948) and west of Priestcliffe (E 54946). Here the rock is transitional to the coarse basalts described above, with groundmass little more than interstitial except in the immediate vicinity of some amygdales (*compare with* Elliot *in* Ramsbottom and others, 1962, pp. 143–163). Plagioclase and olivine megacrysts, the former up to 0.6 mm long, commonly approach the size of crystals in the holocrystalline basalts. Olivines up to 3 mm occur locally in a devitrified glass matrix; this, together with their irregular distribution, favours the interpretation that such crystals are xenocrystic. The amygdales constitute more than half the volume of some of the rocks, and the former voids are then commonly interlinked and partially collapsed. This tends to obscure any evidence of clasticity in thin-section, but macroscopically it is commonly clear that such rocks are autobreccias, probably representing the surface zones of lava flows, either *in situ* or incorporated in an advancing magma stream.

All samples collected from near the recognised tops and bottoms of flows are either variolitic basalts of the types described above or devitrified porphyritic tachylites representing an even more extreme degree of chilling. A thin-section of sample E 52561 from the junction between a hyaloclastite and overlying basalt in the Lathkill

Lodge Lava shows a sharp marginal chilling of the basalt base, with a densely clouded tachylite skin, 0.5 to 1.5 mm thick, grading sharply into a groundmass with plagioclase microlites up to 0.15 mm long. This and some other marginal chilled samples, even from the tops of flows, are not all rich in amygdales. E 52549 and E 52550, for example, from the weathered brecciated top of the Conksbury Bridge Lava contain few amygdales but show evidence of progressive upward chilling and, in E 52550, signs of heterogeneity in the flow, with the intermixing of variolitic textures of differing coarseness.

The less common variant of the fine-grained facies, represented by two samples only (E 48016 and E 52553 from the Conksbury Bridge Lava), is distinctively fine-grained, yet holocrystalline. In both samples, the groundmass olivine pseudomorphs are 0.15 to 0.45 mm across, but the plagioclase and augite of E 48016 are respectively only 0.05 to 0.08 mm long and 0.04 mm across compared with analogues of 0.08 to 0.4 mm and 0.08 mm in E 52553. The finer grained E 48016 contains sparse olivine megacrysts not seen in E 52553. In both rocks the olivine is replaced by 'bowlingite', and the late hydrothermal interstitial phyllosilicate is ?smectite. Both rocks contain sparse amygdales, calcite-filled in E 48016 and ?smectite-filled in E 52553 where there is also general replacement by the same mineral of the whole rock immediately around the amygdales. The two samples are sufficiently similar to one another and distinctive from others to suggest that they originate from the same flow.

(c) Volcaniclastic rocks

A common feature of all the volcaniclastic rocks examined is the total replacement of their basaltic glass by aggregates of olive-green to olive-brown phyllosilicates; most of this material is probably smectite, and in some samples this has been confirmed by X-ray diffraction analysis (analyst R. W. O'B Knox: Plate 9). In most samples the matrix of the clasts is composed largely of similar but spherulitic phyllosilicates; calcite, analcime and quartz are also not uncommon. Pseudomorphs after olivine are preserved in a minority of clasts, as are microphenocrysts of plagioclase. Primary calcic plagioclase was identified in only two samples, E 52559 and E 52569 [Borehole SK 26 NW/25]; elsewhere it is albitised, sericitised or even smectitised. Opaque oxide dust occurs in the majority of clasts and is probably leucoxenised throughout. As is usual in rocks of this type it is commonly concentrated at former surface planes (Plate 9.3), such as clast or amygdale walls and shatter fractures, in otherwise clear former glass (sideromelane); in other cases the opaque clouding is more general (tachylite).

On textural grounds, two main types of volcaniclastic basalt are distinguished in this suite. First, a type (Plate 9.1, 9.2, 9.3) in which the angular clasts contain comparatively large and few amygdales (20% of less of the rock) and are themselves commonly very poorly compacted. This type typifies the central part of the Lathkill Lodge Lava (Plate 9.2, 9.3) in boreholes at both Haddon Fields (E 52557, E 52559–60) and Youlgreave (E 48020), and possibly signifies hyaloclastite lithology in the strictest sense, that is clasticity caused by the spalling-off of shards from the surface of lava pillows:

E 52558 is a tachylitic variolitic basalt from a pillow preserved in this unit. The same lithology (Plate 9.1) also occurs below the Lathkill Lodge Lava in a borehole at Haddon Fields (E 52568–69). Clastic basalt containing more highly amygdaloidal lithoclasts occurs as a relatively thin sheet (4 m thick) in the Shacklow Wood Lava penetrated by a borehole [SK 16 NE/19] near Bolehill Farm, but the general fabric of this rock and the restricted variety of lithoclasts (E 53778A and B) suggest that it too is a true hyaloclastite that has not suffered significant reworking.

The volcaniclastic rocks elsewhere described as tuffs (p. 59; Plate 9.4, 9.5, 9.6) are characterised by more spheroidal (though still angular) clasts typically containing very abundant amygdales and amygdules. The mineralogy is similar to that of the hyaloclastites except that calcite is in some cases much more abundant, not only as matrix but also as clasts. Carbonate enrichment is especially prominent in the Ravensdale Tuff (E 54943–44), but is also present in an otherwise analcime-rich and in part finely bedded lithology (Plate 9.4, 9.5, 9.6) at the base of the Lathkill Lodge Lava (E 52561–64). These rocks cannot be distinguished on petrographical grounds from some types of hyaloclastite.

Petrogenesis

Petrographical evidence indicates that the range in primary chemical composition was small, though not negligible: in particular the presence or absence of subporphyritic plagioclase crystals in otherwise similar rocks is probably significant. The inference already made in the text that olivine is at least in part xenocrystic means that mantle contamination of magma must be considered as a genetic factor just as much as it is in the more obvious case of the peridotite-bearing Calton Hill intrusion (p. 96).

Elliot (in Ramsbottom and others, 1962, p. 157) draws attention to well documented potassium enrichment of many altered Derbyshire basaltic rocks, and Walkden (1972) identifies the related clay wayboards as potassium-bentonites. It is probable that a high proportion of smectite and 'bowlingitic' alteration comprises mixed-layer phyllosilicates, and the potassium enrichment of the altered basalts may be related to this. The primary basalts may have contained only an average proportion of potassium, but during or after alteration the lattice structure of the secondary phyllosilicate would have been an ideal 'sponge' to absorb selectively or retain this element, perhaps by cation exchange with ambient sea water; primary potassium enrichment of the magma cannot, however, be ruled out. The initial formation of the expanding-lattice clays may have been in part autometasomatic and in part pedogenic. Some flows show oxidised altered tops suggesting tropical soil formation, but the tops of many are converted to green argillaceous material similar in appearance to, and locally passing laterally into, clay wayboards. These greenish clays formed under reducing conditions and they grade into the more highly altered types of basalt, characterised by albitised labradorite. Since the secondary plagioclase is commonly sodic andesine (not albite), this alteration is more likely to have occurred in autometasomatic conditions, probably involving some interaction with overlying sea water, than by weathering at normal surface temperatures.   NGB

CHAPTER 5

Namurian

The Namurian outcrop is separated into two unequal parts, to east and west of the Derbyshire Dome (Figure 1), but the rocks were originally deposited in a single basin and consist mainly of sandstones, siltstones and mudstones derived from landmasses to the south and north-east. Variations in thickness and detailed succession between the two outcrop areas are shown in Figure 22.

CLASSIFICATION

The Namurian rocks of the Pennine region are also known as the 'Millstone Grit Series'. 'Millstone Grit' was a term originally used by Whitehurst (1778, p. 147) for a coarse-grained sandstone in the Derwent valley just east of the present district, but later, as 'Millstone Grit Series', the term acquired a time–stratigraphic connotation and was extended to include all strata of Namurian age. The name as used in this sense is inappropriate, for the sequence consists of over 50% mudstone and siltstone, and coarse-grained sandstones or 'grits' are present only in the upper half. Earlier classifications, and those used in adjacent districts, are shown in Figure 23, and compared with the nomenclature used in the present account. Stages and zones based on goniatites have been used as the most convenient way of subdividing the sequence, and on the published maps a long-established scheme of colours has been used to indicate the relative ages of the sandstones. This scheme emphasises the major structures on the face of the map though it does not show differences in sandstone facies. These are represented in a generalised way in Figure 22.

LITHOLOGY

Grey mudstones and argillaceous siltstones make up more than half the sequence in most areas. They are commonly laminated, and many contain grey-brown sideritic ironstone nodules and bands. Subordinate grey to black fissile mudstones occur, and in places contain dispersed pyrite and thin limestone beds and 'bullions' (concretionary bodies with a discoidal form studied in detail by Holdsworth, 1966). The marine marker bands with their goniatite-bivalve faunas belong to this facies.

Arenitic rocks interbedded with the mudstones and siltstones are of three distinctive petrographic types (Holdsworth, 1963b); calcareous siltstone, protoquartzitic sandstone and feldspathic (subarkosic) sandstone. The calcareous siltstones are interpreted as turbidites, and are restricted to the western outcrop where they occur only in beds of E_{1a} to E_{2a} age. The protoquartzites, known traditionally as 'crowstones' (*but see* Holdsworth, 1964), range in age from E_{1b} to R_{1c}; the earliest thus interdigitate with the calcareous siltstones. In the present district almost all are

turbidites, and deposition continued longer in the south than in the north (Evans and others, 1968, fig. 5). These rocks are considered to be derived from a southern landmass (*see below*) and were not deposited in the area east of the Derbyshire Dome. Feldspathic sandstones derived from the north and east first appear in R_{1c} times, and thereafter dominate the sequence in all areas. They include turbidites as well as sheets of very coarse-grained cross-bedded sandstone of shallow-water origin, the original 'Millstone Grits'.

PALAEOGEOGRAPHY AND DEPOSITIONAL HISTORY

In Namurian times the present district lay within the Central Pennine Basin (Craven Basin of Ramsbottom and others, 1978, p. 16), a depositional area which extended from the Craven Faults to the 'Wales–Brabant Island' ('St George's Land'; 'Midland Landmass'). In the west of the district lay the North Staffordshire Basin (Trewin and Holdsworth, 1973) where a thick sequence accumulated, while in the east a thin sequence was laid down on an upstanding area over the stable basement block (p. 109). These palaeogeographical features are shown in Figure 24.

In the earliest part of the Namurian (E_{1a} – E_{1b}) sedimentation tended to continue in the style that was established in the late Dinantian (P_2 Zone), with calcareous argillaceous deposits predominating. In the North Staffordshire Basin, these comprised mudstones and calcareous siltstones (172.9 m thick in the Gun Hill Borehole, p. 69) probably deposited by turbidity currents transporting the sediment from a marine shelf area farther west (Trewin and Holdsworth, 1973). In the east, the equivalent succession is greatly condensed, being only about a metre thick in the Stanton Syncline (p. 74).

The first protoquartzites, the Minn Sandstones, appear in the upper part of the E_{1b} Zone. They were deposited in thin sheets intermixed with silt and mud, all brought in by turbidity currents probably flowing from a source in the deltas flanking the Wales–Brabant Island to the south. Thereafter, until the end of the E_{2a} Zone, this facies alternated with the calcareous siltstones. Subsequent turbiditic influxes of protoquartzitic sand and silt (Figure 24) vary in thickness, extent and direction of flow throughout the E_{2b} to R_{1c} zones. The area of maximum thickness of the Hurdlow Sandstones lies to the south of the district, whereas the Lum Edge Sandstones appear to have their greatest development within the district. The outcrop pattern and limited exposure of the latter suggest the presence of several channel-centred sand bodies with north-eastward flowing palaeocurrents rather than extensive overlapping sheets. Near Thorncliff this unit, according to Bolton (1978), includes the only non-turbidite protoquartzitic sequence in the district (p. 79). The Blackstone Edge Sandstones have a similar but more

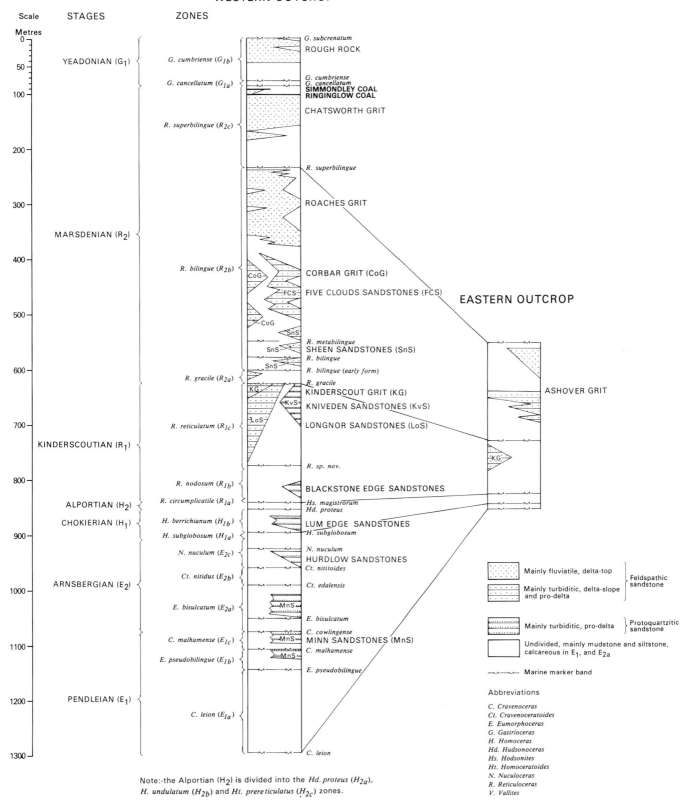

Figure 22   Generalised vertical sections, classification, and major sandstone facies, of the Namurian strata of the district.

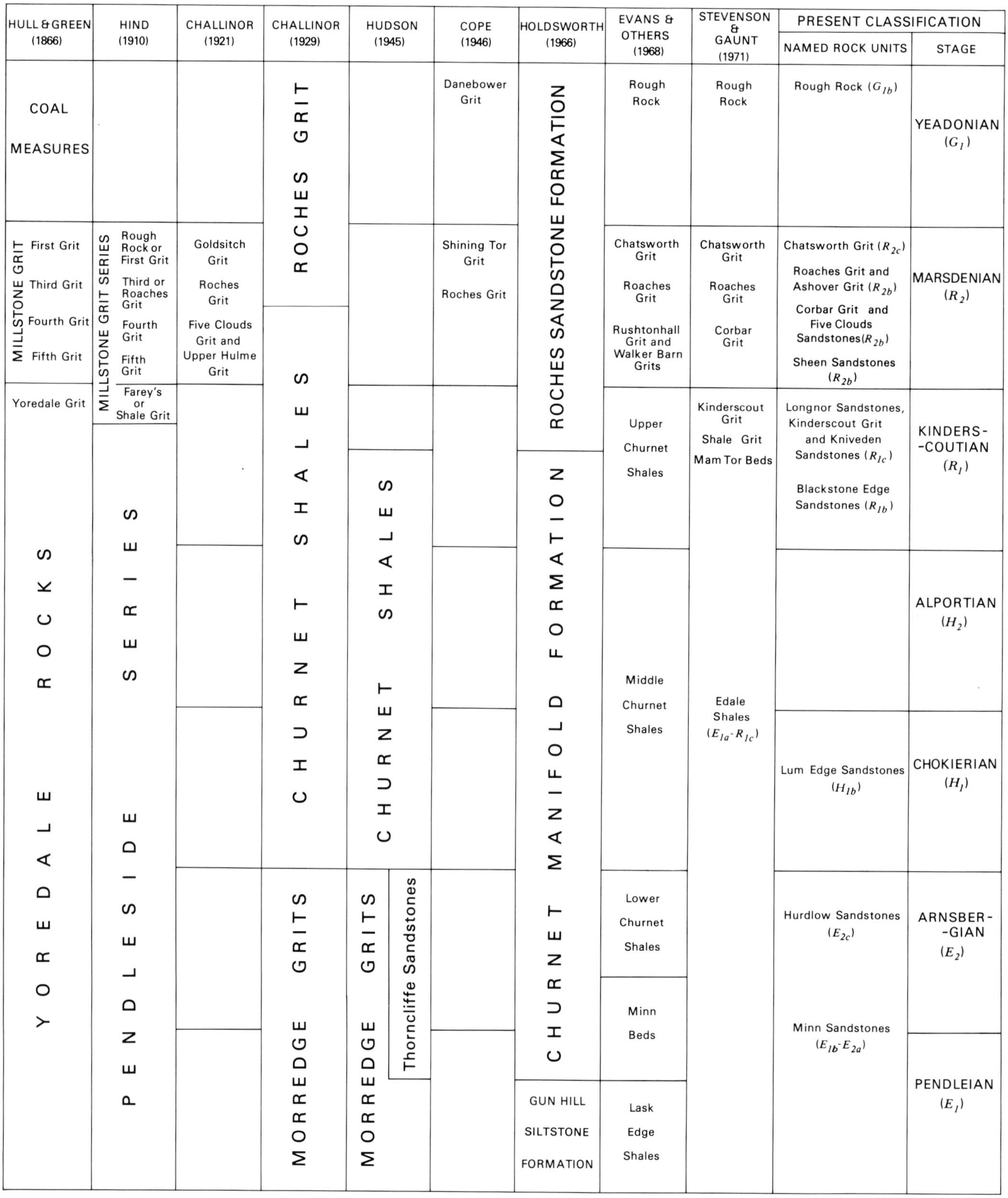

HULL & GREEN (1866)	HIND (1910)	CHALLINOR (1921)	CHALLINOR (1929)	HUDSON (1945)	COPE (1946)	HOLDSWORTH (1966)	EVANS & OTHERS (1968)	STEVENSON & GAUNT (1971)	PRESENT CLASSIFICATION	
									NAMED ROCK UNITS	STAGE
COAL MEASURES			ROCHES GRIT		Danebower Grit	ROCHES SANDSTONE FORMATION	Rough Rock	Rough Rock	Rough Rock (G_{1b})	YEADONIAN (G_1)
MILLSTONE GRIT — First Grit, Third Grit, Fourth Grit, Fifth Grit	MILLSTONE GRIT SERIES — Rough Rock or First Grit, Third or Roaches Grit, Fourth Grit, Fifth Grit	Goldsitch Grit, Roches Grit, Five Clouds Grit and Upper Hulme Grit			Shining Tor Grit, Roches Grit		Chatsworth Grit, Roaches Grit, Rushtonhall Grit and Walker Barn Grits	Chatsworth Grit, Roaches Grit, Corbar Grit	Chatsworth Grit (R_{2c}), Roaches Grit and Ashover Grit (R_{2b}), Corbar Grit and Five Clouds Sandstones (R_{2b}), Sheen Sandstones (R_{2b})	MARSDENIAN (R_2)
Yoredale Grit	Farey's or Shale Grit		CHURNET SHALES			CHURNET MANIFOLD FORMATION	Upper Churnet Shales	Kinderscout Grit, Shale Grit, Mam Tor Beds	Longnor Sandstones, Kinderscout Grit and Kniveden Sandstones (R_{1c}), Blackstone Edge Sandstones (R_{1b})	KINDERS-COUTIAN (R_1)
YOREDALE ROCKS	PENDLESIDE SERIES			CHURNET SHALES			Middle Churnet Shales	Edale Shales (E_{1a}-R_{1c})		ALPORTIAN (H_2)
									Lum Edge Sandstones (H_{1b})	CHOKIERIAN (H_1)
		MORREDGE GRITS		MORREDGE GRITS	Thorncliffe Sandstones		Lower Churnet Shales		Hurdlow Sandstones (E_{2c})	ARNSBER-GIAN (E_2)
							Minn Beds		Minn Sandstones (E_{1b}-E_{2a})	
						GUN HILL SILTSTONE FORMATION	Lask Edge Shales			PENDLEIAN (E_1)

Figure 23   Development of classification of the Namurian rocks in the district and adjacent areas.

Figure 24 Generalised palaeogeography and major phases of fill of the Central Pennine Basin in Namurian times (after Jones, 1980, figs. 1 and 13). Upstanding areas with sedimentation are ornamented. DB = former late Dinantian shelf, WS = western shelf. Rectangle indicates area of Buxton 1:50 000 sheet.

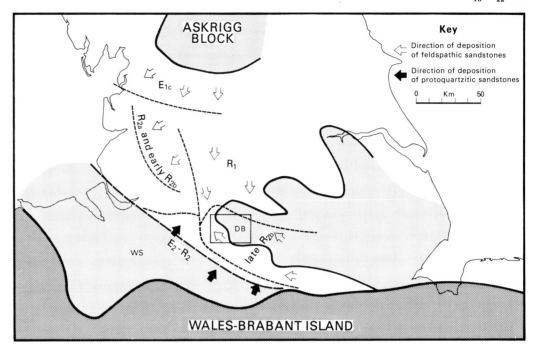

restricted development, while the Kniveden Sandstones are mostly confined to the Leek area.

During R_{1c} times, a turbiditic influx from the north brought in the first sub-arkosic sand, to form the Longnor Sandstones and the distal parts of the Kinderscout Grit. These turbidites heralded the approach of a large delta complex which had been prograding intermittently from the north, probably at times of lower sea level, throughout the Namurian (Ramsbottom, 1969, fig. 5). Great thickness contrasts show the continuing effect of the upstanding area over the stable basement block.

In R_{2b} times, relatively minor sand deposition in the earlier subzones formed the lower and middle leaves of the Sheen Sandstones. A major turbiditic influx from the southeast followed, depositing spreads of sand which for the first time probably extended right across the former upstanding area, to form the lower leaf of the Ashover Grit in the east, and the upper leaf of the Sheen Sandstones, the Five Clouds Sandstones and the Corbar Grit in the west. Finally, in the later part of the R_{2b} cycle, fluviatile delta-top sedimentation was established in this southern part of the Central Pennine Basin (Figure 24) and is represented by the upper leaf of the Ashover Grit and the main part of the Roaches Grit.

After a marine inundation which formed the *R. superbilingue* Band, and after subsequent minor turbidite influxes, fluvial conditions returned from a general north-easterly direction and the main part of the Chatsworth Grit was deposited. The overlying Ringinglow Coal is the lowest Upper Carboniferous coal in the district; it was at one time regarded as the local base of the Coal Measures (Hull and Green, 1866).

The highest Namurian stage, the Yeadonian, commenced with two marine episodes and finished with a fluviatile sandstone, the Rough Rock. This is the most extensive of the Namurian sandstones and has a complex internal sedimentary structure (Heath, 1973).

Evidence for the rhythmical or cyclical nature of Namurian sedimentation in the Central Pennine Basin has been presented by several authors. Both small-scale cycles, such as the sequence of faunal phases in the marine and quasi-marine mudstones (Ramsbottom and others, 1962), and large-scale cycles or mesothems (Ramsbottom, 1969, 1977; Trewin and Holdsworth, 1973; Bolton, 1978), have been recognised. These cycles may reflect the interplay of several factors but eustatic changes in sea level are generally held to be the dominant cause.

STRATIGRAPHY

Cravenoceras leion Band to base of Minn Sandstones ($E_{1a}-E_{1b}$)

In the western area, the sequence (Gun Hill Siltstones of Holdsworth and Trewin, 1968; Trewin and Holdsworth, 1973; Lask Edge Shales of the adjacent Macclesfield district) generally consists of alternating dark grey fissile ('shaly') mudstones and dark grey slightly calcareous siltstones. The latter form sharp-based graded beds, some with thin paler laminae at or near the base. These beds are individually thicker in the basal part of the sequence which reaches a maximum of 4.3 m in the headwaters of the Warslow Brook where it includes the diffuse faunal band of *Cravenoceras leion*. Some of the siltstones are sufficiently calcareous to be argillaceous limestones.

The $E_{1a}-E_{1b}$ sequence is 172.9 m thick in the Gun Hill Borehole [SJ 96 SE/18], and a similar thickness is probably present around the Mixon–Morridge Anticline. Eastwards it decreases markedly to about 60 m around Back of Ecton, and to 27 m north-east of Warslow Hall.

The Namurian–Viséan junction is conformable to the west of Warslow, but the oldest Namurian rocks known to overlie the Viséan limestones along the eastern side of the upper Dove valley are of late Pendleian (E1c) age (*see below*). Therefore the junction, to the east of Warslow, is probably one of gradual overlap, with some overstep onto various parts of the Dinantian sequence. Green and others (1887, p. 28) considered that much of the limestone–shale boundary was faulted, and not until much later did Hudson (1931) recognise its unconformable nature. The boundary is moderately indented along the east side of the Dove valley from Beresford Dale to near Glutton Bridge; from there to near Stoop Farm [064 681], however, it is highly indented with marked changes in altitude (Figures 25, 9). Shale outliers [such as 0837 6738] and embayments, such as those at Glutton [084 670] and Dowel Farm [076 673], partially fill hollows in the high relief topography of the apron-reef limestone (Plates 4 and 5) which, as Hudson (1931) suggested, is probably an exhumed pre-Namurian topography that was progressively buried by sediments during early Namurian times. The Stoop Farm Borehole [SK 06 NE/20] was drilled through 105.87 m of Namurian rocks filling such a hollow (p. 76).

In the eastern outcrop, the basal beds pass down conformably into the Longstone Mudstones, except north of Hassop where there is overstep onto Monsal Dale Limestones and Eyam Limestones. The sequence, probably no more than a metre thick (p. 74), is greatly condensed, which may reflect the continuing influence of the stable basement block on sedimentation.   NA

DETAILS

WESTERN AREA

The basal Namurian (E1a) sequence, and the upward transition from the Mixon Limestone-Shales and Onecote Sandstones, are displayed in numerous scattered exposures around the Mixon inlier, for example near Mixon [0433 5650], Intake Farm [0384 5533], Onecote Lane Head [0324 5580], Warslow Brook [SK 06 SW/12] and at Upper Elkstone [0543 5918]. The beds are also well exposed near Butterton Moor End [0524 5560], where Trewin and Holdsworth (1973) found a fauna including *Cravenoceras sp.* and *Eumorphoceras sp.* (*?E. angustum*); similar beds occur in Warslow Brook south-east of Stoneyfold [0697 5779 to 0748 5760]. However, diagnostic E1a faunas are generally badly preserved and scarce, *Cravenoceras leion* having been recorded from only one exposure, south-west of The Hill [0640 5730] (Holdsworth and others, 1970). In a stream section [0306 5840] near Old Mixon Hay, *Cravenoceras sp.* occurs in dark fissile mudstone intercalations between thin calcareous silty mudstone beds. These occur in a section 0.6 m thick lying 1.2 m above the top of the Onecote Sandstones, and are assumed to represent part of the *C. leion* Band. The only other exposure with an undoubted E1a fauna is in a stream [0877 5970] north-west of Warslow Hall; it yields *Eumorphoceras* cf. *tornquisti*. A prolific conodont fauna, together with *E. tornquisti*, was recorded by Coffey (1969). The best E1a section in the area lies just south of the present district at Bullclough [SK 05 NW/10]; it yields *Cravenoceras sp.* and shows the Namurian–Dinantian boundary.

The basal beds with argillaceous limestones pass upwards into thinly interbedded slightly calcareous silty mudstones or siltstones and mudstones, which are well exposed in tributaries of the River Hamps around Onecote, for example near Intake Farm [0372 5552], in Warslow Brook and its tributaries near Hole [0646 5797], Mount Pleasant [0589 5889], Hob Hay [0467 6021],

Blake Brook and its tributaries near Lower Fleetgreen [0551 6106], and Upper Fleetgreen [0440 6145]. The only goniatite locality in this part of the sequence yields *Eumorphoceras pseudobilingue* from alternating calcareous and fissile mudstone south-west of Warslow Hall [0862 5925] (Coffey, 1969). It is estimated to lie about 14 m above the base of the Namurian.

In the upper Dove valley (Figure 25), 10.9 m of interbedded calcareous siltstones and mudstones are exposed in an old track south-east of Crowdicote [1086 6454]. The base of these lies about 7 m above the top of the Dinantian apron-reef limestones, and the top about 17 m below a fossiliferous argillaceous limestone band 1.1 m thick, exposed on the left bank of the River Dove [1085 6443], which probably represents the *C. cowlingense* (E2a1) horizon. Below this horizon the Glutton Bridge Borehole [SK 06 NE/17] proved 18.75 m of dark grey mudstones with a few silty calcareous bands without reaching the *C. malhamense* Band.

EASTERN AREA

Around Hassop, the basal Namurian beds probably rest unconformably on Dinantian limestones; a section at Back Dale Mine [2319 7343] shows 0.19 m of dark limestone with *C.* cf. *leion*, overlying beds referred to Eyam Limestones. Namurian strata are present in an outlier near Thornbridge Hall where a railway-cutting [1953 7113] shows 0.5 m of dark calcareous mudstone, at the base of the section, with brachiopods and *C. leion*. Near Bakewell, the *C. leion* Band (0.4 m thick) is exposed near Holme Hall [2169 6931] and also, in part, on the southern outskirts of the town [2161 6820]. South of Ashford, hard calcareous mudstone with brachiopods and *C. leion* was found in a trench [2000 6939] overlying Longstone Mudstones; this establishes the existence of small Namurian outliers here. Farther south, the horizon is known from a large open working (now filled in) at Raper Mine [2169 6523], which showed 1.7 m of mudstone with *Cravenoceras sp.* in the lowst 0.1 m, overlying Longstone Mudstones. Farther east, an inclined borehole [SK 26 NW/21] proved 12.82 m of Namurian strata, with the *C. leion* Band between 22.75 and 24.72 m including a dark limestone band 0.15 m thick at the base, overlying Longstone Mudstones.

IPS, JIC

Base of Minn Sandstones to base of *Cravenoceratoides edalensis* Band (E1b–E2a)

In the western area the sequence (Minn Beds of the Macclesfield district; Figure 23) crops out at Gun Hill and around the Mixon–Morridge and Ecton anticlines. The outcrop continues eastwards to the upper Dove valley where these beds are almost wholly argillaceous, as in the lower part of the Edale Shales of the Chapel en le Frith district (Stevenson and Gaunt, 1971).

The overall estimated thickness is 158 m at Gun Hill, 146 m east of Hurdlow, 130 m south of Small Shaw, 98 m near Lower Elkstone and 73 m near Back of Ecton.

The sequence has been described by Trewin and Holdsworth (1973), who noted alternation of two main lithofacies; calcareous siltstones, similar to those forming the underlying E1a–E1b sequence, and protoquartzites (Pettijohn, 1957, p. 316). The latter form the main mappable units in the sequence and are here named the Minn Sandstones; they consist mainly of parallel-sided sharp-based graded sandstone and siltstone beds, commonly with silty mudstone intercalations. Scattered bands and nodules of sideritic ironstone occur as replacements of both rock types. The protoquartzites are interpreted as turbidites and show characteristics that indicate increasing distance from source

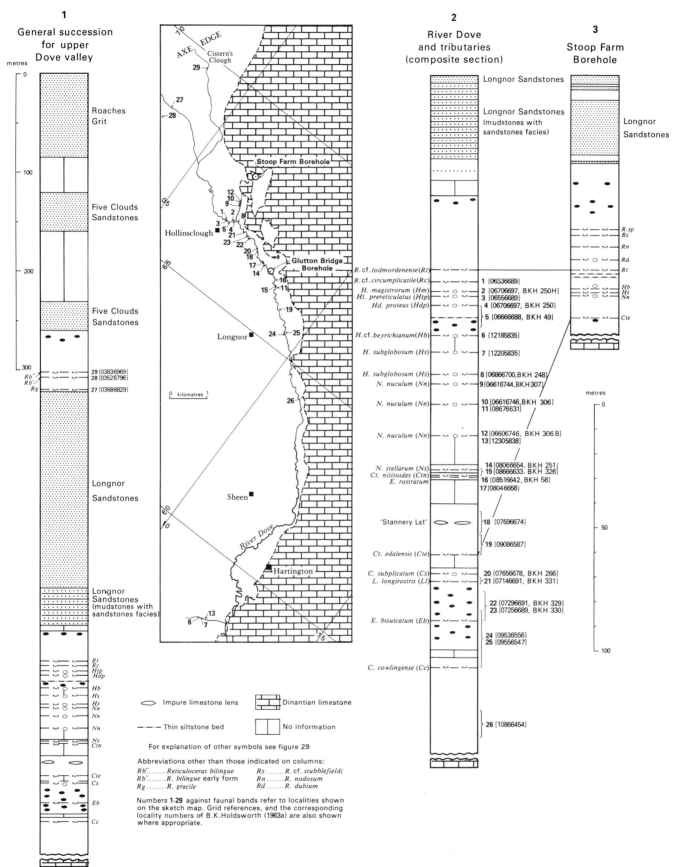

Figure 25 Sections of Namurian strata in the upper Dove valley.

when traced from south-west to north-east; mainly a decreasing overall thickness and diminishing proportion of sandstone to mudstone. Thus within the present district the most proximal sandstones are found in the Gun Hill area, but mudstones predominate over sandstones from Hurdlow eastwards, and the term Minn Sandstones becomes inappropriate. Even these distal turbidites, however, still form mappable units, and the term Minn Mudstones-with-Sandstones has been introduced to distinguish this more argillaceous lithofacies. Farther east, on the flanks of the Ecton Anticline, these strata are not mappable but can be recognised in stream exposures; in the upper Dove valley they have passed laterally into mudstone with ironstone bands and nodules.

The Minn Sandstones comprise three extensive arenaceous units, together with separating beds of calcareous siltstone and mudstone containing the *Cravenoceras malhamense* (E_{1c}), *Cravenoceras cowlingense* (E_{2a1}) and *Eumorphoceras bisulcatum* (E_{2a2}) bands (Figure 26); the last two are the bands referred to by Trewin and Holdsworth as *E. bisulcatum grassingtonense* and *E. bisulcatum erinense*. A fourth mudstones-with-sandstones unit is present locally below the lowest of the three main units; its outcrop extends from just within the boundary of the present district west of Onecote to Waterfall in the adjacent Ashbourne district. In addition, a sequence of mudstones with ironstone nodules is present below the *E. bisulcatum* Band.

The succession also contains seven beds of pale clay (Plate 10), 5 to 20 mm thick, which Trewin (1968) recognised as K-bentonite and considered to be the decomposition product of widely distributed volcanic ash. Three of these occur in the uppermost part of the sequence, which contains a band yielding the bivalve *Leiopteria longirostris* at its base, overlying the upper unit of the Minn Sandstones.

In the eastern part of the district the E_{1c} to E_{2a} sequence consists of calcareous dark grey fossiliferous mudstone and is greatly condensed.   NA

DETAILS

GUN HILL AREA

In the Gun Hill Anticline, exposures of the Minn Sandstones are generally small and scattered. The Gun Hill Borehole [SJ 96 SE/18] started in mudstones, probably near the horizon of the *C. cowlingense* Band, between the middle and uppermost units of the Minn Sandstones. These mudstones form a slack which extends north–south along the length of the hill, and exposures of the *C. cowlingense* and

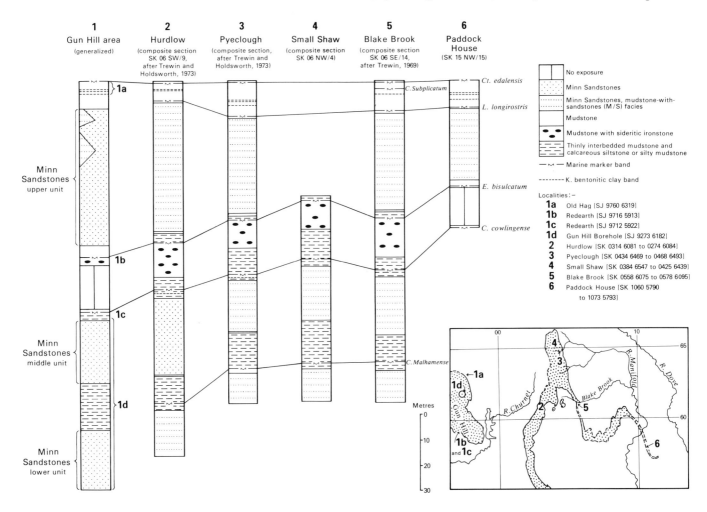

Figure 26 Comparative vertical sections of strata between base of the Minn Sandstones and *Ct. edalensis* Band.

Plate 10 Stream bank near Thorncliff; a pair of white K-bentonite bands 18 cm apart in Namurian (E_{2b}) mudstones. This pair occur over a wide area some 3 to 8 m below the *Cravenoceratoids edalense* band (L 1209).

E. bisulcatum bands have been noted near Redearth [9712 5922 and 9716 5913], near where the slack passes beneath Triassic cover. The middle unit of the Minn Sandstones, 24.4 m thick in the borehole, forms the highest part of Gun Hill, but the only good exposure occurs north-west of Redearth [9705 5915]. This shows about 5.0 m of fine-grained siliceous sandstone in beds ranging in thickness from 7 cm to 1 m with mudstone intercalations. The uppermost unit of the Minn Sandstones has an extensive outcrop on the eastern slopes of Gun Hill and on the isolated hill to the south [984 587]. In this area the top part is split into several leaves. The only sandstone exposures of note are in a quarry north-west of Gunside [9789 6009] and in the stream west of Oxhay [9751 6064]. The sub-drift outcrop north of North Hillswood was proved in site investigation boreholes for Tittesworth Reservoir (*see also* p. 134).

The sequence between the top of the Minn Sandstones and the *Ct. edalensis* Band is partially exposed near Wetwood [9774 6141 to 9777 6139; 9806 6179]. The best section, at Old Hag [9754 6308 to 9760 6319], exposes the *Ct. edalensis* Band and the three underlying K-bentonite bands (Trewin, 1968).

The only record of these beds outside the Gun Hill outcrop is from a borehole [SJ 95 NE/9] at Leek. This proved 14 m of thinly interbedded sandstone and mudstone, now assigned to the upper unit of the Minn Sandstones, overlain by 7.3 m of mudstone. These latter contained *L. longirostris* near the base and extended up to the bottom of the *Ct. edalensis* Band.

MORRIDGE TO PADDOCK HOUSE

All three units of the Minn Sandstones, comprising mainly the mudstones-with-sandstones lithofacies, have been traced around the Mixon–Morridge Anticline. The underlying fourth unit is present locally; it is exposed within the present district only in a stream section [0353 5504] 9.5 m thick south of Intake Farm. The main lowest unit also forms outliers east of Merryton Low [045 610], and south of Ryecroft [058 579]. Locally, these units cannot be traced owing to complexity of the structure.

Composite sections (Figure 26) have been built up from exposures in streams on both sides of Morridge. Two of the best of these, north-east of Hurdlow [SK 06 SW/9] and near Pyeclough Farm [0466 6487 to 0349 6483], were described by Trewin (1968) and Trewin and Holdsworth (1973). Blake Brook and its tributaries south of Lower Fleet Green provide a third composite section of the complete sequence [SK 06 SE/14] (Figure 28). Around Morridge these beds are well seen in a ditch west of Wellington Farm [SK 05 NW/14] and in a stream south of Small Shaw [SK 05 NW/4]. To the south-east, around the southern flank of the Fernyford Syncline, there are good exposures north of Lower Elkstone [0637 5849 to 0639 5860, and 0646 5877], in a gully north-west of Heath House [0712 5845] (localities 507 and 521 of Trewin), and in a stream east of Clough Head [SK 05 NE/11]. A ditch north of Paddock House provides the most easterly and distal section [SK 15 NW/15] of the Minn Mudstones-with-Sandstones.

East of Gun Hill the mudstone-with-sandstones facies predominates except in the Hurdlow section (middle unit) and in a stream section east of Thorncliff [0181 5861] (upper unit), where the sandstone facies is present.

Upper Dove valley

In the upper Dove valley, Namurian mudstone rests with slight discordance on Dinantian limestone at the foot of Chrome Hill [0708 6698]. The junction was passed through in the Stoop Farm Borehole where the lowest goniatite horizon proved in the mudstone sequence was that of *Ct. edalensis* 7.69 m above the limestone contact. At the foot of Chrome Hill the contact is seen:

	m
Limestone, dark grey argillaceous finely laminated with *Posidonia corrugata* and *P. corrugata elongata*	0.30
Mudstone, dark grey with ferruginous bands	1.22
Conglomerate of medium to dark grey rubbly micritic limestone and dark mudstone, irregular top and base	0.25
Limestone, medium grey, massive, fine calcarenite with scattered crinoid debris	0.91

The conglomerate bed is assumed to be of Namurian age, with the plane of unconformity at its base. However, the possibility that it is of high P_1 or P_2 age is not ruled out, in which case the top could also represent an unconformity. The two top beds lie very near the projected position of the *E. bisulcatum* Band (E_{2a2}), which occurs just above the unconformity 298 m to the north-west [0694 6725], and is represented by a small exposure of dark argillaceous limestone with *E. bisulcatum* and *Kazakhoceras sp.* (Figure 9).

The *E. bisulcatum* Band is also seen at several places in the banks of the River Dove [0729 6691, 1066 6452] and north-west of Crowdicote, where it occurs in a section [0953 6556 to 0955 6547] 22.33 m thick, with the *C. cowlingense* Band at the base (Figure 25).

NA

Eastern area

In the stream south of Hassop [2229 7114] (*see also* p. 76) an exposure of 0.32 m of hard dark mudstone yielded *E. sp.* and *L. longirostris*, probably representing the highest faunal band in the sequence (E_{2a3}). A borehole [SK 26 SW/46] in the Stanton Syncline proved a highly condensed sequence (Figure 27), with only 1.02 m of dark calcareous mudstone separating a band with *C.* aff. *leion* from one with *L. longirostris*, which in turn was only 0.89 m below a band containing *Ct. edalensis*. The only surface exposure of note is in a stream [2299 6384] east of Greenfields where 1.5 m of dark mudstone with hard carbonate bands yielded *C. cowlingense* and *E. sp.* (E_{2a}).

JIC, IPS

Cravenoceratoides edalensis Band to base of lowest *Homoceras subglobosum* Band (E_{2b} and E_{2c})

In the western outcrop this part of the sequence is equivalent to the Lower Churnet Shales of the Macclesfield district (Figure 23). The sequence crops out in three separate areas: in the Meerbrook valley, in a small and ill-exposed area near Buxton, and in the main outcrop with extends continuously from the southern margin of the district near Holly Dale [019 556], round the flanks of the Mixon–Morridge and Ecton anticlines, into the upper Dove valley. The thickness ranges from 134.7 m at Thorncliff [015 585] to 72.8 m in the upper Dove valley but is only 10.8 m in Stoop Farm Borehole (Figures 25 and 29) on the adjacent part of the stable block.

The *Ct. edalensis* Band, at the base of the sequence, is exposed in many places, enabling its outcrop to be related to the top of the underlying Minn Sandstones. The band normally comprises hard calcareous and fossiliferous mudstones, the maximum known thickness being 5.63 m in a stream [SK 06 SE/19] at Wiggenstall.

Ct. nitidus, the eponymous goniatite of the next highest marker band (E_{2b2}) has been recovered at only one locality [0902 6078], in a bullion near Wiggenstall, though the same horizon is probably exposed at four other places in the area.

Some 12 to 15 m above the base of the *Ct. edalensis* Band, a thin laminated ankeritic limestone is exposed in several sections; this was called the Stannery Limestone by Holdsworth (1963a) after the farm [077 667] where the limestone was first described. The overlying *C. holmesi* Band is poorly exposed, the fauna having been recorded by Trewin and Holdsworth (1972) only in a stream near Oakenclough Hall, again in a bullion. This band is overlain by distinctive 'cherts' associated with the *Ct. nititoides* Band (E_{2b3}): pale grey-brown dolomitic cherty siltstone 1.0 to 1.5 m thick, showing a fine streaky lamination. The band commonly shows incompetent S-shaped folds in an otherwise evenly dipping sequence. The hardness of the band and its resistance to erosion result in good exposure in many places, enabling its outcrop to be traced over considerable distances. The associated benthonic fauna is both rich and varied, including brachiopods and trilobites in addition to the more usual marine fossils. Benthonic faunas have been associated by Ramsbottom (1969) with well oxygenated shallow-water conditions of deposition. The widespread occurrence of the band in the Pennines, Ireland and Upper Silesia has been taken by Bolton (1978) to indicate a eustatic lowering of sea level.

The basal E_{2c} band of *Nuculoceras stellarum* lies about 3.5 m above the 'cherts' and has been recorded at several localities. It consists of a bed of argillaceous ankeritic limestone, 6 to 15 cm thick, containing abundant crushed *N. stellarum* and *Posidoniella vetusta*.

At all the *N. stellarum* localities, apart from those in the River Dove, the band is closely overlain by protoquartzitic turbidites. Along the western flank of the Mixon–Morridge Anticline these form a mappable unit up to about 46 m thick, the Hurdlow Sandstones. Elsewhere, exposures are poor but only a few thin sandstone beds appear to be present, scattered through a sequence of silty mudstones. In the Dove

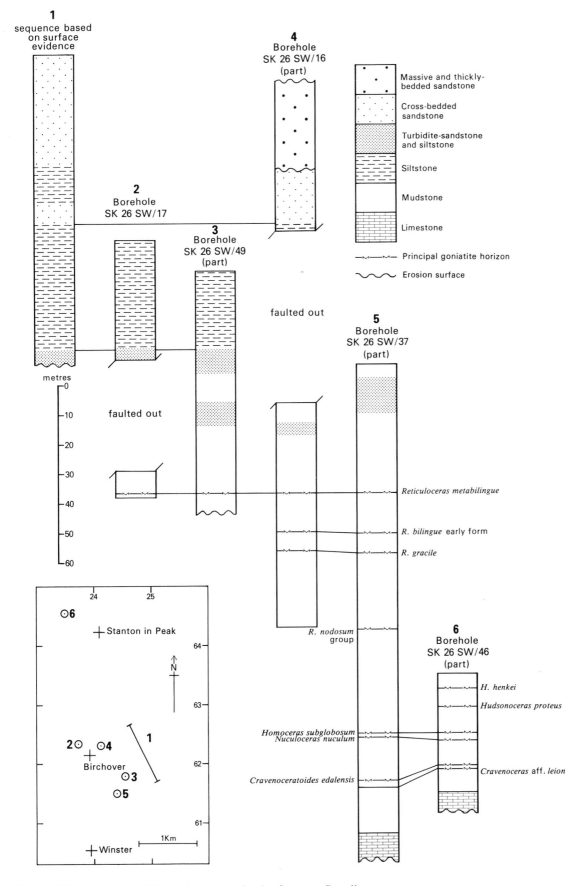

Figure 27   Sections of Namurian strata in the Stanton Syncline.

valley the sequence is thin and probably entirely argillaceous. According to Bolton (1978), the Hurdlow Sandstones consist of mature turbidites in the Thorncliff and Hurdlow sections, while their lateral equivalents in the Oakenclough, Blake and Wiggenstall brook sections are of distal character, the last being the most distal. This conclusion, together with palaeocurrent evidence, suggests a general southerly source for the coarser clastic sediments.

The remainder of the E$_{2c}$ sequence, above the Hurdlow Sandstones or their argillaceous equivalents, consists mainly of silty mudstone. It includes three bands with *N. nuculum*, usually associated with thin ankeritic limestones (Figures 25 and 29). The Swallow Brook section [SK 06 NE/29] in the upper Dove valley is the only place where all three bands have been recorded in sequence (Holdsworth, 1963a; Bolton, 1978); the associated limestone takes the form of bullions rather than continuous beds. The highest *N. nuculum* band is separated from the basal Chokierian *Homoceras subglobosum* band by 4.7 to 5.6 m of mudstone. Sideritic ironstone nodules occur in these beds at certain horizons, particularly between the *Ct. nititoides* and *N. stellarum* bands. Rare K-bentonite bands have also been recorded (Figure 29; Trewin and Holdsworth, 1972).

In the eastern part of the district, the E$_{2b}$–E$_{2c}$ sequence is highly condensed and consists largely of fossiliferous dark calcareous mudstone about 10 m thick, similar to that proved in the Stoop Farm Borehole in the western outcrop.   NA

DETAILS

MEERBROOK VALLEY

Exposure in this area is limited, though two small exposures of the cherty *Ct. nititoides* Band, near Old Hag [9761 6347] and Thornyleigh [9794 6297], have yielded elements of the typical benthonic fauna. At Tittesworth Reservoir, the main cut-off trench showed about 119 m of east-dipping predominantly argillaceous strata, probably ranging in age from E$_{2c}$ to H$_{2b}$. At its western end, the trench revealed a few irregular beds of sandstone dipping sub-parallel to the valley side, and the record of a nearby surface exposure [0922 5867] with *Ct. fragilis* and *N. nuculum* suggests that these are the Hurdlow Sandstones (*see also* p. 74). A thin representative of this unit in the Ballhaye Green Borehole at Leek [SJ 95 NE/9] (Figure 29) is recorded as 'dark grey shale — with a few silty harder bands up to ½ inch and brown ironstone bands with spar-filled joints'.

WEST SIDE OF MORRIDGE TO UPPER DOVE VALLEY

Sections which include the E$_{2b}$ to H$_{1a}$ sequence are near Thorncliff [SK 05 NW/13], Hurdlow Farm [SK 06 SW/13], Oakenclough [SK 06 SE/18], Blake Brook [SK 06 SE/15], Wiggenstall [SK 06 SE/19], and the upper Dove valley (Figure 25). Details of these sections are summarised in Figures 28 and 29.

A ditch at Wellington Farm [SK 05 NW/14] exposes the *Ct. edalensis* Band and 23.75 m of the overlying sequence, including a bullion band with *Dimorphoceras sp.* and three very thin clay bands, apparently K-bentonites. Nearby [0200 5594], a gully exposes one of the best sections of the Hurdlow Sandstones [SK 05 NW/15]. To the north, near Hurdlow Farm, the type section of the Hurdlow Sandstones [SK 06 SW/8] totals 38.42 m in thickness. Individual sandstones range in thickness from a few centimetres to 3.82 m; a few arc sufficiently well exposed to show fining-upwards grading with directional sole structures and laminated tops. The sandstones

are interbedded with grey silty mudstones, the overall sandstone to mudstone ratio being about 1:1. Scattered exposures NNE of Hurdlow are predominantly of silty mudstone with irregular sandstone beds which are usually thin, suggesting that the unit is passing northwards into a more distal facies. The *N. stellarum* Band is exposed immediately below the base of the sandstones at one locality [0304 6182]. Scattered exposures, lacking sandstone, but including several of the faunal bands, are present near Highash [0469 6493 to 0469 6532].

The distribution of exposures in the upper Dove valley is summarised in Figure 25; here the thicknesses between the three *N. nuculum* bands and the lowest *H. subglobosum* band are based on those given by Holdsworth (1963). Bolton (1978) remeasured part of the section [SK 06 NE/29] and found four *N. nuculum* bands in about 34.95 m of strata. This thickness is anomalous when compared with Bolton's figure of 21.5 m for the same interval in the Oakenclough section [SK 06 SE/18], thought to be situated nearer the centre of the basin.

The Stoop Farm Borehole (Figure 25) proved only the highest *N. nuculum* band, comprising 0.24 m of hard calcareous mudstone (?bullion) with *N. nuculum,* at a depth of 87.59 m. *Cravenoceras sp.* occurred in the mudstone 0.19 m below. The top of the *N. nuculum* (E$_{2c}$) Zone, and hence that of the Arnsbergian (E$_2$) Stage, is taken at 87.35 m. As no other goniatite bands were proved below until the *Ct. edalensis* Band at the base of the *Ct. nitidus* Zone, only the combined thickness of E$_{2b}$ and E$_{2c}$ (10.83 m) is known from the borehole; the estimated composite thickness for the River Dove sections is about 73 m (37 m for E$_{2b}$ and 36 m for E$_{2c}$). The difference in thickness illustrates the influence of the steep slope at the edge of the late-Dinantian shelf province against which the Namurian mudstones were deposited (Figure 9). The thin sequence at Stoop Farm accumulated near the margin of this upstanding area, while the thicker Dove valley succession was deposited at the foot of the frontal slope of the apron-reef. Thin sequences also occur in the central part of the upstanding area, for example in boreholes at Stanton (Figure 27) and Ashover (Ramsbottom and others, 1962).
      NA

BUXTON

Rocks of E$_{2b}$ to E$_{2c}$ age are known only from a 2 m section in the banks of the River Wye at Burbage [0434 7316], which shows mudstone with a band, 5 cm thick, containing *N. nuculum* 0.25 m above the base.

EASTERN AREA

In a stream [2216 7099] south of Hassop, 0.46 m of mudstone with *Ct. edalensis* is exposed. The band is also seen downstream [2225 7108]. To the south-east of Bakewell, a 2.9 m section in Buck Pingle [2340 6769] contains a 0.15 m band with *N. nuculum*. In the Stanton Syncline, a borehole [SK 26 SW/46] (Figure 27) proved the *Ct. edalensis* Band to be 0.49 m thick, the remainder of the overlying Arnsbergian Stage consisting of dark mudstone with hard carbonate-rich bands. An 8 cm carbonate band at 58.00 m, containing the trilobite *Paladin* cf. *parilis*, probably represents the cherty dolomitic band at the *Ct. nititoides* horizon in the western area. Two bands containing *N. nuculum* were proved in the top 0.92 m of the E$_{2c}$ sequence in this borehole.   JIC, IPS

Chokierian and Alportian stages (H$_{1a}$ – H$_{2c}$)

In the western area, rocks of this interval are the equivalents of the Middle Churnet Shales of the Macclesfield district (Figure 23). Their outcrop is continuous from the Meerbrook valley, round the Mixon–Morridge and Ecton anticlines to the upper Dove valley and northwards to the

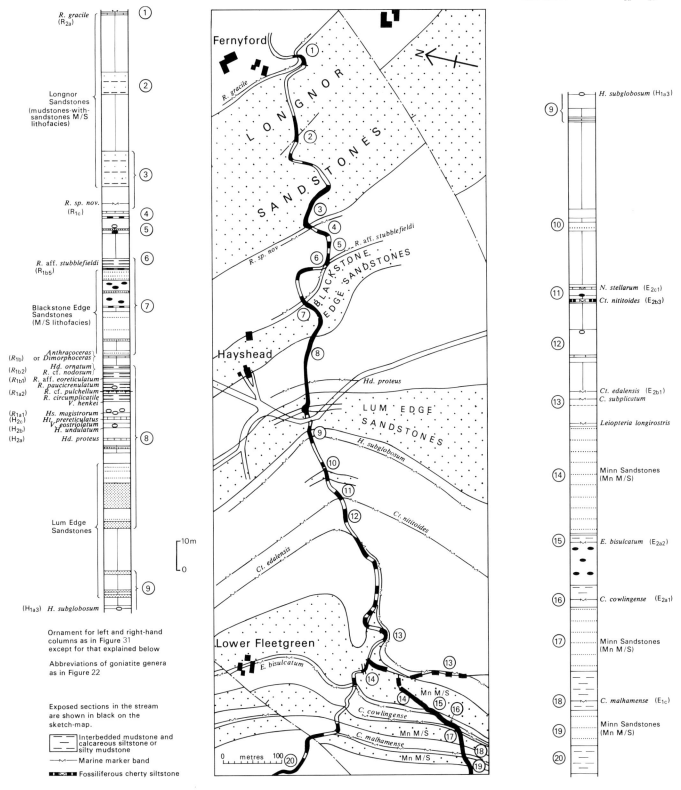

Figure 28   Sketch-map and section of Namurian (E₁c – R₂a) strata in Blake Brook.

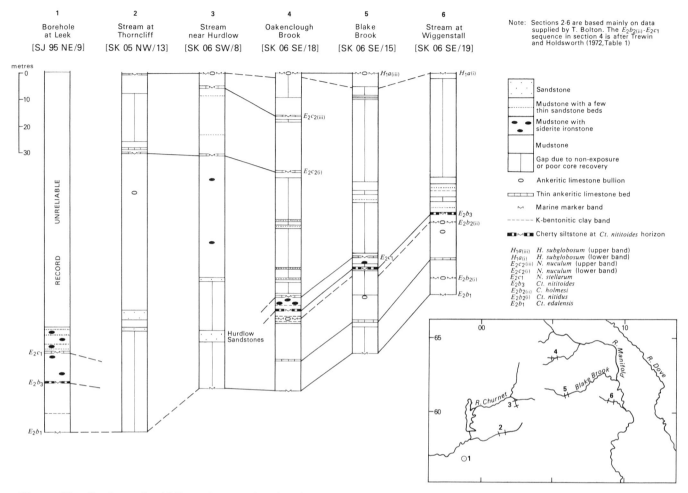

Figure 29   Sections of middle and upper Arnsbergian (E_{2b}–E_{2c}) strata in the western part of the district.

vicinity of Buxton, though no diagnostic fossil bands have been found north of Brand End [051 686] and the nearby inlier around Fairthorn [045 689].

Thicknesses of the Chokierian and Alportian, based partly on Bolton (1978), range respectively from about 27 m and 11 m near Gun End House [964 633] just outside the district, to a maximum of 73 m and 9.3 m in Blake Brook and a minimum combined thickness of only 6.88 m in the Stoop Farm Borehole (Figures 30 and 25). The sequence consists of dark mudstone with an arenaceous unit, the Lum Edge Sandstones, present locally in the middle part. The mudstones of the basal *H. subglobosum* (H_{1a}) Zone generally include three fossiliferous bands with scattered bullions containing *H. subglobosum*, though four bands are known at Tittesworth Reservoir.

Mudstones of H_{1b} age containing faunal bands with *H. beyrichianum* occur in the area but have been found in isolated exposures only. Their apparent absence could be due either to cut-out by the Lum Edge Sandstones or merely to poor exposure; the sandstone unit is inferred to lie mainly within the H_{1b} Zone. The Alportian mudstones are almost continuously fossiliferous with several ankeritic limestone bands or bullion horizons.

In the eastern outcrop the Chokierian and Alportian sequences are again condensed and consist largely of fossiliferous mudstone.

The Lum Edge Sandstones are named after the prominent feature of Lum Edge [061 604], and consist of thin to thick sharp-based turbiditic beds of fine to coarse-grained proto-quartzitic sandstone and silty mudstone. The unit and its lateral mudstones-with-sandstones equivalent form a continuous outcrop on the east flank of the Mixon–Morridge Anticline from Oakenclough Brook [053 636] south-east to Archford Moor. West of the anticline, the outcrop is discontinuous, possibly due to a change in facies in the Thorncliff area (p. 79). The base lies above the highest *H. subglobosum* band and the top below the basal Alportian *Hudsonoceras proteus* (H_{2a}) Band.

DETAILS

MEERBROOK VALLEY

A section [SJ 95 NE/21], 41.35 m thick, on the east bank of Tittesworth Reservoir, includes the highest E_{2c} beds and four bands containing *H. subglobosum* lying 11.7 m, 18.4 m, 23.6 m and 34.8 m above the base. A temporary section [SJ 95 NE/22], near the reservoir dam, revealed 25.01 m of mudstone containing fossiliferous bands with *H. undulatum* (H_{2b}) and *Reticuloceras sp.* [*circumplicatile* group] (R_{1a}).

WEST OF THE MIXON–MORRIDGE ANTICLINE

The best sections are continuations of those at Hurdlow [SK 06 SW/13] and Thorncliff [SK 05 NW/13]; they are shown in Figure 30. The Lum Edge Sandstones are absent at Hurdlow, but are traceable by features farther south and are exposed both in the Thorncliff section and south of Easing Farm. The mainly fine-grained, thinly bedded protoquartzitic sandstones exposed at the former locality differ from the turbidites exposed in the type area to the east, and have been interpreted by Bolton (1978) as deltaic deposits derived from the west. The unit greatly thickens farther south around Easing Farm, where there are exposures of massive or parallel-laminated coarse-grained pebbly sandstones with mudflake cavities [0139 5730; 0151 5716]. The lithology and general lenticular form of these deposits suggest deposition in deltaic channels. Similar outcrops occur farther north between Dry Stones [0307 6263], where 2.74 m of massive pebbly sandstone form a small tor, and Morridge Top [032 654].

EAST OF THE MIXON–MORRIDGE ANTICLINE

The Chokierian and Alportian part of the Blake Brook sequence [SK 06 SE/16] provides the type section through the Lum Edge Sandstones [0610 6116 to 0616 6117], though even here some 15 m are hidden by a culvert beneath a road bridge (Figure 28). In general, three interbedded elements are present: lenses of massive thickly bedded coarse-grained to pebbly sandstones with convex erosive bases and flat tops; thinly bedded fine to coarse-grained mainly discontinuous sandstone beds; and thin to thick beds of silty mudstone and siltstone, the last making up slightly less than 50% of the total succession. Most of the sandstone beds show graded bedding and sharp bases with load structures and directional sole marks characteristic of turbidites. Bolton (1978) interpreted the depositional environment as the proximal turbidite fills of small channels and interchannel areas on a pro-delta slope. Thus, they were probably linked with the channel deposits of the Thorncliff–Easing Farm area to the south-west, as suggested by Bolton, who measured sole structures indicating that palaeocurrents flowed from this direction.

Features suggest that the Lum Edge Sandstones pass laterally into a mudstones-with-sandstones sequence in the vicinity of Reaps Moor [081 616]. In the Wiggenstall stream section [SK 06 SE/19], the sandstone beds, interbedded with predominant silty mudstones, are thin but laterally persistent. Higher up the stream [0875 6097], a mudstone 2.12 m thick contains *H. beyrichianum* (H_{1b}). Field relations indicate that this exposure underlies the Lum Edge Sandstones, confirming the H_{1b} age of the latter. The sandstones are overlain by mudstones containing bullions with *Hd. proteus* (H_{2a}) in the main Wiggenstall section.

UPPER DOVE VALLEY

In four localities a band with bullions containing *H. subglobosum* is exposed, but it is uncertain whether these represent the same or different horizons. The best section in the Chokierian, which also includes part of the underlying Arnsbergian sequence, is in the banks

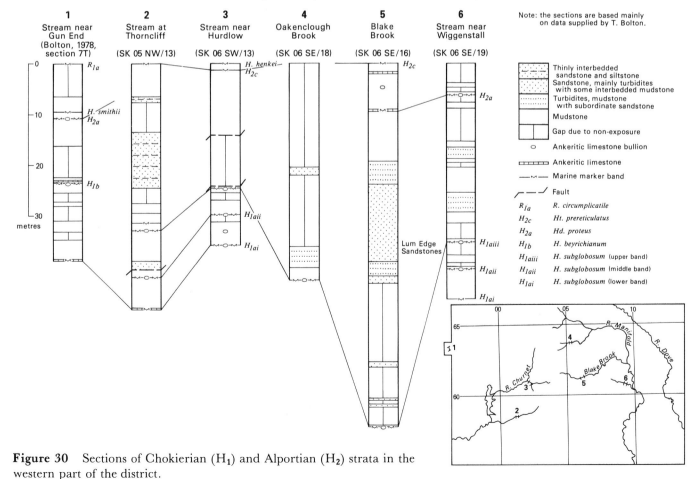

Figure 30 Sections of Chokierian (H_1) and Alportian (H_2) strata in the western part of the district.

of a stream north-east of Field House [SK 15 NW/14]. Faunal bands here contain, in upward sequence, *N. nuculum* (E_{2c}), *H. subglobosum* (H_{1a}), and *H.* cf. *beyrichianum* (H_{1b}). In the Stoop Farm Borehole, *H. beyrichianum* occurs in fissile dark grey mudstone at a depth of 86.22 m, but the position of the top of the H_{1b} Subzone is uncertain. However, in the banks of the River Dove [0677 6698; 0666 6688], Holdsworth (1963a) has recorded sequences of up to 8.08 m of mudstone with protoquartzitic laminae and thin impersistent ironstone bands, which underlie the *Hd. proteus* Band and therefore almost certainly belong to the H_{1b} Subzone. These strata are probably the extreme distal representatives of the Lum Edge Sandstones. The *Hd. proteus* Band is also present (Holdsworth, 1963b) near Brand End [0514 6856], and near Fairthorn [0443 6902; 0439 6907; 0457 6887]. At all of these localities, *Hd. proteus*, usually poorly preserved, occurs as the only fossil in grey finely granular ankeritic limestone that generally forms a string of lenses, ranging from 15 cm to 41 cm in thickness, set in dark grey mudstone. The most extensive exposures are in the River Dove and in a long shaly scar below Fairthorn, where a mudstone section [SK 06 NW/7] of 7.19 m ranges from H_{2a} to R_{1a}, but mudstones of the uppermost Alportian and lowermost Kinderscoutian are absent, probably due to faulting. NA

EASTERN AREA

Exposures in the eastern part of the district are scarce, the best being located north-west of Haddon Park Farm [SK 26 NW/36]. Here some 8 m of mudstone contain *H. subglobosum* in a band 18 cm thick, 3.07 m above the base: this is succeeded closely by a band with *H. beyrichianum* which also contains bullions. In the Stanton Syncline, Chokierian and Alportian strata have been .proved in several boreholes, notably SK 26 SW/46 (Figure 27); they consist largely of fossiliferous mudstone with a few thin carbonate bands and with most of the main goniatite horizons present. The thicknesses of the two stages are: Chokierian 10.33 m; Alportian 5.51 m.   JIC, IPS

Kinderscoutian (R_1) strata

Rocks of this age in the western area are the equivalents of the Upper Churnet Shales of the Macclesfield district (Figure 23). The lower part of the succession is mainly argillaceous, containing numerous mudstone bands with marine faunas, especially in the R_{1a} and R_{1b} zones; protoquartzitic turbidites are locally present and include the Blackstone Edge Sandstones. The upper part is dominated by the Longnor Sandstones, but also includes small outcrops of Kinderscout Grit in the north and Kniveden Sandstones in the south.

The sequence, excluding the Longnor Sandstones, ranges from a maximum of about 71 m in Blake Brook [SK 06 SE/17] to a minimum of 45.43 m in the Stoop Farm Borehole (Figure 25). The equivalent mudstone sequence near Buxton is only about 24 m thick, indicating that the buried late Dinantian shelf area was still upstanding in Kinderscoutian times.

Over much of the district, the R_{1a} and R_{1b} succession consists mainly of dark mudstone, associated with a few thin argillaceous limestone beds and bullions. Marine fossils are present throughout the R_{1a} mudstones, but are interspersed with thinner barren mudstone bands in R_{1b}. The R_{1b} sequence in the western area has been shown by Ashton (1974) to contain five marine bands usually separated by barren mudstones (Figure 31).

Thin pale clay bands occur in the mudstones; these, the 'kaolinised ash bands' of Ashton (1974), are similar to the K-bentonites described by Trewin (1968) from the Arnsbergian. Ashton noted a total of eight bands, six in the R_{1a} Zone, and two near the base of the R_{1b} Zone. The bands range in thickness from mere partings to 1.5 cm, and consist of pale buff, orange, or greenish-grey clay, locally with abundant pyrite. The clay comprises mainly either mixed-layer clay minerals or well ordered kaolinite.

Locally, around the northern culmination and east flank of the Mixon–Morridge Anticline, protoquartzitic sandstones occur in the middle part of the R_{1b} sequence forming a recognisable unit, here named the Blackstone Edge Sandstones, the equivalent of the Ballbank Sandstone Member of Holdsworth (1963a). The sandstones and their predominant mudstones-with-sandstones equivalent form Blackstone Edge [051 644] and Lady Edge to the south, and are of turbidite facies; they lie in the middle part of the R_{1b} Zone (Figure 22), but exposure is too poor to allow more precise definition. Another sequence of mudstones with protoquartzitic sandstones and siltstones, about 9 m thick, occurs in the uppermost part of the R_{1b} Zone, but it has been found in only the Brund boreholes.

Fossil bands probably representing the R_{1c} Zone have been found at many localities in the area but correlations are tentative owing to the poor preservation and sparsity of the faunas. Two separate bands are probably represented, an upper one with *R. sp. nov.* and a lower one with *R. reticulatum*, but there is no record of both bands being present in any one section in the area.

Over much of the Kinderscoutian outcrop in the western part of the district the change in turbidite deposition from southerly-derived protoquartzites to northerly-derived feldspathic sandstones (Holdsworth, 1963b) took place in R_{1c} times. The Longnor Sandstones are the earliest representatives of the latter. They were first described by Hull and Green (1866, p. 59) as 'Yoredale Grit' or 'Longnor Sandstone', and later as 'Longnor Grit' by Bisat and Hudson (1943, p. 390). Holdsworth (1963a) showed that they overlie the highest R_{1c} marine band containing *R. sp. nov.*, now regarded by N. J. Riley (written communication, 1984) as a correlative of the *R. coreticulatum* or Butterly Marine Band and are therefore equivalent to the Upper Kinderscout Grit of the ground farther north. During the present survey, the sandstones were found to be immediately overlain by the *R. gracile* Band and to have an extensive outcrop from near Stanley Moor Reservoir [042 710] in the north, to near Upper Hulme [014 604] in the south-west and near Hulme End [106 590] in the south-east; they are now known to extend westwards into the adjacent Macclesfield district around Swythamley (Ashton, 1974).

The thickness of the Longnor Sandstones reaches an estimated maximum of 200 to 240 m near Brand Top [0444 6817 to 0384 6833], but there is little trace of the unit north of Stanley Moor Reservoir, only 2.7 km to the north. Figure 32 shows isopachytes, outcrop distribution, the gradual lithofacies change from proximal turbidites in the north-west to distal turbidites in the south and east, and the palaeocurrent evidence summarised by Ashton (1974, appendix IV); the data all suggest that the Longnor Sandstones were deposited mainly from the north-west in a major lobe that was thickest some 5 km south-west of Buxton. A second

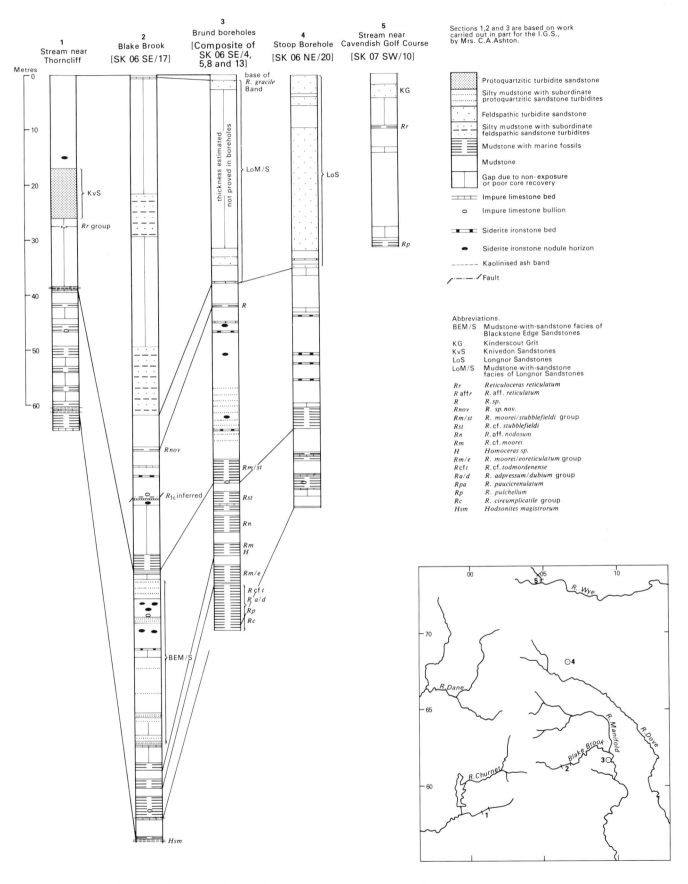

Figure 31 Sections of Kinderscoutian (R₁) strata in the western part of the district.

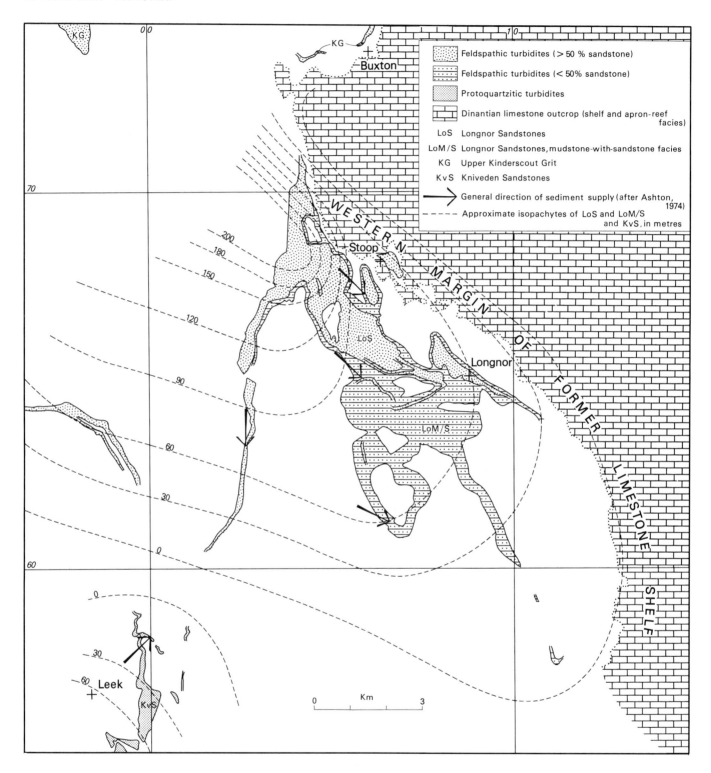

Figure 32 Outcrop and facies of the Longnor Sandstones and correlatives.

lobe was probably located west of the district (Ashton, 1974). The thickness variations of the Longnor Sandstones (Figure 32) suggest that the late Dinantian shelf still formed an upstanding feature on the sea bed, but sand was being deposited on its margins for the first time, as shown at Stoop Farm Borehole. The sand did not extend far onto the positive area, for, in a Pocket Deposit at Hindlow (p. 106), 2 km to the north-east, a foundered mass of mudstone of E_2 to R_{2a} age contained no sandstone. The turbidite lobe probably accumulated at the foot of a submarine pro-delta slope, sediment being transported via feeder channels from the delta-top where the Upper Kinderscout Grit (Collinson, 1968) was deposited. At the north-western extremity of the district, an outcrop of sandstone in the core of the Todd Brook Anticline is shown on the map as Upper Kinderscout Grit, although it is of turbidite facies; it is assumed to belong to the same lobe of sediment as the Longnor Sandstones.

In the eastern area, the Kinderscout Grit thins southwards from its major fluviatile facies in the district to the north to a limit [224 689] near Bakewell. Such evidence of facies as is available suggests that the grit here is also turbiditic, as would be expected at the southerly and most distal part of its outcrop.

For at least part of the time when the Longnor Sandstones were being deposited, protoquartzitic sands were entering the district from the south-west to form a unit here named the Kniveden Sandstones, after a farm immediately east of Leek [000 560]. The sandstones are mainly fine-grained and in thin beds, with erosive bases locally channelling into the underlying bed or amalgamating two or three beds into thicker units. Several features, including the sequence of sedimentary structures within beds and high sandstone:mudstone ratios, suggest that the sediments are proximal turbidites (Ashton, 1974). The outcrop cannot be traced northwards beyond the vicinity of Thorncliff [010 585] and Troutsdale Farm [995 589], the unit being assumed to die out at least 1 km short of the southern limit of the Longnor Sandstones (Figure 32). No interfingering of the protoquartzitic and feldspathic lithofacies has been found.

To the south, in the Combes valley (Ashbourne district memoir, *in preparation*), the Kniveden Sandstones are of shallow-water type, but details of the transition from the turbidite facies are lacking.   NA

DETAILS

MEERBROOK VALLEY

The only extensive exposures are in a stream south-west of Hillyless* [9667 6395 to 9705 6407] where several goniatite bands in the R_{1a} and R_{1b} marine mudstone sequence are exposed; a diagrammatic composite section is given by Ashton (1974). To the north-west of Roachside [994 627] the Longnor Sandstones form a prominent double scarp which dies out to the south-east. The few streams cutting through this feature show scattered exposures of interbedded flaggy sandstone, siltstone and silty mudstone, and small exposures of minor sandstone interbedded with mudstone occur as far south as the River Churnet [9978 6024].

---

* The name of the well-known Swythamley Hall does not appear on the topographic base of the 1:50 000 geological map of the district.

LEEK AREA

Much of the are a south of the River Churnet and below Tittesworth dam is probably underlain by Kinderscoutian mudtones, but few of the scattered exposures have yielded faunas well enough preserved to be diagnostic of zone or subzone. A basal R_{1a} band containing *R. sp.* [*circumplicatile* group] was found in a temporary section [SJ 95 NE/22] near Tittesworth dam (p. 79), while in the valley 280 m to the south-east [9976 5829], near the top of the mudstone sequence, some 4 to 5 m of weathered mudstone, just below the inferred position of the base of the Kniveden Sandstones, yielded a probable R_{1c} fauna. Although much mudstone is exposed in the valleys [988 561; 994 556] immediately south-east of Leek, it is mostly weathered and of a very silty ferruginous nature, probably indicating closer proximity to the deltaic environment to the south. The sandstones between Cornhill Cross and Lowe Hill are assigned to the Kniveden Sandstones mainly by inference. Exposures in the wooded ridge east of Cornhill Cross [988 556] reveal up to 12.64 m of turbiditic sandstones in parallel beds with subordinate weathered mudstone intercalations. More irregularly bedded sandstones, including some with a channel-like form, were recorded with interbedded siltstones and mudstones in a 10 m section in a road-cutting at Lowe Hill [9979 5553]. To the north, the Kniveden Sandstones cap the hill on which Kniveden stands, but there is little exposure of note. The best Kinderscoutian section in the Leek area is in a stream running WSW from Thorncliff that cuts across the southern closure of the Goyt Syncline and exposes the Kniveden Sandstones in both limbs of the fold. The section is complicated by landslipping of the mudstones, making thicknesses difficult to estimate. A composite section after Ashton (1974) is given in Figure 31 for the east limb of the syncline [0122 5845 to 0080 5839]. The Kniveden Sandstones are estimated to thicken from 9 m on the east limb to 15 m on the west [9983 5828 to 0014 5811], where the unit is best exposed. A 6.1 m section of the sandstones is also seen in the stream south-east of Thorncliff [0100 5816 to 0102 5815] where an R_{1c} marine band, 1.52 m to 2.44 m below the base, contains *R. sp.* [*reticulatum* group].   NA

MORRIDGE TO BUXTON

Farther north, in tributaries of the River Churnet near Hurdlow [0218 6092] and Swainsmoor [0265 6172; 0263 6176], mudstones with bullions of R_{1a} and R_{1b} age are exposed; here Ashton has found a total of seven interbedded 'kaolinised ash bands', some correlated with similar bands in sections in Blake Brook, at Thorncliff (see Figure 31), and in the River Manifold [SK 06 NW/8]. The most complete record of R_{1a} to R_{1b} marine mudstones was obtained from site investigation boreholes in the Manifold valley near Brund [SK 06 SE/4, 5, 8 and 13] (Figure 31). Parts of the sequence are exposed at many other localities such as the bank of the River Dove near Hollinsclough (R_{1a}) [0653 6689], a bank south of Booth (?R_{1a} to mid R_{1b}) [0562 6785 to 0568 7793], the streams south-east of Fairthorn (R_{1a}) [SK 06 NW/7] and (R_{1b}) [SK 06 NW/9], and near Cavendish Golf Course, Buxton [SK 07 SW/10] (Figure 31).

An apparently discontinuous unit of thinly bedded protoquartzitic sandstones forms a feature at Swainsmoor [024 613], and is tentatively assigned to the Blackstone Edge Sandstones because of the position of its outcrop in relation to the Longnor Sandstones. The best exposures of the Blackstone Edge Sandstones (mainly mudstones-with-sandstones facies) occur in the right bank of the River Manifold near Thick Withins [0494 6566] and Hole Carr [0625 6530 to 0562 6515]. The protoquartzitic sandstones are mainly in beds less than 30 cm thick, with sharp bases and some directional sole structures. Many beds are laterally impersistent and lenticular beds, interpreted as channel fills, are common. Thin ironstone beds and nodules are also present. The member is also well exposed in Blake Brook where it includes isolated lenticular channel-fill sandstones showing contortion and overfolding due to penecontemporaneous movement. Similar slumped beds are exposed in the brook at Oakenclough Hall [0553 6366]. Thick lenses of fine to coarse-grained sandstones form strong laterally impersistent features on Lady Edge [056 621; 058 618] and Swallow Moss [073 605], and are interpreted as large-scale channel-fills.

Sections exposing the *R. sp. nov.* Band and the transition to the overlying Longnor Sandstones occur at Blake Brook [0644 6123] (Figure 28), the left bank of the River Manifold [0586 6510] near Ballbank House, and near Fawside [0775 6466]. The Blake Brook section and its significance in fixing the R_{1c} age of the base of the Longnor Sandstones were first noted by Holdsworth (1963a). The section near Ballbank House, which is similar to that near Fawside, shows 8.08 m of grey-brown soft argillaceous sandstone and grey silty mudstone in sharp-based graded beds up to 1.22 m thick, overlying 2.93 m of grey to dark grey silty mudstone with a 38 mm soft brown decalcified band containing *Reticuloceras sp.* 0.76 m above the base.

Other sections exposing the more argillaceous basal part of the Longnor Sandstones occur in the banks of the River Manifold and near Hollinsclough, in a cliff 22 m high, in the left bank of the River Dove [SK 06 NE/30]. In a track south-west of Hollinsclough, about 52 m of these beds are exposed [SK 06 NE/31]. Sandstones in thicker beds up to 1.5 m, but probably at the same stratigraphic level as the upper part of the last section, are exposed in a landslip scar [0613 6682]; some poor flute casts and shale intercalations up to 13 cm thick are present in a crag of 6.1 m. Another landslip scar near Fough [0694 6756] exposes 15 m of sandstones, including one bed 6.1 m thick, again at a similar stratigraphic level. Many of these thick compound beds were formed by the erosive amalgamation of one or more individual proximal turbidite flows. Similar thick beds have been worked in a small drift mine near Longnor, said to have been in operation about 1887 [SK 06 NE/32], and in two quarries at Newtown [0595 6310; 0598 6300].

The Longnor Sandstones reach their maximum thickness in the Dove above Hollinsclough and in its tributary above Fairthorn. In the latter, scattered exposures of sandstone indicate a total thickness of about 200 m (including the mudstones-with-sandstones lithofacies at the base); the formation thickens to 240 m in the River Dove to the south (Figure 32). The top 70.7 m of the unit was penetrated by a borehole [SK 06 NE/23] near Moorside. In the Stoop Farm Borehole the basal 34.04 m were drilled, interbedded mudstones making up about 18% of the sequence. The sandstones in this hole were mostly medium-grained and unlaminated, a characteristic noted in many surface exposures and an indication of the proximal nature of these turbidites. The sandstone at the northern boundary of the outlier, near Stoop, appears to rest directly on Dinantian limestones but the junction is not exposed and juxtaposition due to faulting cannot be ruled out.

The northernmost exposure is in a disused quarry [0428 7073] south of Stanley Moor Reservoir:

	m
Sandstone, massive	1.52
Mudstone passing laterally into sandstone, with a slump breccia in upper part	4.27
Sandstone, mainly well bedded with mudstone pellets in places and mudstone partings near base	21.34
Sandstone, massive, coarse to medium-grained	2.74

In a typical exposure [0523 6688], near Moorside, the topmost bed of the Longnor Sandstones is overlain by 2.13 m of mudstone which separate it from the base of the *R. gracile* Band.   NA, IPS

Kinderscoutian mudstones are largely unexposed and known mainly from boreholes in the Stanton Syncline, for example SK 26 SW/37 and SK 26 SW/46 (Figure 27). In the former, the limits of the stage are delineated on the evidence of poorly preserved goniatites; the total thickness is 46.88 m. The lowest 25.24 m, up to a band with *R. sp.* [*nodosum* group] (that is, most of the R_{1a} and R_{1b} zones), are almost continuously fossiliferous, but above this band the facies largely comprises bioturbated mudstone, with sideritic ironstone bands and nodules, most of which is probably in the R_{1c} Zone.

In the Calver area the Kinderscout Grit comprises some 90 m of sandstone with an underlying thin lower leaf. South of here, the sandstone thins progressively, the lower leaf dying out east of Hassop, and the remaining part of the unit near Castle Hill, Bakewell. The Kinderscout Grit forms a well marked scarp over much of its outcrop but there are few exposures. Its probable turbidite character in this most distal part of its development is indicated by an exposure [2323 7239] of 2.74 m, near Oxpasture, which consists of mainly parallel-laminated medium-grained sandstone in sharp-based graded beds with sole structures. Some 640 m to the south the unexposed top of the grit lies a short distance below an exposure [2326 7175] of the *R. gracile* Band.   NA, IPS

Marsdenian (R_2) strata below *R. superbilingue* Band (R_{2a} and R_{2b})

In the western half of the district, strata of R_{2a} and R_{2b} age crop out on both flanks of the Goyt Syncline and in outliers around Sheen [113 614], Moseley [959 666], Bank House [066 633] and Fernyford [067 618]. In the east, the outcrop extends from Baslow to Birchover.

The succession consists of a lower, mainly mudstone, sequence, and an upper deltaic sandstone sequence including the Roaches Grit in the western outcrop and the Ashover Grit in the east. The lower sequence includes at its base the *R. gracile* Band, the best exposed of the Marsdenian goniatite bands in the district. The composite nature of this band around Ashover (Ramsbottom and others, 1962) has been confirmed in the western part of the present district by Ashton (1974) who found that it can usually be subdivided into several fossiliferous mudstone leaves separated by barren beds. When complete, the following subdivisions are present: a lower leaf with *R. sp. nov.* A; a middle leaf, which is further subdivided into a lower part with *R. gracile* and *R. gracile* early form, and an upper part with *R. gracile* late form and *R. sp. nov.* B; and an upper leaf with *R. sp. nov.* A. Locally, the top two subdivisions merge, and in places the lower leaf is absent where the Longnor Sandstones are present beneath.

The *R. gracile* Band is succeeded by the *R. bilingue* early form Band and then the *R. bilingue* Band. The latter has not been found in the eastern area but occurs nearby in the Chesterfield district (Smith and others, 1967). It is succeeded by bands with *R. bilingue* late form in the Macclesfield Forest area, *R. eometabilingue* in the upper Churnet valley, and a composite band with both *R. eometabilingue* and *R. metabilingue* in the Stanton area. The precise relationship of these highest bands is uncertain but they probably all occur at about the same stratigraphic level. Some sandstones of turbidite type are present locally between the *R. bilingue* and *R. eometabilingue* bands in the upper Churnet valley [0248 6295], and in the Sheen area where they are named the Sheen Sandstones. These include the Lower Sheen Grit and Upper Sheen Grit of Bisat and Hudson (1943, p. 390) but, in detail, three sandstone leaves are present, the two lowest being separated by mudstones containing the *R. bilingue* Band (Figure 33, column 6). The strata between the middle and upper leaves are unexposed, but it is likely that they include the *R. eometabilingue*–*R. metabilingue* horizon (the upper leaf is, therefore, a correlative of the Five Clouds Sandstones). The estimated thickness of 111 m from the base of the *R. gracile* Band to the inferred position of this horizon compares with 70 m in the upper Churnet valley and only 20 m in the Birchover Borehole [SK 26 SW/16] in the eastern area (Figure 27). Thus, the area of maximum sedimentation continued to lie immediately west of the late Dinantian shelf area, though somewhat to the south-east of its position in late Kinderscoutian times (Figure 32).

The lower, mainly mudstone, sequence forms only a small part of the beds between the *R. gracile* and *R. superbilingue* bands, the remainder consisting largely of sandstone, mainly the Roaches and Ashover grits. In the western outcrop, three other sandstone units are locally present below the Roaches Grit; the Corbar Grit, the Five Clouds Sandstones, and the Sheen Sandstones (see Figure 33). Marker bands are lacking, and all these sandstones tend to map out as laterally discontinuous leaves separated by silty mudstones and siltstones (Figures 22, 33). These names are of local significance only, and the sandstone sequence as a whole represents a single regressive cycle (Jones, 1980).

The upper sandstone sequence is estimated to reach a maximum thickness of about 460 m at Macclesfield Forest and to be over 300 m thick in the western outcrop generally. In the eastern outcrop, the full thickness is not present within the district, but nearby it ranges from about 162 m east of Stanton to about 50 m east of Pilsley. These general thickness variations tend to confirm the regional isopachyte diagram drawn by Jones (1980, fig. 11).

The full successions in both western and eastern outcrops represent the advance of a delta from the south-east (Mayhew, 1966, 1967; Jones, 1980) in a way that is analogous to the earlier deposition of the Shale Grit – Kinderscout Grit sequence (in this case from a different direction) in the Edale – Alport area (Walker, 1966; Collinson, 1970). Jones (1980) found the Roaches Grit – Corbar Grit – Five Clouds Sandstones sequence to consist of three facies associations representing deep-water turbidite fan, delta-slope and delta-top environments. The Five Clouds Sandstones, much of the Corbar Grit and the lower leaves of the Ashover Grit (Chisholm, 1977) are the main representatives of the deep-water turbidite fan association. The sandstones occur in graded sharp-based beds, intercalated with silty mudstones and siltstones. Beds over 0.5 m thick tend to be structureless, while the thinner beds tend to show the Bouma (1962) sequence of internal sedimentary structures and to have bottom structures which indicate a general direction of transport towards the north-west (Jones, 1980, fig. 2).

Strata which Jones assigns to the delta-slope association include parallel-laminated siltstones, ripple-laminated and bioturbated sandstones, and lenticular-bedded structureless or parallel-laminated sandstones. The latter are turbidites, in some cases occupying channels up to 20 m deep.

The delta-top association of Jones is made up of coarse-grained sandstone in cross-bedded sets up to 20 m thick, in cosets of tabular cross-bedding with individual sets up to 3 m thick, and in channel forms with faint internal lamination and strongly erosive bases. The latter type in particular is very coarse with pebbles up to 3 cm in diameter. The delta-top sandstones form the main escarpments of the Roaches Grit, including the spectacular crags of The Roaches, Hen Cloud and Ramshaw Rocks (Plate 1); they also form the upper leaf of the Ashover Grit in the Stanton Syncline in the east, where facies are similar except for the absence of very large solitary sets of cross-bedding and a tendency for sharp changes in thickness and lithology to take place across well defined lines that are interpreted as growth faults (Chisholm, 1977). These are listric (concave-up) faults that developed during the deposition of the sandstone and allowed the accumulation of increased thicknesses of sandstone on their downthrow sides. The best known example is at Birchover, but all the faults that affect the Ashover Grit of the Stanton Syncline are believed to be of this type. They are generally somewhat arcuate in plan, with downthrows to the north-west (the same direction as the palaeocurrent flow). The fault planes flatten downwards into the mudstone and do not appear to affect the underlying Viséan limestones. Siltstones and cross-bedded sandstones laid down after the fault movement had ceased form the top of Stanton Moor, and are the youngest deposits preserved in the area. According to Jones (1980) some growth faulting may also have occurred in the Roaches Grit of the Dane valley.

The coarse fluvial sandstones of the delta top are succeeded in places by parallel-laminated fine-grained sandstones interpreted by Jones as a post-abandonment facies of the delta. These in turn are overlain by silty mudstones and siltstones which continue to the base of the *R. superbilingue* Band.

DETAILS

The lowermost leaf of the *R. gracile* Band is exposed in a stream bank [0004 5874] north-west of Tittesworth Farm, where some 4.6 m of fossiliferous mudstones are exposed, with goniatites particularly common in the top 1.8 m. This exposure also contain both middle leaves according to Ashton (localities 118 and 119). Good exposures of the middle leaves occur in the banks of the River Churnet [0241 6190; 0245 6208] north of Swainsmoor, and can be related in a composite section [SK 06 SW/7] to the R$_{2b}$ bands of *R. bilingue* early form and *R. bilingue* exposed higher up the main stream, and to the *R. eometabilingue* Band seen in a tributary [0243 6298] south of Stake Gutter (Figure 34; Ashton, personal communication, 1971). The lower part of the middle leaf at the second of these localities contains limestone bullions and the 7.32 m section through the *R. bilingue* early form Band [0257 6320] showed this, too, to be split into several leaves with a barren interval, 2.51 m thick, in the middle with ironstone ribs. The band with *R. eometabilingue* at the top of this composite section comprises some 30 cm of dark grey mudstone containing the eponymous goniatite. A second composite section [SK 06 SW/6] in the Churnet and its tributary west of Hurdlow includes exposures of the *R. gracile*, *R. bilingue* early form and *R. bilingue* bands. The last named is exposed on the left bank of the Churnet [0145 6065] and has a total thickness of 5.6 m; it contains leaves with evolutionary changes in the Reticuloceratid fauna which have been described in detail for the area by Ashton (1974; *see* p. 84).

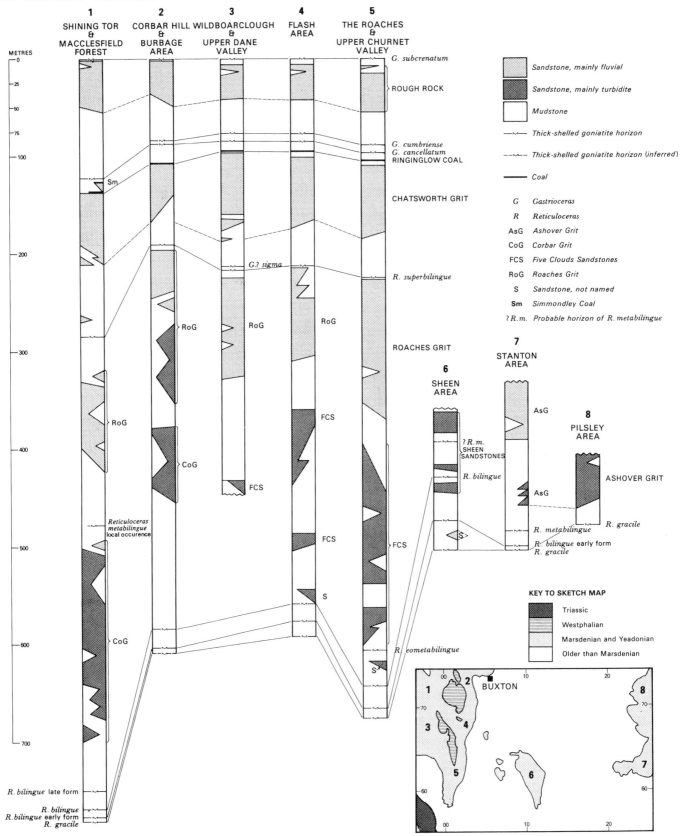

Figure 33 Generalised sections of Marsdenian (R$_2$) and Yeadonian (G$_1$) strata. Sketch-map shows outcrop and location of sections.

Two site investigation boreholes near Brund [SK 06 SE/8 and SK 06 SE/12] have proved the strata between the *R. gracile* and *R. bilingue* early form bands, and between the latter and the *R. bilingue*-Band (Figure 34). The *R. gracile* Band shows an exceptional thickness of 16.39 m; it contains three leaves identified by Ashton as the upper *R. sp. nov.* A leaf from depths of 26.82 to 27.43 m, and the upper and lower parts of the middle leaf from 30.56 to 34.19 m and from 39.07 to 43.38 m respectively. The middle leaf contains a bullion of 5 cm at 32.11 m. The mudstones between these leaves appear barren, though traces of bioturbation are present. The *R. bilingue* early form Band was found from 28.80 m to the bottom of borehole SK 06 SE/12, the total thickness of 5.95 m probably representing almost the full extent of the band. It comprised nearly continuously fossiliferous dark grey mudstone. The basal part of this band was also penetrated near the top of the borehole SK 06 SE/8 allowing a continuous composite section to be drawn. A feature of the succession between the *R. bilingue* early form and *R. bilingue* bands in SK 06 SE/12 is the presence of a fossiliferous mudstone band, 25 cm thick, at 17.93 containing *Posidoniella sp.*, *Anthracoceras* or *Dimorphoceras sp.*, and mollusc spat. Even though the *R. bilingue* Band is incomplete, the total proved thickness of about 4.65 m is the greatest recorded for this band in the district.

The *R. bilingue* early form Band is exposed [1128 6302; 1211 6228; 1196 6097] below the escarpment of the lower leaf of the Sheen Sandstones east of Sheen Hill and at several localities southeast of Hulme End [1077 5916; 1120 5887]. A 15.1 m sequence in a stream [1128 6302] south of Broadmeadow Hall is similar to that in the Brund boreholes. The sandstones forming the lower leaf of the Sheen Sandstones are of turbidite facies and are best exposed in old quarries east of Under White [1003 6388], at Townend [1090 6057], and in a stream gully east of Ridge End [1024 6264 to 1030 6269] where 16.46 m of thinly to thickly bedded grey to pale brown fine-grained sandstone and interbedded silty mudstone with sideritic ironstone are present. The lowest 5 m of the beds between the lower and middle leaves of the Sheen Sandstones is partially exposed in the stream south of Moorhouse [1054 6147]:

	m
Mudstone, dark grey with a 13 m dark argillaceous limestone band, 15 cm above the base; *Caneyella rugata, Dunbarella speciosa, Posidonia sp., R. bilingue*	0.43
Not exposed	3.05
Mudstone, dark grey, silty	1.52
Sandstone (Sheen Sandstones, lower leaf)	—

Numerous small exposures of the overlying middle leaf of the Sheen Sandstones are seen higher up the stream [1086 6160 to 1141 6220]; they comprise mainly thinly bedded ripple-laminated grey-brown fine to medium-grained micaceous sandstone. The overlying argillaceous beds are not exposed and the steep higher slopes of Sheen Hill are formed by the upper leaf of the Sheen Sandstones. The unit is seen in a few quarried blocks of thickly bedded red-stained coarse-grained sandstone in a landslip on the south-west slope of the hill: some parallel lamination is visible in these blocks but, in an *in situ* exposure, 4.57 m thick, at the summit, the sandstone is massive and coarse-grained with bands of mudflake cavities. It probably represents a turbidite feeder channel similar to those of the delta-slope association in the Five Clouds Sandstones noted by Jones (1980) (p. 85).

At Wildmoorbank Hollow [9803 7391], 3.35 m of grey silty mudstone separate the *R. bilingue* early form Band, which is 1.83 m thick, from the *R. gracile* Band, 0.66 m thick; this latter band in turn lies 4.5 m above the top of the Kinderscout Grit. The *R. bilingue* Band, some 8 m above the *R. bilingue* early form Band is exposed [9822 7376] farther up stream; it comprises 0.66 m of mudstone with *R. bilingue* and *Homoceratoides sp.*. In Hogshaw Brook

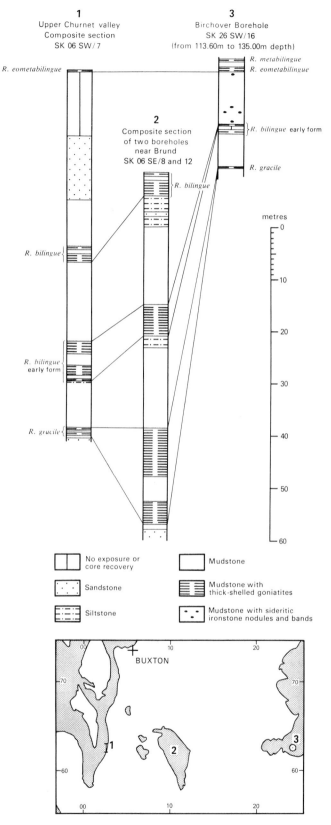

Figure 34 Comparative sections of early Marsdenian strata in areas of maximum and minimum thickness.

[0599 7423] north of Buxton, the *R. bilingue* early form Band consists of 0.84 m of mudstone, grey in the top 0.38 m and blocky and pyritous below, with *R. bilingue* early form, *Anthracoceras sp.* and *Dunbarella sp.*; it lies some 10 m above the top of the Kinderscout Grit. About 18 m higher in the sequence, the *R. bilingue* Band [0596 7434] comprises 1.22 m of dark decalcified mudstone with *R. bilingue* and *Dunbarella sp.*.

An exposure in a ditch near Brownsett [9916 6377] shows 0.80 m of dark mudstone with *R. bilingue* and *Dunbarella sp.*, on 0.40 m of soft grey sandstone with mudstone flakes overlying a few centimetres of dark poorly fossiliferous mudstone. This section seems to represent part of the *R. bilingue* Band split by a turbiditic sandstone incursion. In the same ditch, small exposures of sandstone and mudstone include one showing 0.25 m of dark grey mudstone with *R. metabilingue* [9925 6369], not far below the lowest leaf of the Five Clouds Sandstones.

In the eastern area, north-east of Pilsley [2326 7175], the *R. gracile* Band is seen as 0.75 m of black platy mudstone with bullions, containing *R. gracile* late form; the band lies apparently a short distance above the Kinderscout Grit. Nearby [2321 7209], and probably only slightly higher in the succession, an exposure of the *R. bilingue* early form Band, 0.99 m thick, has yielded *C. rugata*, *Anthracoceratites sp.*, *Hd. ornatum* and *R. sp.*.

In the Stanton Syncline, boreholes and surface exposures prove a mudstone sequence below the Ashover Grit broadly similar to that in the Ashover boreholes (Ramsbottom and others, 1962) (Figure 27). The lowest 44 m consists of dark grey mudstone with thin goniatite bands. The two lowest of these, the *R. gracile* and the *R. bilingue* early form bands, are present both in the Stanton and Ashover boreholes, but at Stanton these are followed by a pair of bands containing *R. eometabilingue* (in the lower) and *R. metabilingue* (in the upper), whereas at Ashover these were not recorded: the *R. bilingue* Band, present at Ashover at about this level, was not found at Stanton. The *R. bilingue* late form Band, used by Smith and others (1967, p. 68) to define the position of the Ashover Grit, is absent in the Stanton and Ashover boreholes, but is recorded at outcrop in the intervening ground near Beeley, where it lies 1.8 m above the *R. bilingue* Band (Smith and others, 1967, p. 69).

The Five Clouds Sandstones are named after the prominent escarpment below the Roaches on the western flank of the Goyt Syncline. Here, a line of discrete crags [0002 6284 to 0020 6255], up to 22 m high, consists of pale brown to pink coarse-grained sandstone without visible bedding. The crags may possibly represent a continuous but variably cemented bed in which the well cemented parts form the crags and the more poorly cemented sandstone the intervening unexposed ground. Old quarries on the same escarpment to the south-east [0031 6242 to 0041 6226] expose a total of 10.7 m of thickly bedded medium to coarse-grained sandstone in mainly massive beds with some indistinct parallel lamination, ferruginous concretionary patches, and scattered mudstone clasts.

An 81.6 m stream section at Upper Hulme [SK 06 SW/14] partially exposes thickly bedded sandstones (Five Clouds Sandstones), which here, as elsewhere, belong to the deep-water turbidite fan association of Jones (1980). The top 35.5 m of this section, between the Five Clouds Sandstones and the base of the Roaches Grit, consist of partially exposed grey-brown ripple-laminated sandstones and thinly interbedded siltstones showing gradational interbed contacts. These beds are referred by Jones to the delta-slope association.

On the east flank of the Goyt Syncline, good sections in the Five Clouds Sandstones are few. At Cistern's Clough, the lower leaf [0375 6979 to 0369 6979] consists of blocky grey fine-grained sandstones interbedded with mudstone, while the upper leaf, now largely obscured by the main road, includes thin beds of ripple-laminated fine-grained sandstone. The lower leaf belongs to the deep-water turbidite fan association, and the upper leaf to the delta-slope association (Jones, 1980) (*see* p. 85). Father north, a section 10.98 m high is exposed at a roadside quarry [0377 7109] and a section 20.12 m high at The Terret [0368 7201]. Both are in the turbidite facies.

The Corbar Grit crops out most extensively in the axial areas of the Todd Brook and Macclesfield Forest anticlines between Longclough [987 737] and Whitehills [973 728]. It forms a variable complex of up to four leaves of sandstone, extending through an estimated thickness of 200 m and is mainly of turbidite facies. The sandstones are commonly flaggy and contain siltstone and silty mudstone partings. Farther south, much of the sandstone disappears, though a thin leaf is still present around the southern culmination of the Macclesfield Forest Anticline.

On the eastern limit of the Goyt Syncline the Corbar Grit is well developed around Edgemoor, where a section [0343 7345] of sandstone, 33.53 m thick, is of turbidite facies. Farther east, similar beds are present in a section, 20.28 m thick, in an old quarry on Corbar Hill [0497 7405].

The Roaches Grit is exposed in crags at The Roaches extending from near Roach End [9964 6449] to Hen Cloud [009 616] and around the southern closure of the Goyt Syncline to Ramshaw Rocks [020 629]. The most extensive sections are at Rockhall [0061 6221] and Hen Cloud where the total thicknesses of the sections are 62.5 m and 34.2 m respectively. The deposits of three discrete superimposed fluvial channels have been recognised in the Rockhall section (Jones, 1980; Jones and McCabe, 1980). The channels tend to have highly irregular erosive bases, overlain firstly by thin faintly laminated coarse sandstone, with pebbles up to 3 cm, and then by large crossbedded sets up to 20 m in height. These are overlain in turn by sets of tabular cross-beds, 0.5 to 2 m thick, in cosets 10 to 35 m thick. Eight similar channels have been distinguished by Jones (1980, fig. 10) laterally superimposed along the length of The Roaches escarpment, and at Ramshaw Rocks four superimposed channels are present. Below The Roaches escarpment, a lenticular outcrop with a crag, 13.5 m high, of massive red-stained coarse-grained sandstone [9985 6353] is interpreted by Jones (1980, plate 3) as a turbidite-filled channel.

Good exposures in the Roaches Grit occur in the valleys of the Dane and its tributary Clough Brook, around Allgreave and Back Dane; a section [9706 6621 to 9703 6641] 74 m high has been described in detail by Jones (1980, fig. 5). Of particular interest [9706 6621] are thick beds of structureless coarse-grained sandstone, with rows of mudflake cavities, interpreted by Jones as turbiditic channel fills within the delta-slope association. The sharp erosive base of the tabular crossbedded sandstones near the top of this section [9703 6641] marks the transgressive base of the fluvial channel sandstones of the delta-top association. Farther up Clough Brook [9698 6740] a section of 6.2 m in the highest beds of the Roaches Grit includes thinly interbedded dark fine-grained sandstones, siltstones and sandy mudstones with low-amplitude ripple-lamination. In the Cumberland Cottage section [SJ 96 NE/9] the equivalent beds are strongly bioturbated and are thought by Jones to represent the post-abandonment phase of the delta.

Farther north the grit outcrop underlies much of the lower slopes of Shutlingsloe and Yarnshaw Hill, where its total thickness is about 100 m. A roadside quarry [9841 7058] shows a typical section of about 15 m of massive pink medium-grained sandstone.

On the east limb of the Goyt Syncline, the grit forms the prominent escarpment of Axe Edge. Here, Cistern's Clough shows a 50 m section [SK 06 NW/10], the top 22.5 m consisting of coarse-grained cross-bedded sandstone of delta-top type while the underlying sandstones show ripple-lamination and bioturbation characteristic of the delta slope.

An indication of the thickness of the Roaches Grit north of Axe Edge is provided by the Portobello Bar Borehole [SK 07 SW/3], which proved head to 8.23 m, on a variable sequence of sandstone and mudstone to 150.57 m. The total calculated thickness here is about 190 m. Nearby [0327 7207 to 0331 7209], the coarse-grained upper part of the grit is well exposed in a section 27 m thick.

In the eastern part of the district, all the sandstones are named Ashover Grit, regardless of facies or relative level in the sequence. The delta-slope association is represented by grey well bedded siltstones with fine to medium-grained sandstones, mainly of turbidite facies. In the Stanton Syncline, sandstones of this type are present only locally, and at the base of the sequence; they have been interpreted as a submarine fan deposited at the foot of the delta slope (Chisholm, 1977, p. 311). They were proved in boreholes around Birchover (Figure 27), where individual beds ranged up to 5 m in thickness. Well marked features east and west of Upper Town [240 617] suggest the presence of three lenticular sets of sandstones. Elsewhere the turbidites have been recognised, mainly by features, at Anthony Hill [220 618] and above Beech Wood [206 627], but appear to be absent along the northern flank of the Stanton Syncline. North of the River Wye they are common, however, and occur throughout the slope association, forming an extensive outcrop around Pilsley (Figure 33, column 8). Exposures are few, the best being at Lees Moor Wood [2477 6720], where a section, 7.6 m thick, consists of sharp-based sandstone beds, from 0.40 m to over 3.0 m thick, largely lacking internal lamination. North of Baslow the grit thins markedly prior to its disappearance in the ground to the north (Stevenson and Gaunt, 1971, p. 234).

In the Stanton Syncline the main leaf of the Ashover Grit is generally about 60 m thick consisting of coarse-grained feldspathic cross-bedded sandstone with impersistent siltstone bands. its thickness is locally increased, however, on the downthrow sides of syndepositional listric faults (growth faults), the extra sandstone being massive or thickly bedded.

The best example of the structures associated with a growth fault is at Birchover [240 622], where good exposures have been supplemented by two boreholes [SK 26 SW/16, SK 26 SW/17]. The fault plane was penetrated by both holes and was shown to flatten off in the mudstones beneath the lower (turbidite) leaves of the grit (Chisholm, 1977, fig. 4). The cross-bedded sandstones have been faulted down by about 70 m, as can be inferred from an exposure at Rowtor Rocks [2349 6211], and the total thickness of sandstone has been increased on the downthrow side by the addition of a similar thickness of massive and thickly bedded sandstone. The latter facies is also well seen at Rowtor Rocks, and has been much worked for freestone in quarries [241 624] above Birchover, where the bedding planes dip towards the fault plane at angles up to 24°. The faulted material is succeeded on Stanton Moor [243 627] by flat-lying cross-bedded sandstones apparently laid down after the fault movement had ceased.

Other examples of growth faulting have been described by Chisholm (1977, pp. 319–321), for instance at Stoney Ley Wood [232 635] and Bury Cliff Wood [217 615], where cross-bedded sandstones are tilted towards the fault planes at angles up to 45°, and at Robin Hood's Stride [224 623] and Lees Cross Quarry [2505 6369], where there are exposures of massive and thickly bedded sandstones analogous to those worked at Birchover.   NA, JIC, IPS

Marsdenian strata above *R. superbilingue* Band (R_{2c})

Strata of R_{2c} age are found only in the western part of the district, in the Goyt Syncline and round the southern culmination of the Macclesfield Forest Anticline (Figure 40). The estimated total thickness is about 102 m around Burbage Edge, some 125 to 131 m in the southern part of the Goyt Syncline, and 162 m around Shining Tor.

The sequence consists of a coarsening-upwards unit, with the *R. superbilingue* Band at the base, overlain by barren mudstones and siltstones that grade up into the Chatsworth Grit. Locally the *Donetzoceras sigma* Band occurs a few metres above the *R. superbilingue* Band. The Chatsworth Grit includes a relatively thin (up to 15 m) and impersistent lower leaf. The main leaf is present everywhere and is estimated to range in thickness from 60 to 74 m. The Chatsworth Grit is overlain by the Ringinglow Coal, which is generally 1.3 to 1.8 m thick, and this by 7 to 18 m of silty mudstone, with sporadic ironstone nodules, up to the *Gastrioceras cancellatum* Marine Band. Locally, a thin sandstone with the Simmondley Coal above is present beneath this marine band (Stevenson and Gaunt, 1971, p. 182).

An unpublished sedimentological study of the Chatsworth Grit has been made by Kerey (1978) in an area extending from the southern culmination of the Goyt Syncline as far north as Glossop, and including part of the outcrop in the present district. This work confirmed the general conclusion (Harrison *in* Stevenson and Gaunt, 1971, p. 249) that the sandstones are mainly subarkosic in composition (p. 91). Kerey concluded that these rocks were deposited from low to medium sinuosity rivers with a dominant westward palaeocurrent direction.

DETAILS

The *R. superbilingue* Band generally consists of dark grey to black calcareous mudstone from 0.23 to 0.76 m thick. Locally [as at 0272 7375] *Lingula sp.* occurs in the immediately underlying mudstone, and thin 'contorted beds' (Cope, 1946; *see* p. 113) are also present in places just above or below the band.

The best sections showing the band, together with the underlying and overlying strata, are south and east of Cumberland Cottage [SJ 96 NE/9], near Berrybank [9810 6801 to 9813 6798], south of Wicken Walls [0154 6706] and north-west of Burbage Edge [SK 07 SW/11]. In the section near Berrybank, a separate band, 25 cm thick, with poorly preserved goniatites, and a 'contorted bed' lie respectively 3.21 m and 0.38 m above the top of the *R. superbilingue* Band. The higher marine horizon may correlate with the *Donetzoceras sigma* Band exposed beyond the western boundary of the district in the right bank of the stream south-west of Lower Nabbs [9663 9798] (loc. 170 of Evans and others, 1968). This is also the locality where Cope (1946) reported the presence of the *R. superbilingue* Band, overlain by a 'contorted bed'. The Cumberland Cottage section shows a sequence of mudstone, 30.24 m thick, coarsening upwards to siltstone, between the *R. superbilingue* Band and the lowest leaf of the Chatsworth Grit. The siltstones, which predominate in the top 20 m, show a fine parallel lamination with alternating dark grey and brown colour banding. On the right bank of the River Dane at Gradbach [9944 6618], the highest 6 m of these strata comprise grey mudstone and siltstone with a few beds of ripple-laminated and strongly bioturbated sandstone.

The lower leaf of the Chatsworth Grit has been mapped continuously for about 4 km south of Shining Tor [993 733]. Elsewhere, it can be traced only where is is exposed, being difficult to detect in the steep slopes below the main escarpment. In the Cumberland Cottage section, 11.5 m of turbiditic sandstone and subordinate siltstone are exposed below the lower leaf in beds up to 0.58 m thick. The lower leaf consists of 12.1 m of cross-bedded buff medium-grained sandstone, and a similar facies is present in the section at Gradbach (*see above*). South-east of Lower Nabbs [9688 6787; 9710 6786], up to 2.5 m of massive medium-grained turbiditic sandstone with large ferruginous concretions are exposed.

The best section through the main leaf of the Chatsworth Grit is in the Dane valley around Three Shire Heads [0094 6830 to 0074 6892]. The lowest bed is of turbidite character and consists of 2.74 m of speckled pale grey fine- to medium-grained sandstone, forming a waterfall. It is overlain by about 6 m of thinly interbedded shaly siltstone and current ripple-laminated fine-grained sandstone. A gap of 4.57 m separates these pro-delta slope sediments from beds representing the delta-top facies. The lower part of these beds consists of 9.14 m of medium- to coarse-grained sandstone in poorly defined ?trough cross-bedded sets 13 to 61 cm thick. This sub-facies passes up into coarse-grained sandstone in large-scale crossbedded sets, mainly over 3 m thick, exposed in the upper part of the section; large-scale crossbedding is characteristic of the sandstone generally. The following (with thicknesses shown) are the best exposures: crags in the east face of Shutlingsloe [9765 6956], 11.6 m; Heild Rocks near Allgreave [9790 6734], about 17 m; old roadside quarry south of Shining Tor [9958 7270], 7.6 m; old quarry north-west of Burbage Edge [0270 7343], 8.5 m; crag south-west of Flash Bottom [0185 6573], about 10 m; and Gib Tor [0182 6473] about 22 m. The sandstone in these exposures is generally pink or red-stained and contains small pebbles up to 2.5 cm in diameter. The greatest proved thickness is in the upper level of Burbage Collieries, where 43 m of 'coarse red grit', were recorded.

The succession from the top of the Chatsworth Grit to the *G. cancellatum* Marine Band is best seen in a composite section [SK 06 NW/5] in a stream near Orchard Farm. The total thickness of these beds is 15.33 m. The Ringinglow Coal is exposed at two places [0179 6971; 0216 6883]. At the former locality, 36 cm of coal and carbonaceous shale on 20 cm of seatearth mudstone were noted; at the latter, 1.52 m of coaly shale were seen lying about 1.5 m above the pinkish grey fine-grained sandstone forming the topmost bed of the Chatsworth Grit. Exploratory boreholes for coal north-east of Orchard Farm proved up to 2.13 m of coal on sandstone, at the top of the Chatsworth Grit. Elsewhere, its presence is marked by numerous old bell-pits, and in a few places by collapsed adits, for example NNE of Holt [009 697] and ENE of Burntcliff Top [995 664]. The latter workings are referred to (Hull and Green, 1866, p. 52) as 'Green Hills Colliery', and the coal ('Feather Edge Coal') was 1.83 m thick. The most extensive workings were probably in Burbage Collieries, where the upper level intersected the seam [0285 7202]: coal 0.33 m on 'bat' 0.15 m on coal 0.89 m. Deep shafts around Thatch Marsh [0273 7067], near Moss Chain, probably connect with these workings.

The outcrop of the Simmondley Coal is traceable into the district for a short distance in the valley north-east of Shining Tor. In workings to the north, near Errwood Hall, it was 0.38 m thick (Stevenson and Gaunt, 1971, p. 238). The top of a thin sandstone with rootlets exposed in a stream near Dane Bower [9992 7037] may represent the horizon of this coal.

The highest 4.48 m of the sequence exposed in the Orchard Farm stream section contains sporadic bands of ironstone nodules and a 'contorted bed', 18 cm thick, at the base.

Yeadonian (G₁) strata

The outcrop of Yeadonian strata is confined to the Goyt Syncline from its southern culmination, near Harpersend [012 627], to the upper part of the Goyt valley at the northern margin of the district. The estimated thickness ranges from 83 m around the Dane valley to a maximum of about 121 m near Shining Tor.

The succession (Figure 33) is predominantly argillaceous in the lower part, with the *G. cancellatum* Marine Band at the base and the *G. cumbriense* Marine Band 5 to 8 m higher. The association of these bands with thin 'contorted beds' in the adjacent mudstones was noted by Cope (1946) (*see also* p. 113). The upper part comprises a coarse-grained sandstone unit, the Rough Rock, which passes up into silty seatearth or ganister, overlain by a thin coal, the Six-Inch Mine. The

coal is succeeded by the *G. subcrenatum* Marine Band, at the base of the overlying Westphalian.

Estimates of the thickness of the Rough Rock range from 35 to 52 m. Typically the sandstones are cross-bedded in sets up to 3 m thick, pink to buff in colour, and medium- to coarse-grained. They are generally finer grained and less thickly bedded in this district than the Roaches and Chatsworth grits, and do not form such prominent escarpments. Mudstone or siltstone beds are present near the top and base.

The Yeadonian sequence represents a transition from transgressive marine to fluviatile conditions, as in the case of the cycle containing the Chatsworth Grit.

Heath (1974) gave details of palaeocurrent measurements in tabular cross-bedding in the Rough Rock in the district as part of a wider study. Around the southern end of the Goyt Syncline, south of Quarnford, palaeocurrents flowed south-westwards and westwards, but in the adjacent area, as far north as Errwood Reservoir, the directions were highly variable, while at the north end of the syncline the flow was towards the south-east or south-west, more in accord with the overall south-westward trend established by Shackleton (1962).

DETAILS

The best exposure of the lowest part of the sequence is in a stream to the east and north-east of Orchard Farm in the upper part of a section [SK 06 NW/5] extending down to the Chatsworth Grit. This section is taken as typical and compares closely with that proved about 9 km to the north in the Fernilee No. 1 Borehole (Stevenson and Gaunt, 1971), except that the *G. cumbriense* Marine Band is only 0.12 m thick in the stream section compared with 0.89 m in the borehole. The marine bands are separated by grey to dark grey ferruginous mudstone with sporadic ironstone bands and nodules, and are closely associated with thin 'contorted beds' (Cope, 1946) (p. 113), standing out as ribs in the mudstone walls of the stream gorge.

The overlying sequence, up to the base of the Rough Rock, is poorly exposed but includes thin sandstone beds in its upper part, as in a 3.66 m exposure, in a stream bank [0158 6434] east of Goldsitch House. In Burbage Colliery upper level (Hull and Green, 1866, p. 46) approximately 60 m of strata are present between the probable position of the *G. cancellatum* Marine Band and the Rough Rock. The upper half of these strata is described as 'stone bind' with a sandstone 9 m thick, while the lower half is 'black shale'. However, the only detailed records of this part of the sequence in the Goyt Syncline are those of the Fernilee boreholes (Stevenson and Gaunt, 1971, pp. 240–241).

The best exposures of the Rough Rock occur in several disused building stone quarries in the Goyt Syncline: Goytsclough Quarry [0115 7333], 12.20 m; roadside quarry [9983 7222], north-west of the Cat and Fiddle Inn, 4.27 m; Danebower Quarries [0125 7005], 3.05 m; and Reeve Edge Quarries, [0126 6967], 14 m. The main rock type exposed is a medium to coarse-grained tabular cross-bedded sandstone, except at Goytsclough where sandstone with only faint lamination predominates. Trough cross-bedded and parallel-laminated sandstone also occurs in addition to thin beds of shale pellet conglomerate. The only direct measurement of thickness was from Burbage Collieries upper level where 42.7 m of 'flaggy sandstone' were reported.

The lowest beds of the Rough Rock occur in the Dane valley north-west of Greens [0042 6696 to 0047 6743]. They consist of thinly bedded, ripple-laminated fine-grained sandstones with interbedded siltstones and are spectacularly folded (Francis, 1970).

These beds pass up into pink and red cross-bedded medium-grained sandstone.

Near Greens, a stream section [SK 06 NW/11] shows the top 14 m of the Rough Rock which includes several beds of sandstone with rootlets, each overlain by matted carbonaceous plant fragments or coaly partings, the highest forming a seatearth, 1.65 m thick, overlain by an 8 cm coal, the Six-Inch Mine. This horizon appears to lie about 9.36 m below the base of the *G. subcrenatum* Marine Band, the intervening strata being poorly exposed. Some 680 m to the south-east, in the stream near The Wash [0131 6627], Cope (1946, p. 145) described a section in which the coal, 4 to 5 cm thick, lies only 0.74 m below the *G. subcrenatum* Marine Band. In Stake Clough [0046 7259; 0068 7298] the coal is 2 to 5 cm thick and lies 0.53 m below the marine band.   NA, IPS

Petrography

The following account is based on the petrographical analyses of selected samples representing the main mapped sandstone formations; it must be emphasised that some of the general conclusions are not supported by a statistically valid spread of samples. The conclusions therefore are not incorrect but their substance would undoubtedly be strengthened by further sampling.

Thin-sections and whole-rock X-ray diffractometry (by K. S. Siddiqui) confirm the known main characteristics of the 'Millstone Grit' suite: the marked distinction between 'protoquartzitic' and 'feldspathic' sandstones; the high degree of diagenetic compaction; the common presence of authigenic kaolinitic clay; the dominance of fresh microcline among detrital feldspars; the commonly subrounded to subangular grain form; the poor to moderate sorting and general variation in sediment coarseness (*see, especially,* Trewin and Holdsworth, 1973; Gilligan, 1920).

As detailed in Table 2, the distinction between the two main suites is even greater in this district than has been recorded in neighbouring areas (Evans and others, 1968, pp. 83–92; Stevenson and Gaunt, 1971, pp. 247–263). Here the protoquartzites approximate to orthoquartzites because their lithoclast content, amounting to 4% or more of the rocks, is itself mainly quartzose and the feldspar content is less than 1% (3% if authigenic clay is included as feldspar). In contrast, the feldspathic sandstones contain a minimum of 8% feldspar and less than 1% non-plutonic lithoclasts. The high degree of compaction in these rocks, signified by the redistribution of quartz (sutured junctions and syntaxial overgrowth) and extreme distortion of some feldspar and mica clasts, must be responsible for a proportion of analogous distortion in quartz grains. Therefore the presence of strained extinction cannot be equated with a metamorphic origin, even though some grains certainly do have such an origin.

The fabric of the sandstones is controlled partly by their mode of deposition and partly by the extreme nature of diagenetic alteration. In most cases the proportion of detrital matrix, for example, is equalled or surpassed by proportions of authigenic clay and the assessment of primary grain-matrix ratios is therefore subjective. Even so, it can be deduced that some of the rocks that are technically grainstones may once have been packstones if not quite wackestones (Knox *in* Nockolds and others, 1978, p. 229). E46782 (Chatsworth Grit), for example, contains much void pore space from which detrital matrix may have been leach-

ed since compaction, because its presence is otherwise difficult to reconcile with the intense compaction implied by distorted feldspar and mica; authigenic kaolinite here, as elsewhere, does not show this distortion.

Features distinguishing the protoquartzitic sandstones in particular, include the presence of notable minor proportions of detrital chert and foliated quartzite, and the accessory amounts (only) of plagioclase and microcline, with white mica, opaque oxides, tourmaline and zircon. One sample (E46770: Kniveden Sandstones) is in part cemented by poikilotopic aggregates of goethite.

The feldspathic sandstones, though diverse in composition, are all either arkosic or sub-arkosic 'grainstones' (*see* comments on matrix, *above*). Apparent trends in their distribution and composition include a predictable tendency for less resistant components, such as feldspar and mica, to decrease in proportion with sediment maturity, such as grain rounding and to some degree grain sorting and grading. This contrasts strongly with the protoquartzitic sandstones in which even the least mature rock (E46775; Lum Edge Sandstones) contains little other than quartzose detritus — an obvious indication of a restricted variety of source material.

A second trend, less predictable and therefore a clear case for testing by additional sampling, is for the ratio of potassic feldspar to plagioclase to increase through the Namurian and possibly on into the Westphalian; detrital biotite concentrations tend to parallel those of plagioclase. These phenomena would correlate best with a progressive change from tonalitic to potassic granite source material since sandstones of comparable fabric occur across the span of samples, eliminating possible correlation of feldspar proportions with host grain-size availability or with differential hydraulic or attrition characteristics between the feldspars. Validity for naming the main host rocks in igneous, rather than metamorphic, terms is provided by the occurrence in coarse sandstones of composite clasts containing the main detrital minerals in igneous, rather than metamorphic, intergrowth (such as 46777: Five Clouds Sandstones). The classic exhaustive treatment of the Yorkshire Millstone Grit by Gilligan (1920, pp. 254–259) substantiates the dominance of microcline–microperthite-bearing igneous rocks among the pebbles in the series.

Detrital potassic feldspar is mainly fresh microcline and at least a proportion of it is perthitic. The plagioclase is mostly clouded by alteration, and that which is determinable is sodic rather than calcic. Both types of feldspar show some degree of diagenetic deformation in a proportion of grains of virtually every sample. In the most extreme cases some former grains are reduced by cataclasis to bundles of shards (for example, E46778: Five Clouds Sandstones). There is no obvious correlation between degree of deformation and deposit age.

Detrital biotite, or its partial decomposition products, is abnormally abundant in several samples collected from low in the succession (E46763, E46764, Longnor Sandstones; E46767, Sheen Sandstones; E51342, Ashover Grit), and in each case its abundance corresponds with a high detrital plagioclase content and a high plagioclase to potassic feldspar ratio. Except where protectively engulfed in lithoclasts the biotite is more or less altered by bleaching, or more thorough decomposition, to aggregates variously containing

Table 2 Sandstone modal analyses

	1	2	3	4	5	6	7	8	9	10	11	12	13
Quartz	73.1	82.7	58.2	57.3	71.4	70.4	67.9	71.7	82.2	61.9	73.3	69.0	72.4
Potassic feldspar	—	—	5.5	10.1	15.4	12.1	24.8	13.5	8.5	3.8	17.5	19.4	14.0
Plagioclase	0.1	0.7	14.1	12.1	3.9	7.6	1.7	1.9	0.1	7.1	2.2	4.7	0.5
Biotite	—	0.5	12.9	11.3	0.7	0.1	—	0.4	0.1	3.8	1.1	0.9	0.3
Muscovite	—	0.4	2.7	0.4	—	0.2	0.4	0.3	0.1	1.2	—	0.5	0.4
Authigenic clay	2.4	1.2	1.2	4.7	4.0	5.1	2.1	3.6	2.2	8.0	3.6	2.4	5.9
Detrital clay	6.0	4.3	3.7	2.2	1.8	1.6	1.8	6.5	4.7	8.7	0.4	1.6	5.0
Lithoclasts	17.2	7.0	0.3	—	0.4	0.1	0.4	0.2	0.9	0.6	0.7	0.3	0.2
Opaque minerals	1.1	2.8	1.2	1.8	2.4	2.4	0.9	1.7	1.0	0.7	1.2	1.2	1.3
Calcite	—	—	—	—	—	—	—	—	—	3.8	—	—	—
Others	0.1	0.4	0.2	0.1	—	0.4	—	0.2	0.2	0.4	—	—	—

1	E 46775	Lum Edge Sandstones	6	E 46778	Five Clouds Sandstones	11 E 51341 Ashover Grit (Upper)
2	E 46769	Kniveden Sandstones	7	E 46779	Roaches Grit	12 E 51340 Ashover Grit (Upper)
3	E 46763	Longnor Sandstones	8	E 46782	Chatsworth Grit	13 E 46786 Woodhead Hill Rock (the data
4	E 46767	Sheen Sandstones (Lower)	9	E 46783	Rough Rock	for this Westphalian sandstone are included
5	E 46766	Sheen Sandstones (Upper)	10	E 51342	Ashover Grit (Lower)	for convenience)

Sample localities are given in Appendix 3

The definition of terms is as follows:

Quartz refers to monocrystalline grains of the mineral together with large composite grains of probable plutonic origin. *Potassic feldspar* includes all feldspar stained by sodium cobaltinitrite; it is mostly fresh microcline, much of it being microperthitic. *Plagioclase* tends to be more or less clouded by alteration products but is probably generally sodic. *Biotite* is taken to include altered relics as well as the fresh mineral. *Muscovite* is taken to include all white mica above clay grade particle size. *Authigenic clay* refers to well crystallised interstitial aggregates of kaolinitic clay, including kaolinite itself and dickite. *Detrital clay* incorporates all heterogeneous clay-grade matrix and may therefore include unrecognised highly deformed and altered plagioclase etc. *Lithoclasts* are in all cases mainly metamorphic and sedimentary polycrystalline quartz varieties including chert and quartzite; siltstone, acidic volcanics etc are comparatively rare. *Opaque minerals* include detrital grains and in some cases matrix-forming ferric oxide. *Calcite* occurs as a poikilotopic cement in one sample only. *Others* generally means detrital accessory minerals such as zircon, tourmaline, rutile and garnet.

assemblages of chlorite, clay minerals, siderite, white mica and ferric oxide. Whole-rock X-ray diffraction analyses indicate significant proportions of chlorite in the biotite-rich rocks but no measureable chlorite in other samples; it is inferred that chlorite, rather than vermiculite or smectite is the dominant biotite degradation product. Biotite clasts also show strong degrees of physical deformation and in extreme cases (such as E46767: Sheen Sandstones) locally appear to engulf neighbouring felsic grains. Probably the proportion of detrital biotite in many sandstones was much greater than is now apparent: in E46779 (Roaches Grit) for example, traces of the mineral are so heavily altered to ferric oxide and clay that their identity can be inferred only by analogy with less altered samples.

As is clear from Table 2, detrital grains other than quartz, feldspar and biotite rarely exceed accessory proportions in the feldspathic sandstones. The accessories include chert, metamorphic lithoclasts, white mica and opaque oxide grains, with probably less abundant heavy minerals such as tourmaline (schorlite), epidote, zircon and (in E46777) garnet.

Detrital silty clay matrix is commonly a very minor constituent although it may have been reduced in proportion by diagenetic solution or replacement in at least some rocks. The clay component is probably mainly kaolinite, but mixed-layer clay occurs in E46763 (Longnor Sandstones) and E46779 (Roaches Grit), and montmorillonite is present in E46782 (Chatsworth Grit). The X-ray diffractograms indicate that illite is generally absent or present in only very small proportions.

Authigenic clay, occurring in interstitial clots of undeformed well crystallised 'books', is disseminated through all samples in varying degrees of prominence. X-ray diffraction analysis indicates that kaolinite is most commonly the mineral involved, but its polymorph dickite is sometimes dominant (E46778).

Although cementation is normally confined to redistributed quartz and authigenic kaolinite, calcite is sometimes present as a rare accessory proportion, and locally contributes more significantly to the cement (E51342, Ashover Grit).

NGB

CHAPTER 6

Westphalian

Rocks of Westphalian age are confined to the western part of the district, where they occur in a number of outliers along the axis of the Goyt Syncline. The outliers partly bridge the gap between the Goyt Coalfield (Stevenson and Gaunt, 1971, pp. 267–292)) and the North Staffordshire Coalfield (Figure 35).

As elsewhere in the Pennine area, the rocks are arranged in a number of cyclothems. These, where complete, consist of coals, overlain by dark mudstones with marine fossils, mudstones with non-marine 'mussels', silty mudstones and siltstones, and fluviatile sandstones capped by seatearths or ganisters. Sandstones are generally thinner and finer-grained than in the underlying Namurian rocks, and coals are more numerous.

The strata present comprise the lowest 150 m or so of the Westphalian A Stage and lie wholly within the *Anthraconaia lenisulcata* Zone. Four coals are present, the Ganister (or Cannel), White Ash (or Thick), Red Ash (or Thin) and Big Smut (or Silver). There are two main outliers. The southern one, at Goldsitch Moss and Robin's Clough, contains beds ranging from the *Gastrioceras subcrenatum* Marine Band to the Big Smut Coal (Figure 35). In the northern outlier, at Goyt's Moss, the beds present extend up to just below the Ganister Coal.

DETAILS

Beds below Woodhead Hill Rock

These strata are some 18 to 36 m thick, the lower figure applying to areas where the Woodhead Hill Rock is thick. They are dominantly argillaceous with the *Gastrioceras subcrenatum* Band at the base, closely overlain by a 'contorted bed' (Cope, 1946; p. 113).

In the Goldsitch Moss outlier, a section [SK 06 SW/10] in Black Brook shows the full thickness of these beds: the marine band, 0.45 m thick, is overlain by 26.19 m of mudstone with a 'contorted bed', 5 cm thick, at the base. These strata are also exposed, though less well, in a stream south-east of Shaw Bottom [013 635] and near The Wash [0132 6627] (Cope, 1946, pp. 145–149).

At Goyt's Moss, the section in Burbage Colliery upper level (Hull and Green, 1866, p. 46) [025 719] proved a total of 30.48 m of mudstone between the Rough Rock and Woodhead Hill Rock, with the lowest 6.10 m described as 'clunch'.

The lowest beds are well exposed in a stream section on Axe Edge Moor [0241 7029], where the marine band, 0.23 m thick, with *G. subcrenatum* and bivalves, is separated from the overlying 'contorted bed' by 0.10 m of dark platy mudstone with plant debris. Higher beds are in general poorly exposed; to the north of Raven's Low, they contain a thin sandstone, close below the Woodhead Hill Rock.

On the north-west side of Goyt's Moss, a section in Stake Clough [0046 7259], shows mudstone with ironstone bands 6.10 m, on 'contorted' mudstone 0.15 m, on platy calcareous mudstone with marine fossils. Farther north [0066 7291], the marine band, 0.23 m, yields a fauna including *G. subcrenatum*.

In Deep Clough there are small inliers showing basal Westphalian strata. Both marine band [0100 7240] and 'contorted bed' [0100 7266] are present: the most complete section showing both is exposed [0120 7286] in the northern part of the valley.

In the outlier north of Raven's Low, a section [0161 7428] shows the 'contorted bed' overlying the marine band.

Woodhead Hill Rock

This is a sandstone between 20 and 37 m thick. It is clearly visible at Goldsitch Moss, where a section [SK 06 SW/10] 31.9 m thick, shows reddened poorly cemented and friable sandstone, fine-grained in the upper part and coarser and crossbedded below. Near The Wash [0100 6831 to 0116 6833] it is a red-stained coarse sandstone.

The best exposures of the sandstone are along the River Goyt, one at Foxhole Hollow [0159 7235] showing coarse crossbedded feldspathic sandstone 22.86 m, on thin-bedded flaggy sandstone 1.83 m. Farther upstream [0156 7150], the ganister top of the sandstone is 3.96 m thick. In the vicinity of Cheshire Knowl, the sandstone is pebbly in places [0162 7055 and 0183 7003].

On the eastern limb of the Goyt Syncline, pink-staining is not common; it is, however, present in 2.44 m of coarse feldspathic sandstone [0265 7095]. Hull and Green (1866, p. 46) record 30.48 m of sandstone at this horizon in Burbage Colliery upper level and (ibid., p. 48) 36.58 m of sandstone in a well at the Cat and Fiddle [001 719].

Yard Coal and its seatearth

The main outcrop of this coal is at Goyt's Moss, where it is some 1.2 m thick.

At Goldsitch Moss, the seam contains many shale partings which have rendered it unworkable. A stream section [SK 06 SW/11], near Goldsitch House, shows 3.30 m of silty mudstone with thin coals at this horizon.

At Goyt's Moss, the coal was formerly worked in Burbage Colliery and in extensive bell-pits along much of the outcrop east and south-east of the River Goyt. The abandonment plan records: coal 0.30 m, bat (thinly interlaminated coal and mudstone) 0.08 m, coal 1.09 m, on 'soft clunch' (seatearth mudstone). An identical section is recorded in the upper level (cross measures) of the colliery by Hull and Green (1866, p. 46), though here the coal is stated to rest directly on sandstone.

A section at Derbyshire Bridge [0171 7189] shows the lower part of the coal and the underlying strata: coal 0.76 m, soft grey clay 0.05 m, grey micaceous seatearth with rootlets (more silty below) 0.23 m, ganister with *Stigmaria* and rootlets 0.30 m, grey silty seatearth with rootlets 0.76 m, on grey mudstone. The base of the section lies close to the top of the Woodhead Hill Rock. Farther upstream [0168 7155] the 1.27 m of coal exposed in the left bank of the Goyt is nearly the full thickness.

The Yard Coal is also seen in Tinkerpit Gutter [0144 7080], where 1.12 m of dirty coal rests on grey clay 0.91 m, representing the seatearth seen 30 m to the north-west.

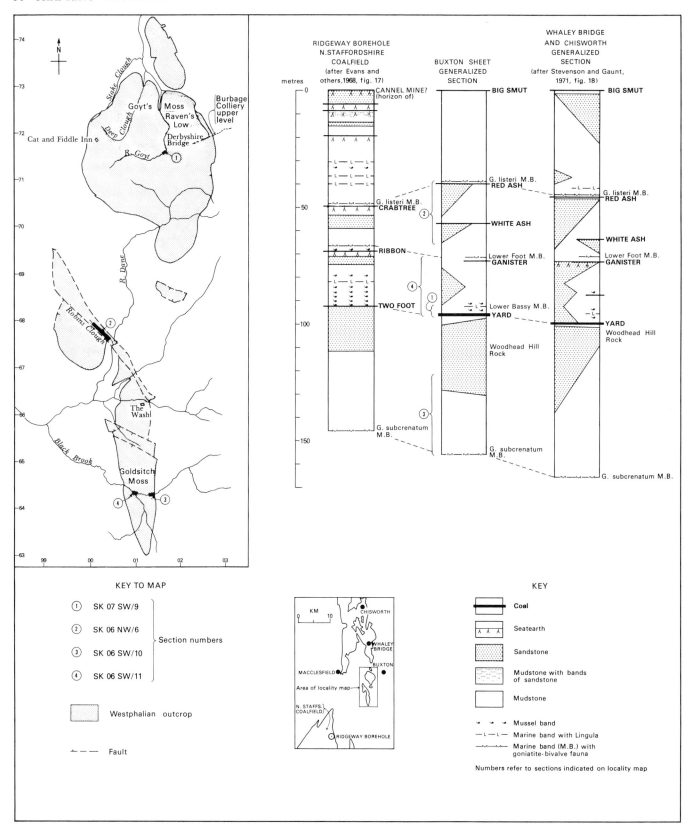

Figure 35   Sketch-map of Westphalian outcrops and vertical sections, showing correlations with adjacent districts.

Beds between Yard Coal and Ganister Coal

These are some 13 to 20 m in thickness. For the most part they comprise mudstones which contain a non-marine 'mussel' fauna above the Yard Coal, overlain by the Lower Bassy Marine Band. At the top, a thin sandstone is common beneath the Ganister Coal.

At Goldsitch Moss, a section [SK 06 SW/11], near Goldsitch House, shows 13.1 m of mudstones and siltstones, with a thin sandstone near the top. Cope (1949b, p. 477) notes a dark band 5 cm thick, of marine character, and a *Lingula* band, about 6.10 m and 12.20 m respectively above the Yard Coal; these may represent the Lower Bassy and Upper Bassy marine bands. In a stream near Blackbank [0119 6405], dark grey mudstone yields a mussel fauna including *Carbonicola sp.* (*protea* group) and *C.* aff. *declinata*; the fauna represents the middle division of the Bassy Mine sequence (Eagar, 1947). Mudstones exposed downstream [0113 6411] have yielded *L. mytilloides* and *Caneyella* cf. *multirugata* [juv.] in a band, 2 m thick, possibly representing the Lower Foot Marine Band.

At Goyt's Moss, three separate outcrops of a sandstone occur in the northern half of the outlier in the lower half of this division. The two most north-easterly, interpreted as a channel fill, show a diachronous base and disappear southwards, while the western outcrop is more regular and sheet-like, but is thought to lie at the same horizon. An exposure [0256 7144] shows 5.18 m of coarse feldspathic sandstone. In Burbage Colliery upper level, Hull and Green (1966, p. 46) record 6.10 m of sandstone on 4.57 m of shale in this position.

South of Derbyshire Bridge, the beds above the Yard Coal are well exposed. The best section [SK 07 SW/9] shows 2.74 m of mudstones with mussels including *Carbonicola* aff. *fallax* and *C.* aff. *protea* between the coal and the Lower Bassy Marine Band, in platy mudstone 5 cm thick which yields *Lingula sp.* and fish debris. The overlying 3.6 m of mudstone yield *C.* aff. *extenuata, C. fallax* and *C.* aff. *doliqua*.

A section [0168 7174] farther north shows the marine band to lie 6.78 m below the sandstone noted above. Nearby [0173 7169 to 0178 7161] an angular discordance in the beds above the marine band has been noted by Cope (1949b, p. 475–476). It may be due to the syndepositional slipping of strata bordering a channel. Cope has given a detailed description of these beds (now part obscured) in a section 21.84 m thick which shows the Lower Bassy Marine Band 3.96 m above the Yard Coal. The channel base lay some 6.7 m above the marine band and there is a band 6 cm thick with *Posidonomya* cf. *gibsoni, Posidoniella sp.* and indeterminate goniatites at the top of the section; this band is considered to represent the Lower Foot Marine Band, the Ganister Coal and its seatearth being absent.

Ganister Coal and overlying strata

Some 87 m of these beds are present at Goldsitch Moss. They comprise the Ganister Coal and Lower Foot Marine Band at the base, the White Ash Coal, the Red Ash Coal with overlying *Gastrioceras listeri* Marine Band, and the Big Smut Coal (Figure 35).

Near Goldsitch House, a section [SK 06 SW/11] shows the Lower Foot Marine Band 3 cm, overlying sandstone 1.5 m, above the Ganister Coal (coal and dirt 0.35 m). Higher beds are best seen in Robin's Clough [SK 06 NW/6], where 41.18 m of strata are visible below the Red Ash Coal. A sandstone at the base of the section probably lies below the horizon of the Ganister Coal: it is overlain by mudstone with pyrite-filled burrows, which probably represents the Lower Foot Marine Band. Only the lowest 2 cm of the White Ash Coal are seen; the seam lies 20.25 m below the Red Ash Coal and is underlain by a thin sandstone with rootlets. Over half of the intervening strata are sandstone, which bears rootlets at the top. The sandstone is overlain by mudstone and a seatearth, with the Red Ash Coal, 0.25 m thick above. The coal is succeeded by 10.1 m of strata, with the *Gastrioceras listeri* Marine Band at the base; this yields a fauna including *G. listeri, Dunbarella papyracea* and bivalves. There is evidence of old workings in the White Ash Coal nearby. The thickness of 0.61 to 0.69 m was given for the seam by Hull and Green (1866, p. 26).

The highest beds at Goldsitch Moss are poorly exposed. The Big Smut Coal is known from bell-pits near outcrop; its thickness was given by Hull and Green (1866, p. 26) as 0.41 to 0.46 m.   NA, IPS

CHAPTER 7

Intrusive igneous rocks

The intrusive igneous rocks of the district comprise dolerite sills and pyroclastic rocks, the latter in volcanic necks. The dolerites are significantly later in age than the rocks into which they are intrusided (*see below*), but the necks form part of the contemporaneous vulcanism and are of Viséan age.

DOLERITE SILLS

The term dolerite has been used on the map for the rocks forming the igneous intrusions. Though some of these are true olivine-dolerites, others are ultrabasic and may result from the hybridisation of the original magma. They are intruded into the upper part of the Bee Low Limestones and the lowest Monsal Dale Limestones. Fitch and Miller (1964, pp. 164, 165) give an age of 295 ± 14 Ma (late Westphalian) for the Calton Hill Intrusion, though the Waterswallows Intrusion, lying just beyond the northern limit of the district, has been dated at 311 ± 6 Ma (close to the Namurian–Westphalian boundary) using the K–Ar method (Stevenson and others, 1970). The Tideswell Dale Sill has been dated by Fitch and others (1970, p. 775) as 287 ± 13 Ma; the Ible Sill was dated as 309 ± 21 Ma. Both determinations were based on the K–Ar method. The accuracy spans of these ages largely overlap one another and the intrusions may therefore be contemporaneous.

The outcrops of all three intrusions present in the district are closely associated with those of the lavas within the limestone sequence; all have lavas in their roof, and these may have exercised some control on emplacement. At Calton Hill, the presence of irregular masses of tuff and agglomerate, otherwise unexposed, within the intrusion points strongly to it having followed the line of a volcanic neck.

The extent of thermal metamorphism due to the intrusions is uncertain in many places due to the absence of critical exposures; in the adjacent Chapel en le Frith district (Stevenson and Gaunt, 1971, p. 293) it was found to be highly variable. Metamorisation is pronounced, however, in the limestones underlying the Tideswell Dale Sill (*see below*).

The three intrusions described below show considerable differences in character despite their probable consanguineous orgin within the Carboniferous basic igneous suite. The Calton Hill Intrusion has a fine-grained heterogeneous ultrabasic phase that could well be the product of the assimilation of much mantle lherzolite by a basaltic 'host' magma. The Ible Sill is, by contrast, a coarse ophitic olivine-dolerite remarkably uniform in its primary composition. The Tideswell Dale Sill is intermediate between the Calton Hill and Ible intrusions in almost every respect; the composition mainly resembles that of the Ible Sill but the rock as a whole is less homogeneous and more fine-grained and in part contains traces of xenocrystic material comparable with that of Calton Hill. All three intrusions show evidence of partial hydrothermal alteration,

perhaps most widespread at Ible and probably concentrated at or restricted to the margins in the case of Calton Hill and Tideswell Dale. Marginal hydrothermal alteration of Carboniferous sills is a common feature. NGB

CALTON HILL INTRUSION†

Description is complicated by the incomplete nature of the exposures and their variation during the period of quarrying at the site.

Tomkeieff (1928) published details of the quarry in the years 1922–1925 (north-east part of the intrusion) and described the irregular distribution of stratified tuff, amygdaloidal lava and analcite-basalt (*sic*). He noted the occurrence of fragments of limestone as well as lava in the tuff, and attributed it to the presence of a vent which subsequently controlled the emplacement of the intrusion. Sections quoted (Tomkeieff, 1928, p. 707) show:

i 'Southern Bay and Island-Stack': 2.3 m of baked tuff, laminated calcareous clay, stiff plastic clay, green tuff and agglomerate, below dolerite replaced laterally by vesicular lava.

ii 'Level-Cutting': 9.14 m of agglomerate and tuff.

At the time of survey (1971) much of the quarried area was obscured by tip. In the north-east of the outcrop, a section [1185 7148] shows intrusive rock 1.2 m on 4.2 m of tuff with bombs up to 0.35 m. Nearby [1189 7149], an exposure of indurated fine-grained rock is probably the 'baked clay' of Tomkeieff. This part of the quarry is now filled-in.

Workings on the western side of the quarry area show large irregular masses of amygdaloidal lava within the intrusive rock. The junction between the two lithologies is associated with the presence of geodes containing quartz or calcite and pockets of hematite also occur [1163 7139]. Columnar structure is present locally [1161 7139] in the intrusive rock.

On the southern side of the intrusion, a small area of bedded tuff has been mapped. A section [1197 7122] shows olive-green tuff with some coarser bands of pumice fragments (up to 5 mm) and rare bombs up to 6 cm. Nearby [1205 7115], a section shows 6 m of vesicular lava on 10.5 m of dolerite with peridotite xenoliths up to 0.11 m (*see below*). IPS

The petrographic character of the intrusive rock (E 7332, 8340, 8341, 54949–56, 56429–31, 56434–35) is variable but the commonest type is probably that illustrated by Tomkeieff (1928, fig. 4B, p. 711). It consists of a groundmass fabric of granular augite, slender laths of plagioclase and grains of iron-titanium oxide set with subhedral phenocrysts of olivine and augite, ultrabasic xenoliths and xenocrysts, amygdales containing zeolite and/or chloritic phyllosilicate, and 'nests' of fine-grained prismatic augite. The groundmass constituents and olivine phenocrysts occur throughout but other components are variable in abundance or absent in places. The feldspar fabric commonly shows a preferred flow orientation.

The plagioclase laths are fresh; they range in length from about 0.1 to 0.66 mm but do not exceed 0.6 mm in thickness. According

---

† An alternative view, that the igneous rocks at Calton Hill are all extrusive, has recently been published by Macdonald and others (1984).

to Donaldson (1978, p. 365) the composition is An_{64} but there is some degree of zoning, probably from cores of about An_{70} in some cases, and there may be gradation to andesine at the extreme margins. The groundmass augite is granular, verging towards prismatic, with a maximum dimension of 0.15 mm; it tends to be more abundant than plagioclase. The composition is $Wo_{50} En_{36} Fs_{14}$ (Donaldson, 1978, p. 365). Nest-like aggregates of prismatic fine-grained augite are of widespread, but relatively sparse, distribution. Phenocrysts of augite tend to be euhedral and about 0.15 to 0.3 mm in diameter; they are rarely as abundant as olivine phenocrysts and are sometimes absent. Crystals tend to show a strongly coloured marginal zone resembling the groundmass type, surrounding a paler core containing fine-grained vermicular inclusions. Donaldson gives an analysis as $Wo_{45} En_{45} Fs_{10}$. Olivine is porphyritic, microporphyritic and xenocrystic, with cognate subhedra rarely exceeding 0.3 mm in diameter. It is everywhere at least partially altered along fractures and at the margins; in some samples it is totally replaced. The composition of the cognate olivine is $Fo_{84.5-79.5}$, (Donaldson, 1978, p. 365). Its main alteration product is an olivine green 'bowlingite', but serpentine may predominate locally. Xenocrystic olivine is not readily distinguishable from cognate except in large examples which occur in some sections. Donaldson gives a xenocryst composition of $Fo_{87.5}$. Fresh titaniferous iron oxide is abundant and occurs in angular grains of 0.06 to 0.15 mm diameter that have a slight tendency to engulf neighbouring plagioclase. Zeolitic minerals, mainly analcime, are of late hydrothermal origin and occur in amygdales and also interstitially in association with spherulitic chlorite.

The peridotite xenoliths have been described by Hamad (1963) and Donaldson (1978). They are coarse spinel-bearing lherzolites and harzburgites deduced by Donaldson to have originated in the mantle at a depth of about 45 to 48 km. The xenolith minerals show marginal transformation to phases compatible with the host magma; olivine, for example, here becomes less magnesian and resembles that of the xenocrysts (Donaldson, 1978, p. 366).

The heterogeneity of the Calton Hill Intrusion is exemplified by published accounts: Tomkeieff records a figure of 23.38% plagioclase (An_{60}) whereas Hamad gives 63.2% of plagioclase (An_{64}). The present work suggests that the former composition is the commoner. The less basic rocks resemble more 'normal' Carboniferous dolerites in both fabric and mineralogy; they tend to lack porphyritic augite, and some are coarser in texture, with linear opaque oxide structure and total alteration of olivine. The late hydrothermal mineral component of the rock is mainly analcime, but in many samples its place is entirely or partially taken by spherulitic chlorite.

Chilled marginal facies display a range of mineralogical variation comparable with that of the main facies. One sample shows severe hydrothermal alteration, with the total replacement of olivine and partial replacement of augite by phyllosilicates, and the albitisation of plagioclase to oligoclase. NGB

TIDESWELL DALE SILL

This sill is intruded mainly into the lower part of the Lower Miller's Dale Lava; it is only slighly discordant. The best section [1551 7383] is in the quarry on the eastern side of the dale, which shows 25 m of dolerite, the top 2.6 m bearing scattered calcite amygdales. The rock shows spheroidal weathering. Variation in the intrusion in the quarry was noted by Arnold-Bemrose (1907, pp. 273–274), and includes the presence of a fine-grained lithology in zones at the top and base, each 4.67 m thick.

Thermal metamorphism resulting from the intrusion is extremely variable. It includes the baking of a clay at the base of the Lower Miller's Dale Lava. An exposure [1547 7378] shows 0.61 m of clay, which here has a prismatic structure, the prisms reaching 6 cm in diameter; the underlying limestone is not seen. On the west side of

Tideswell Dale, an exposure [1539 7410] shows 0.45 m of marmorised limestone beneath the dolerite. The present state of exposure makes the alteration difficult to observe, though Arnold-Bemrose (1899, fig. 2) shows it extending 0–3.9 m below the contact with the overlying clay and dolerite. IPS

A thin-section (E 54998) from about 1 m above the base of the sill comprises a random intergrowth of plagioclase laths (49% of the rock), granular augite (20%) and accessory opaque oxide (3.7%), with phenocrysts and perhaps xenocrysts of olivine (12%) and augite (0.8%), abundant interstitial devitrified glass (9%), and amygdular phyllosilicate (5.5%). There are also traces of an assimilated xenolith.

The plagioclase laths range from 0.15 mm to over 2 mm in length and commonly show a very fine oscillatory zonation around a composition of An_{70} passing to a normal sodic gradation at extreme margins. The augite grains range from only 0.06 mm to 0.15 mm in diameter but olivine, entirely pseudomorphed by opaque oxide and a 'bowlingite' closely resembling brown biotite in optical character, is subhedral to rounded and 0.06 mm to 1.2 mm across. The largest crystals are possibly xenocrystic and some are associated with clots of coarse-grained granular augite, 0.3 mm to 0.45 mm across and containing vermicular inclusions (?devitrified glass). Iron-titanium oxide is fresh and mainly concentrated in rectilinear crystal growths. In places it is penetrated by plagioclase laths in 'ophitic' relationship. The section contains an arcuate lithic xenolith 4 mm long, composed of microcrystalline plagioclase clouded by opaque dust. Interstitial material includes abundant microcrystalline asbestiform brown bowlingite and isotropic olive-coloured devitrified glass, more or less clouded with crystallites among which apatite needles are prominent.

A sample (E 54997), from 2 m lower in the sill, has plagioclase laths 0.3 to 1.35 mm long, though the augite (0.1 to 0.3 mm) is coarser. The ophitic facies described by Arnold-Bemrose (1899) and illustrated by Teall (1888) was not identified and may have been removed in subsequent quarrying.

Three samples (E 54994 to E 54996) confirm progressive chilling towards the base of the sill, although the constituents do not decrease in size. At about 6.5 m from the base (E 54996), plagioclase laths are up to 0.6 mm long and augite grains up to 0.18 mm across, but there are 'ophitic' plates of opaque oxide up to 0.45 mm in diameter. At about 3.5 m above the base (E 54995), augite grains only reach 0.04 mm but plagioclase laths are as large as in E 54996; olivine megacrysts also show a reduction in size to 1 mm maximum, and opaque oxide shows a linear form (up to 0.36 mm). E 54994, from very near the (now hidden) contact, shows no further significant reduction in grain-size, but a very thorough argillisation of all silicates except for a minor proportion of plagioclase; the rock also shows a slight flow foliation and contains small quartz-filled amygdales. There is a progressive tendency in the three samples for quartz also to become a major replacement product of olivine megacrysts.

A sample of 'prismatic clay' (E 56749) from beneath the intrusion (see this page) shows an extremely fine 'net-veined' fabric in thin-section, with veinlets of phyllosilicate only a few microns across permeating a clay-grade microcrystalline matrix. The dominant mineral is an expanding-lattice mixed-layer clay (identified by X-ray diffraction; analyst K. S. Siddiqui). NGB

IBLE SILL

This sill is emplaced within the Bee Low Limestones and appears to be strongly transgressive. The best section [2530 5680] is in a quarry at Ible, which shows some 30 m of well jointed dolerite. Arnold-Bemrose (1907, p. 275) noted the presence of marmorisation in the adjacent limestones, while Gibson and Wedd (1913, p. 28) observed thin quartz veins within the intrusion. IPS

Thin sections (E 9052, 54988–54990) indicate that the rock is an ophitic olivine-dolerite which has suffered a variable and locally severe amount of hydrothermal alteration. There is a close resemblance to the ophitic facies of the Tideswell Dale Sill (Teall, 1888, plate IX, fig. 1; possibly no longer exposed). In E 54989 a random intergrowth of 0.3 to 1 mm long plagioclase laths (53% of the rock) is set with poikilitic titaniferous augite (20%) up to 5 mm across and subhedral to anhedral olivine (15%) of comparable dimensions. Olivine and accessory fresh poikilitic (to plagioclase) titaniferous iron oxide (3%) tend to be concentrated between the augite crystals. The rock also contains interstitial devitrified glass (6%) and interstitial to amygdular hydrothermal phyllosilicates (3%).

Plagioclase is typically fresh, and very strongly and simply zoned from cores of sodic bytownite (An_{77} in some cases) to calcic oligoclase, but the bulk of the mineral is probably labradorite. The titaniferous augite, pleochroic in thick sections, is optically homogeneous, and in typical samples (E 54988, 54989) shows only minimal chloritisation. Olivine shows the usual 'bowlingite' replacement, though relics are locally preserved unaltered (E 9052) or replaced by calcite and quartz (E 54988) as in parts of the Tideswell Dale Sill. Intersertal former glass is replaced by crypto-crystalline and microcrystalline intergrowths of olive-coloured phyllosilicate, opaque oxide and felsic minerals; unobtrusive filamentous accessory apatite also tends to be concentrated in this association. The minor interstitial to amygdular segregations contain asbestiform 'bowlingite' optically akin to that of the olivine pseudomorphs. NGB

VOLCANIC NECKS

These necks are a feature of the Dinantian extrusive episode. Where not modified by subsequent intrusion, the vents are filled by relatively fine-grained pyroclastic deposits, evidently representing an explosive phase or phases not directly connected with the extrusion of lava; this is borne out by the association of the Shothouse Spring Tuff with the Grangemill necks.

CALTON HILL

The presence of a vent, now much modified by the subsequent intrusion of dolerite, has been noted on p. 96.

BAKEWELL AREA

To the west of Bakewell a borehole [SK 26 NW/8] proved pyroclastic rock to 18.90 m. No material of this type is otherwise known in the vicinity, and the record is interpreted as indicating the presence of a neck of unknown dimensions (but presumed to be small) at this site.

At Ditch Cliff [211 670] an oval outcrop of igneous rock (160 m maximum diameter), proved by auger, is interpreted as indicating the presence of another neck. This structure lies at the south-eastern end of a faulted monocline. A similar association between folds and vents has been noted in the Chapel en le Frith district (Stevenson and Gaunt, 1971, p. 300), and both instances suggest a relationship between contemporaneous movements and vent formation.

GRANGEMILL

At this locality the presence of two necks has been known since the work of Geikie (1897) and Arnold-Bemrose (1907). The necks are roughly oval, the larger having a maximum diameter of about 700 m.

The larger, southern neck, in general poorly exposed, is best seen in a quarry [2442 5783], which shows 5.5 m of unbedded spheroidal-weathering tuff with rare limestone lapilli up to 5 cm. Arnold-Bemrose (1907, pp. 262–263) noted the presence of three dykes of east–west trend within the neck 'on the roadside opposite the old Mill'. Recent excavation [2438 5782] near the western edge of the neck have shown the presence of a dolerite dyke some 7.5 m wide and trending at 165°. The rock is intruded into tuff, here very weathered. The smaller, northern neck is ill-exposed, though small outcrops of purple and green tuff are seen in places [such as 2441 5836].

The necks are bordered on east, north and west by the outcrop of the Shothouse Spring Tuff which is interpreted as their extrusive product. The large size of outcrop of the necks is unusual for Derbyshire, and encourages comparison with occurrences in Fife where Francis (1968, p. 168) noted that the vent walls show an inward inclination in their upper part so that the vents are funnel-shaped. The relationship at Grangemill between the necks and the Shothouse Spring Tuff is interpreted as indicating that the outcrops are at a high level in the vents, and a similar funnel shape possibly accounts for their relatively large extent. IPS

A thin section (E 52370) shows that the rock comprises ragged clasts of devitrified amygdaloidal basalt up to 5 mm diameter, replaced mainly by clay material after palagonite. Amygdales up to 0.36 mm contain ?smectite (commonly spherulitic), calcite or analcime; they are surrounded by opaque dust. There are some serpentinised olivine megacrysts up to 3 mm.

The dyke rock (E 56403) is formed of a meshwork of fresh zoned plagioclase laths (0.15–0.6 mm), with labradorite cores, interstitial granular to prismatic augite (0.04–0.16 mm), partly altered to calcite and a green phyllosilicate, and accessory titaniferous iron oxide; scattered olivine megacrysts, with some bowlingite replacement, are also present. A pyroxene-rich clot may represent a partly assimilated xenolith. There is a strong petrographic, though not necessarily genetic, similarity to the marginal facies of the Tideswell Dale Sill. NGB

CHAPTER 8

Triassic

Red pebbly sandstones of presumed Triassic age are present around Leek, in the south-west of the district, where they form an elongated outlier several kilometres to the north of the main Staffordshire outcrop. The sandstone is well known as a water source, and has been exploited by several boreholes (Figure 36). The underlying Carboniferous rocks are stained red or purple, the mudstones to a depth of about a metre, the sandstones more so. Staining of this type affects the coarser Carboniferous sandstones irregularly over much of the western Pennines (Evans and others, 1968, p. 57; Stevenson and Gaunt, 1971, p. 5); the effect is usually attributed to a former, more widespread Triassic cover. A notable example can be seen in the Goyt Syncline, where the Woodhead Hill Rock is strongly stained, as near The Wash [010 663] and Goldsitch House [013 642]. Adjacent mudstones and siltstones here are similarly affected. Neptunean dykes of red sandstone and siltstone, of presumed Triassic age, are present in the Carboniferous Limestone at localities far removed from the Triassic outcrop: for example, on the steep side of the Dove valley [1454 5673] west of Coldeaton where irregular joints up to 0.26 m wide in the Woo Dale Limestones are filled with red calcareous sandstone. Similar occurrences were recorded at depths between 73.0 and 86.4 m in a borehole [SK 15 NW/10] drilled in a knoll-reef in the Milldale Limestones near Alstonefield.

The Triassic rocks of the Leek outlier rest with marked unconformity on folded and faulted Carboniferous rocks, but are themselves only slightly deformed; tectonic dips are generally low (Figure 36) and only one fault is known (Evans and others, 1968, p. 150). Around the margins of the outlier the basal contact dips inward at angles of up to 34°, and borehole provings nearer the centre confirm that the structure is an elongate closed hollow, with about 200 m of sandstone preserved in the deepest part. Approximate contours on the basal surface are shown in Figure 36. The structure is not a simple north–south syncline, for the dips observed at surface are much less than those inferred, or seen, at the basal contact; the steeply-dipping east and west flanks of the hollow must represent the sides of a buried valley in the pre-Triassic landscape, as was recognised at an early date (Wardle, *in* Sleigh, 1862, pp. 232, 288; Hull and Green, 1866, p. 62). The gentle closure to the north probably requires a tectonic explanation, however, for the sedimentological evidence (below) suggests that during deposition the hollow was open at both ends, allowing rivers to flow freely through the area from south to north. The simplest explanation of this, and the one favoured here, is that the pre-Triassic valley was modified by gentle post-Triassic warping, so separating an originally continuous valley-fill into closed hollows, the eroded remnants of which are preserved at Leek and, farther north, at Rushton Marsh (Evans and others, 1968, p. 149).

The deposits were originally classified as Pebble Beds (Hull and Green, 1866, p. 62), the middle division of the Bunter Sandstone (Hull, 1869, p. 10), and this nomenclature was retained (for example, Evans and others, 1968, p. 126) until superseded by that of Warrington and others (1980), in which all the Triassic sandstones are assigned to the Sherwood Sandstone Group. This name was used on the Buxton 1:50 000 map (1978) in anticipation of the change. More recently, formation names have been given to divisions of the Group in the district to the south (Charsley, 1982) and these have been applied to the rocks of the Leek outlier in the present account. The bulk of the sandstones at outcrop belong to the Hawksmoor Formation, with a thin representative of the basal Huntley Formation in Longsdon Wood (*see below*). No other exposures of the Huntley Formation are known, but it may be thicker and more widespread at depth though the two formations cannot be distinguished separately in old borehole records.

HUNTLEY FORMATION

The basal sandstones at Longsdon Wood contain angular clasts of locally-derived Carboniferous sandstone in addition to rounded quartzite pebbles, and are for this reason assigned to the Huntley Formation (Charsley, 1982, p. 2). In the Cheadle area, to the south of the Leek outlier, these beds are considered to have been deposited as alluvial fans by flash floods (Charsley, 1982, p. 4).

HAWKSMOOR FORMATION

Most of the exposures are of soft red sandstone showing trough cross-bedding and parallel lamination. Measurements of the dip directions of foresets indicate a general south to north palaeocurrent flow (Figure 36). The pebbles are mainly of well rounded quartzite and are found scattered through the sandstones or, less commonly, in discrete conglomeratic bands. Subordinate lenticular bands of red silty mudstone ('marl' of older literature) up to 9 m thick are also present. The lithologies have been described in greater detail from the Bunter Pebble Beds of the Cheshire Basin by Thompson (1970). On a regional scale, the deposition of the Hawksmoor Formation and its correlatives has been attributed to braided streams or rivers flowing north-westwards across the Needwood and Cheshire basins (Thompson, 1970; Charsley, 1982, p. 5). All the sediments exposed in the Leek outlier are of this alluvial-plain facies: no aeolian deposits have been found.   JIC

DETAILS

The northern part of the outlier lies in the Macclesfield (110) district and details of sections there have been published (Evans and others,

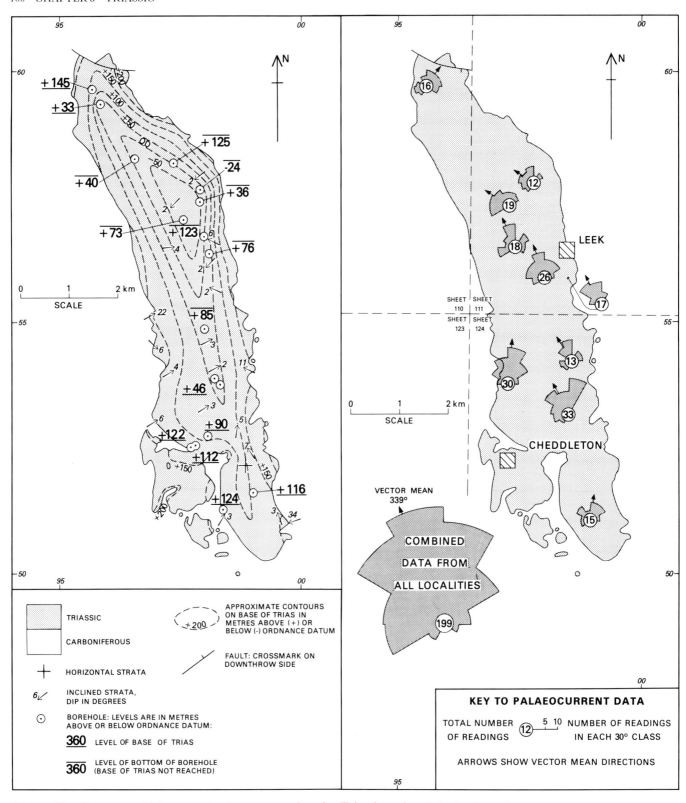

Figure 36   Structure, thickness and palaeocurrent data for Triassic rocks of the Leek outlier.

1968, p. 150). The southern part lies in the Ashbourne (124) district (Ashbourne Sheet Memoir, *in preparation*). The central parts of the outlier, which fall within the present district, are described below.

REDEARTH TO RIVER CHURNET

The ground is relatively free of drift, and the base of the Triassic rocks can be traced with fair accuracy from surface indications. At Redearth [9721 5909] a small exposure of soft red sandstone with red mudstone flakes lies very close to the base, and 0.9 m of reddened Carboniferous mudstone is exposed [9773 5859] at Upper Foker immediately below the base of the Trias. At Abbey Green, red cross-bedded sandstone with scattered pebbles is well exposed in road cuttings [977 577] and old quarries [978 576]. Road cuttings at Bridge End [973 573; 972 572] show up to 6.1 m of the same lithology, and up to 7.6 m are visible in the banks of the River Churnet nearby [9659 5704; 9681 5719; 9706 5720]. A borehole in this area [SJ 95 NE/17B] proved 181 m of red sandstone with pebbly bands and some red mudstone; the pebbly beds account for 57% of the sequence and the mudstone for 11%: the thickest mudstone band, at 78 m, is 9 m thick.

LEEK

In the town there are good exposures of almost flat-lying red pebbly sandstone in road and railway cuttings. In places, as at Mill Street [981 566] the soft sandstone faces have had to be supported with retaining walls. Near the north portal of an abandoned railway tunnel [973 565] 15.2 m of partly cross-bedded sandstone are exposed. Pebbles are rare in the upper part, and absent below. Near the south portal [976 561] 19.8 m of crossbedded pebbly sandstone were visible formerly, with a 1.7 m lens of conglomerate 3.2 m above the base. In a long road cutting at Broad Street [980 560], up to 3.7 m of sparsely pebbly sandstone are exposed. The trough-shaped nature of the cross-bedding is well displayed here. A road cutting [9849 5586] by the cemetery shows broad shallow trough cross-bedding with scattered pebbles. Around Barnfield [980 557] there are small exposures of sandstone with scattered pebbles by the railway and road sides; the rocks show flat bedding and low-angle depositional dips. A section [9795 5568] low in the sequence showed 0.6 m of red silty mudstone with a few green bands up to 4 cm thick, on 1.4 m of red cross-bedded sandstone with mudstone clasts. A band of red and green siltstone 0.66 m thick was recorded among red sandstones nearby [9819 5543].

The base of the Trias was easily traced to the north and south of the town but is conjectural in the built-up area. The basal beds are clearly banked against pre-Triassic topography in a small escarpment [989 551] at the district boundary south of Cornhill Cross.

Three boreholes have been drilled into the Triassic rocks in Leek, but none has penetrated the base. In one [SJ 95 NE/11], red mudstone and pebbly sandstone each account for 8% of the sequence, the total depth being 79.86 m. In another [SJ 95 NE/13], 106.68 m deep, no mudstone was recorded and pebbly beds accounted for 10% of the sequence. The record of the third borehole [SJ 95 NE/10] gives 'red sandstone' to 59.74 m but no further details of the sequence.

WESTWOOD HALL TO WALL BRIDGE FARM

The western margin of the outlier crosses the corner of the district at Westwood Hall [966 562]. The boundary can be mapped with fair accuracy from surface indications, and there is an exposure [9697 5520] of the basal beds west of Wall Bridge Farm, above the canal feeder in Longsdon Wood. Grey Carboniferous mudstone can be seen at water level, and above lies a section 3 m thick, of conglomerate of rounded quartzitic pebbles mixed with subordinate angular blocks, up to 0.4 m across, of siliceous sandstone of Namurian type. Rough bedding planes dipping at 22° to the northeast are presumed to be concordant with the pre-Triassic topography. The conglomerate, which is the only known representative of the Huntley Formation in the district, is overlain by loose orange pebbly sand derived from the Hawksmoor Formation.

JIC, NA

CHAPTER 9

Secondary dolomitisation of Dinantian limestones

The Carboniferous limestones have been altered locally to dolomite in the southern part of the Derbyshire Dome (Figure 37), probably as a result of exposure to magnesian brines percolating down from the Zechstein sea in late Permian times (Parsons, 1922, p. 63; Dunham, 1952a, p. 415). There are no proved Permian deposits nearby, however, and it is possible that Triassic groundwaters were responsible for the alteration. The patchy distribution of the dolomite may reflect the extent to which the impermeable cover of Namurian mudstone had been stripped from the region when the limestone was dolomitised.

The secondary origin of the dolomite was established by Parsons (1922) on the basis of the field relations; the margins in detail are irregular and transgress the bedding planes in the limestone. In its broader distribuion, the dolomite forms a surface zone up to about 200 m thick with a markedly ir-

regular base; it is quite distinct from the earlier-formed dolomite present in the lowest parts of the limestone sequence (p. 10). Its distribution is restricted to the area of limestone outcrop, except north of Winster, where it extends laterally for about a kilometre beneath the mudstone cover. The dolomite is clearly later than the main Hercynian folding, but is cut by the Bonsall Fault and has been mineralised in many places. These relationships do not give an unequivocal date for the dolomitisation, but they are compatible, at least, with the view that it took place during Permian or Triassic times.

Chemical analyses carried out by Parsons (1922, pp. 61–114) showed great variation in the proportion of magnesium carbonate present, and this variability is confirmed by the numerous analyses now available as a result of the industrial minerals assessment programme of the

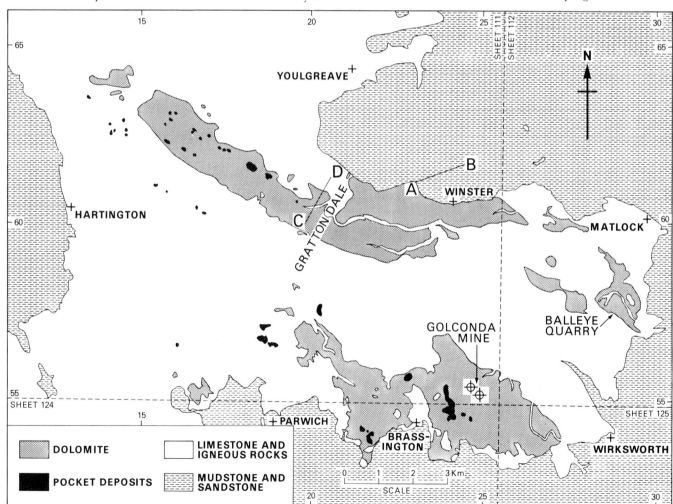

Figure 37 Outcrop distribution of secondary dolomite and Pocket Deposits in the southern part of the Derbyshire Dome.

Survey. A summary of the data for the whole region (Bridge and Gozzard, 1981, p. 14) shows that pure dolomite is unknown, but that rocks of commercial grade are present locally. Most of the dolomites contain some calcite, and it is evident that iron and manganese have been introduced in addition to magnesium. In the field the variation is equally clear; for mapping purposes partially dolomitised limestones have not been distinguished from more completely altered rocks.

At outcrop the dolomite is porous and weathers easily to a buff-coloured sand of individual crystals. Isolated tors have survived, however, and some form conspicuous landscape features: their origin has been discussed by Ford (1963). The porosity was investigated by Parsons (1922, pp. 60, 114), who found that it was due to preferential solution of the calcite component, and by Munn and Jackson (1980), who detected textural evidence for growth of calcite in the weathering zone at the expense of dolomite; later solution of the calcite is believed to result in an enhanced porosity in near-surface rocks.

There is a remarkable similarity between the regional distribution of the Pocket Deposits (p. 105) and that of the dolomite, though the Pocket Deposits are scattered over a somewhat wider area (Figure 37). The association has been explained by Kent (1957) in terms of a common cause, the distribution of both being related to the former margin of the Namurian mudstone cover. According to this view, the area of dolomite was determined by the extent to which the limestone has been stripped of its impermeable mudstone cover by Permian times, and the sinkholes – now filled with Pocket Deposits – were initiated as swallow holes at the margin of the mudstone. According to Ford and King (1969), however, the association is more probably due to the preferential formation of caverns in the dolomite, followed by collapse of overlying sediments as the caverns penetrated to the surface. Pockets lying outside the dolomitised area are thought by these authors to have formed at local downward extensions of the dolomite base below its former regional position, which lay above the present level of erosion.

DETAILS

NORTHERN OUTCROP

The more northerly outcrop is a discontinuous strip 1 to 2 km wide by about 17 km long, between Custard Field [136 639] and Matlock Bath [293 583] (Sheet 112). For much of its length the southern margin of the strip lies close to the Cronkston–Bonsall Fault Zone, and in places appears to be limited by individual faults within the zone. The relationship is exposed at Balleye Quarry [288 574] (Sheet 112), where dolomite has been thrust over unaltered limestone by movement of the Bonsall Fault.

Boreholes show that the dolomite forms an irregular layer with its base subparallel to the present surface and transgressing the limestone sequence. A section (Figure 38, AB), drawn across the margin of the dolomite north of Winster, is based on boreholes drilled along Coast Rake. The alteration here has affected limestone beneath the mudstone cover for a lateral distance of about 1 km from the outcrop, a phenomenon not known elsewhere and perhaps to be explained as a consequence of the easy access afforded to magnesian brines by the fractured rock along the rake. Another section (Figure 38, CD), constructed from surface and borehole information near Gratton Dale, shows the dolomite as a subhorizontal layer broadly transgressive to the bedding of the tilted limestone formations. The base of the dolomite in both sets of boreholes is gradational, through a zone of alternation, or patchy dolomitisation, of very variable thickness. The base is likewise gradational in a borehole [SK 25 NW/21] near Elton: the record shows dolomite 37 m, on alternating limestone and dolomite 9 m, on limestone with patchy dolomitisation 21 m, on unaltered limestone 12 m. A borehole [SK 16 SE/4], sited only about 70 m from the dolomite margin near Oldham's Farm, proved 48.2 m of dolomite above partly altered limestone.

Outside the dolomite outcrop, partly altered limestone has been proved at depth beneath unaltered limestone in boreholes at Parsley Hay Station [SK 16 SW/5] and Pikehall [SK 15 NE/3] and is doubtless present elsewhere. At Parsley Hay Station an alternation of limestone and dolomite was present to a depth of 54.46 m below the surface, and at Pikehall patchy dolomitisation of the limestone was recorded between 35 m and 80 m below the surface; the limestones above and below were unaltered.

There are prominent isolated crags of dolomite at Wyn's Tor [2406 6028] and Grey Tor [2349 6040], and smaller crags along the sides of Long Dale [184 609 to 196 598] and Gratton Dale [201 599]. Good exposures of dolomite, including partly altered limestone, can be seen in an old railway-cutting [151 629] on Blake Moor.

SOUTHERN OUTCROP

In broad outline the outcrop consists of two north-westerly trending patches, although in detail the margins are highly irregular. At Roystone Rocks [197 568] the dolomite has weathered into crags and small tors, and there are many similar features between here and Longcliffe [227 557], notably at Rainster Rocks [218 549], Black Rocks [214 548] and Pinder's Rock [220 555]. At Longcliffe Station Quarry [2267 5563], 19 m of pale brown and buff dolomite are exposed, with impersistent bands of grey mudstone on widely spaced bedding planes; a pillar of pale grey limestone remains at one place. About 20 m of dolomite, with residual masses of limestone up to 4 m thick, are exposed at High Peak Quarry [231 555], above 12 m of limestone which is variably dolomitised on joints. To the east of here, tors and crags of dolomite crop out on rising ground, and at Manystones [236 551] a large mass of variably altered limestone surrounded by dolomite has been quarried.

At Harboro' Rocks [242 553] the dolomite forms a prominent craggy escarpment, and its base has been proved nearby at Golconda Mine, where caverns along the dolomite–limestone contact were the site of a rich lead-zinc deposit (p. 121; Ford and King, 1965). The shaft at Upper Golconda Mine [2492 5514] showed 95 m of dolomite above the limestone, but the thickness of dolomite is locally greater, for the base undulates by as much as 36 m (Ford and King, 1965, p. 1689). The greatest recorded thickness of dolomite, 200 m, is an approximate figure indicated by chipping samples from a borehole [SK 25 SE/37] sited about 1.3 km east of Upper Golconda Mine and a short distance beyond the district boundary.   JIC

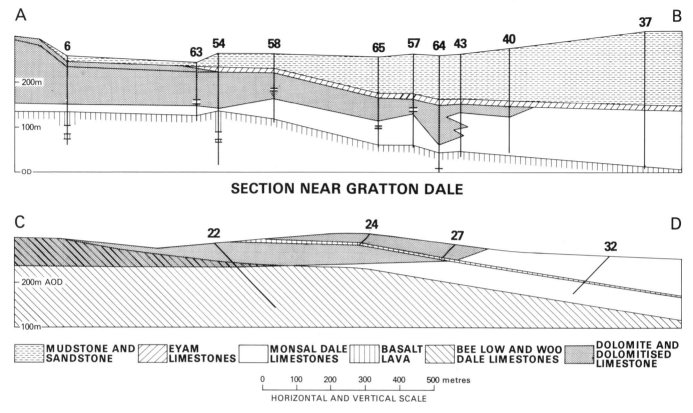

Figure 38   Scale sections showing vertical and lateral extent of secondary dolomitisation based on selected data from confidential boreholes site on or close to lines of section.

CHAPTER 10

Pocket Deposits

Deposits trapped and preserved by collapse into solution cavities in the limestone and dolomite have been collectively termed 'pocket deposits'. They include materials of various ages from Carboniferous to Pliocene. Pleistocene deposits commonly overlie them but have not, in general, been affected by collapse.

The geological importance of the Pocket Deposits is out of proportion to their size, for they provide information on the otherwise little-known Mesozoic and Tertiary history of the region. They owe their economic importance mainly to their content of quartz-rich refractory sand (Boswell, 1918, p. 127; Howe, 1920, pp. 168–175), though the clays have also been used (Green and others, 1887, p. 163). Most of the workings are now abandoned, and many have been filled in.

The distribution of the pockets noted during the present survey, or known from earlier records, is shown in Figure 37; some additional occurrences inferred from surface indications or recorded underground are listed by Ford and King (1969, p. 53). A close, but not exclusive, association with the area of dolomite outcrop is evident.

STRATIGRAPHY

The sequence of deposits at the Bees Nest Pit [241 545] just south of the district (*see below*) appears to be of general application throughout the region, with the exception of a few pits at Ribden, also in the district to the south. The sequence (modified from Walsh and others, 1972, pp. 532–533) is:

5 Pleistocene deposits (till and head)
4 Brassington Formation
3 Black Namurian mudstone,   } *Pocket*
   altered to lilac at top   } *Deposits*
2 Chert breccia and clay
1 Wall rocks

1 Wall rocks

The walls, of limestone or dolomite, are commonly steep, and detached masses are locally incorporated in the Pocket Deposits.

2 Chert breccias

The earliest view of the origin of the chert breccias (Yorke, 1961, pp. 32–60) is that they are glacial deposits 'intruded' into the earlier-formed materials of the pockets by ice pressure: this belief is no longer widely held and was probably based on the similarity of the breccias to the superficial glacial or periglacial deposits, which also contain chert. It is now generally accepted that the position of these deposits, between the wall rocks and the slabs of Namurian mudstone,

indicates that they formed as solution residues. The date of this solution is in dispute. One view is that it took place during a period of pre-Namurian emergence and weathering. No similar deposit is preserved at the Namurian–Dinantian junction elsewhere in the region, however, and this interpretation, if correct, would indicate that the breccias were preserved only where solution hollows were already developing in early Namurian times (Evans, *in* Walsh and others, 1972, p. 552). An alternative view, put forward by Walsh and others (1972, pp. 552–553), is that the stratigraphical position of the breccias is no indication of their age and that they developed by solution of limestone and dolomite beneath the mudstone cover in Tertiary times.

3 Namurian mudstones

Slabs up to 6 m thick of steeply dipping black mudstone have been recorded near the margins of several of the pockets, resting against the chert breccias or the wall rocks. Where overlain by the sands of the Brassington Formation the mudstone is commonly altered to a lilac colour. The slabs are now believed to represent remnants of the surface on which the Brassington Formation was laid down (Walsh and others, 1972, p. 525), rather than ice-transported masses in the till (Yorke, 1961, p. 64). Goniatites, bivalves, and plant miospores present in the mudstones indicate ages ranging from E_1 to R_{2a}: details are given in descriptions of individual pockets (below) and by Walsh and others (1972, pp. 530–535).

4 Brassington Formation

White, pink and yellow siliceous pebbly sands, with some brightly coloured silts and clays, make up the main fill of the pockets. Where bedding is present it dips steeply or is otherwise disturbed, and a basin-like structure has been detected in some of the pockets, notably at Bees Nest and Kenslow Top. The youngest deposits, those at the centres of the basins, are plant-bearing clays of Tertiary age, but the underlying deposits are unfossiliferous and their age has not been established with certainty. The pebbly sands are similar to the Triassic sandstones (p. 99) that crop out widely around the southern Pennines, and some authors have inferred from this that they are of Triassic age (Kent, 1957; Yorke, 1961). More recent authors (Ford and King, 1969, p. 60; Walsh and others, 1972; 1980) have suggested, however, that the deposits are all of Tertiary age, and that the pebbly sands are made up of derived Triassic material.

Detailed studies by Walsh and others (1972, p. 522) have led to the recognition that a tripartite sequence of deposits is consistently present in the larger pockets, and that this probably represents the remains of a widespread sheet of sediment. The deposits are believed to have accumulated in a terrestrial environment, in which sands and gravels laid down by rivers flowing mainly from the south were overlain by silts and clays of lacustrine or swamp origin (Walsh and others, 1972, p. 523; 1980, p. 57).

The name Brassington Formation is applied to the whole sequence and the three members have been named as follows (Boulter and others, 1971):

3   Kenslow Member (grey clays)
2   Bees Nest Member (coloured clays)
1   Kirkham Member (sands and gravels)

The type locality of the formation lies a short distance south of the district boundary, at Bees Nest Pit [241 545]. The thickest sequence known (67 m: Walsh and others, 1980, fig. 2) is at Kenslow Top Pit [182 616].

The **Kirkham Member** comprises siliceous pebbly sands. These are described by Walsh and others (1980, p. 50) as poorly sorted medium or fine-grained sands and silty sands, all of orthoquartzitic composition. The quartz is normally angular to subangular, with rounded grains common only at a few localities. Howe (1920, p. 168) noted that each grain is coated with kaolinitic material. In most exposures the sand is poorly bedded or structureless but locally it is crossbedded; the palaeocurrent flow at Brassington was generally from south to north (Walsh and others, 1980, p. 52).

The pebbles are mainly of quartzite and are well rounded: they are very similar to those in the Triassic sandstones (p. 9) of neighbouring areas. Pebble orientation studies by Walsh and others (1980, pp. 52–56) suggest that, in the Brassington area, long axes of pebbles tend to be aligned parallel to the palaeocurrent flow deduced from the crossbedding. Farther north, however, and within the present district, the pebble orientations are variable and in the absence of crossbedding data do not give a clear indication of the palaeogeography.

The **Bees Nest Member** consists of poorly bedded red, green and yellow, partly mottled, silt and clay. It is only about 6 m thick in the exposure south of the district, but up to 40 m have been recorded at Kenslow Top (*see below*). The base and top are apparently conformable in all sections (Walsh and others, 1972, p. 523; Ford, 1972).

The **Kenslow Member**, up to 6 m thick, is a grey clay with plant remains. Extensive studies of these (Boulter and Chaloner, 1970; Boulter, 1971) indicate a Miocene to Pliocene age for the member.

5   Pleistocene deposits

The oldest are tills consisting of clay with ice-scratched boulders, some from north-western (Lake District) sources. They are of Wolstonian age or older, for the Devensian ice-sheets did not extend into the area where Pocket Deposits are preserved (p. 133). Till has been recorded in only a few of the pockets, but overlying head is present almost everywhere. It consists of red-brown unstratified silt or clay, generally with chert fragments: its origin is discussed on p. 136.

Yorke (1961, pp. 32–60) did not separate till from head in his descriptions, but grouped them together as 'superficial' drift. This was, however, distinguished from 'intruded' drift (now named chert breccias).

ORIGIN OF THE POCKETS

The age and origin of the solution hollows in which the deposits are found is still in some doubt. Kent (1957) suggested that the pockets were initiated in Permian or Triassic times as swallow-holes at the margin of the Namurian mudstone cover, and that the deposits subsided into them as they gradually enlarged during regional uplifts of Mesozoic and Tertiary age. Mesozoic emplacement is clearly impossible if, as is now thought likely, the Brassington Formation is entirely of Tertiary age, but it is still possible that the hollows were initiated as swallow-holes at the mudstone margin: this would account for the association observed between the pockets and the dolomite outcrop (Figure 37). Yorke (1961, p. 22) believed that the Pocket Deposits had been trapped in pre-existing surface hollows, some of them water-courses and others solution cavities, but this seems unlikely in view of the collapse-structure found in the larger deposits. Ford and King (1969) have shown that caverns form preferentially where dolomite rests on limestone, and have suggested that some of the Pocket Deposits have been trapped and preserved where caverns of this type have worked upwards to the surface on which the Brassington Formation was laid down.

Walsh and others (1972, pp. 524–530) have summarised the evidence and conclude, in the light of experimental data of their own, that the majority of the deposits have been preserved in cavities that formed, after the deposition of the Brassington Formation, by gradual solution rather than by sudden collapse of cavern roofs. This view is favoured here.

JIC

DETAILS

The pockets and their contents are described in approximte geographical order from north-west to south-east. Only those that fall within the Buxton district are included.

HINDLOW

An outlying pocket revealed in a working limestone face at Buxton Quarry [0832 6907; not included in the area covered by Figure 37] was about 35 by 20 m in diameter. It contained grey clay of the Kenslow Member with fossil wood identified by W. G. Chaloner as that of a conifer belonging probably to the family Taxodiaceae. Pieces of Namurian mudstone also present contained goniatite-bivalve faunas of E_2, H_1 and R_{2a} ages (Aitkenhead and Holdsworth, 1974, p. 62).

NA

HIGH PEAK PITS

Several small pockets, the largest about 100 m across, have been worked for refractory sand on the limestone plateau south and west of Parsley Hay Station [147 636]. Howe (1920, p. 170) gives details of the workings at one of the pits, where the deposit was mainly white sand. Photographs of an abandoned working are included in the earlier account by Yorke (1954, plates 10, 11), and a general view across the area is given by Ford and King (1969, plate 4). Jackson and Charlesworth (1920, p. 488) recorded black shale with *Posidoniella sp.* in the Pocket Deposits here, and boulders of Eskdale granite and Borrowdale volcanic rocks in the overlying drift.

The abandoned pits have near-vertical walls of Bee Low or Woo Dale limestones, with remnants of pebbly sand in crevices. The excavations vary between 5 and 16 m in depth, but it is not known whether any of the deposits was bottomed. Ponds in some of the pits are perched well above the water table in the surrounding limestone, and the presence of deeper clay deposits can be inferred in these instances. A mass of black mudstone, presumed Namurian, is present in the side of one pit [1455 6247]. Recent excavations have provided details at a pit [1341 6351] north-west of Vincent House, where 10 m of white and yellow sand, partly pebbly, contain irregular masses of pale grey silty clay; at another pit [1405 6278] south-east of the farm 5 m of mixed sand, pebbles and grey clay with chert fragments can be seen. A surface covering of up to 1.5 m of brown head is present in both exposures.

BLAKE MOOR

Four old pits, together with surface indications of two unworked pockets, are located here in dolomite. None exceeds 150 m in width. Their existence was first noted by Howe (1897, p. 145), at 'Parsley Hay'. The deposit in one pit (Howe, 1920, p. 172) was pebbly sand with small amounts of black and greenish clay. Pebble orientations measured in one of the pits by Walsh and others (1980, fig. 12) show a north–south trend.

One of the pits not yet filled [1595 6310] is about 13 m deep, with a pond at the bottom. It contains red, yellow and white sand with bands of red clay and silt, and there are siliceous pebbles up to 3 cm across. The dip is 45° to the south-east. Blocks of Namurian mudstone are present on the north-west side, and the dolomite wall rock can be seen on the south-east. A deposit of head (brown silty clay with chert fragments) is thickest, 3 m, against rising ground to the north-east. A shallower excavation [1565 6280] contains similar pebbly sand and a breccia of chert blocks up to 0.3 m across, with some masses of pink stellate baryte. The superficial covering of head here reaches 4 m in places. The area occupied by Pocket Deposits may exceed that shown on the map, for the wall rock is visible only at the south corner of the pit.

GREEN LANE TO OLDHAMS FARM

Nine pockets, all in dolomite, are known in this area. Six have been excavated and the remainder have been proved by boreholes or can be inferred from surface indications. Only three of those excavated remain open, the largest [1655 6262] being 22 m deep, 150 m long and up to 50 m wide. The wall rocks here are well exposed, but of the deposits only traces of pebbly sand remain, except at the south-east, where 5 m of disturbed red and white sand with clay and ochreous cherty bands dip at 45° to the north-east. Patches of head are present at the surface. A photograph of a smaller pit nearby [167 624] is included in the account by Ford and King (1969, plate 5).

HEATHCOTE

Two pockets in Bee Low Limestones are located north-east of Heathcote. The smaller is known from traces of variegated clay and pebbly sand in the sides of an old excavation [157 608]; the larger [154 610] is still worked from time to time. The horseshoe-shaped excavation, known as Heathcote Pit, is 21 m deep by about 110 m wide, with near-vertical walls of partly decalcified limestone. Dolomitisation is localised along joints. The deposits are mainly white sand with siliceous pebbles, although there are also small masses of mottled red, grey and yellow silt and clay (Bees Nest Member?) against the wall at the east side of the pit [1547 6104], together with greenish grey silt and clay from which a Kenslow-type flora has been obtained (Walsh and others, 1980, p. 49). At the west side a mass of brown till with scratched and facetted pebbles of limestone has lodged in a widened joint of the wall-rock. A larger mass of the same material, lying apparently among sand, was illustrated by Yorke (1961, p. 54).

NEWHAVEN

Howe (1897, p. 145) noted the existence of six pits and a disused mine shaft here. The deposits were recorded as 'clean white sand and yellowish sand and clay', with few pebbles. The Washmere Pit [1683 6025], now filled in, was at 48 m the deepest in the area (Yorke, 1961, p. 27), and a mine shaft sunk into the deposit reached 64 m below surface without encountering the wall rock (Howe, 1920, p. 169). Jackson and Charlesworth (1920, p. 488) recorded the presence of drift containing ice-scratched boulders of dolerite, limestone and sandstone at this pit. The Friden Pit [1661 6042], the only one hereabouts that remains open, is about 100 m across. Pebbly sand attached to the vertical walls of partly decalcified Bee Low Limestones is all that now remains of the fill, which Howe (1920, p. 172) described as grey and white sand and clay. Ford and King (1969, p. 61) recorded the presence of dark blue clay with plant remains (the Kenslow Member) in 'the core of a synclinal sag'.

KENSLOW

Two large pits have been excavated in complex areas of Pocket Deposits near Kenslow Knoll [184 617]. The wall rocks, where seen, are dolomite.

The Kenslow Top Pit [182 616] or greater Friden pit (Yorke, 1961) contains the best exposures in the district (Kent, 1957, p. 5; Walsh and others, 1972; 1980). In 1971 the excavation showed an irregular sheet of chert breccia lying against the dolomite wall and forming projections into the central areas. Against the breccia at the north side of the pit lay two masses of black mudstone. The age of these has not been determined, but goniatites, bivalves and miospores obtained from slabs of mudstone previously exposed at Kenslow Top indicate early Namurian (E_1 or E_2) ages (Kent, 1957, p. 5; Yorke, 1961, p. 65; Neves, in Walsh and others, 1972, p. 534). Inside the lining of chert breccias and Namurian mudstone, steeply dipping sediments of the Brassington Formation occupied the central parts of the pocket. All three members, apparently in conformable sequence, were present, the succession being similar to that recorded by Walsh and others (1980, fig. 2) in sections more recently exposed. The Kirkham Member, comprising red pebbly sand, was about 35 m thick, with a few bands of red mudstone. At the base a thin ferruginous-cemented bed rested on Namurian mudstone. Walsh and others (1980, fig. 8) have recorded pebble orientations in the Kirkham Member, with a peak towards the ENE. The Bees Nest Member consisted of up to 40 m of stiff, poorly bedded red and yellow mottled clay and silt, with a few sandy and pebbly layers. A preferred orientation of pebbles in a north-easterly direction was found by Walsh and others (1980, fig. 9) in a band of gravelly clay about 11 m above the base of the member. The type section of the Kenslow Member consisted of a few metres of soft grey clay and silt with pieces of wood and other plant fossils of Miocene–Pliocene age (Boulter and Chaloner, 1970; Boulter, 1971). A deposit, 6 m thick, of boulder clay with striated stones was at one time exposed above the Kenslow Member (Yorke, 1961, pp. 82, 84), but this has since been removed, and the Pocket Deposits exposed in 1971 were overlain by a layer of head up to 3 m thick, consisting of red-brown cherty silt.

The Kenslow Lesser Pit [187 613] (Walsh and others, 1980) was known to Yorke (1961, p. 14) as the lesser Friden pit. It is about 10 m deep with masses of dolomite standing up from the floor in places. A body of chert breccia near the north side contains blocks of pink stellate baryte. The Brassington Formation is represented only by the pebbly white sands of the Kirkham Member.

ALSOP MOOR

A pit [160 565] about 100 m across at Alsop Moor Lime Works shows degraded sections of yellow, white and brown pebbly sand with some silty clay. The wall rock, of Bee Low Limestones, is visible only at the south-east. Howe (1920, pp. 171–172) states that the deposit was white sand, stained red and yellow in places, dipping inwards at 60°.

LOW MOOR

A group of irregularly shaped pockets in Bee Low Limestones is situated west and north-west of Lowmoor Farm [191 565]. Yorke (1961) includes a sketch map showing six pits, and several photographs taken during the period of working; in one of the pits the sand deposit was free of drift cover and contained little clay and only a few pebbles. Some of the pits have since been filled, but one still open [1857 5668] shows several masses of black Namurian mudstone near the edges and degraded exposures of red clay, sand and pebbles elsewhere. Another pit [1867 5710] shows partly overgrown sections of red and brown sand, with red, grey and cream clay. These belong to the Brassington Formation but the status of some cherty clays, which are also common here, is uncertain.

Details of one of the bodies of Namurian mudstone at Low Moor are given by Walsh and others (1972, p. 533). It is about 5 m thick, with near vertical bedding, and lies directly against the limestone wall of the pocket. The black colour has been altered to lilac at the contact with the overlying sands of the Kirkham Member. Goniatites collected from Namurian mudstones during the present survey include cf. *Homoceras subglobosum* (H_1), and a fauna recorded earlier (Yorke, 1961; p. 65) includes goniatites and bivalves of E_{2b} and E_{2c} age.

MINNINGLOW

A group of old pits excavated in clay and sand is located in a drift-filled hollow south-east of Minninglow Grange [200 578]. The full extent of the pocket, or pockets, is unknown. The largest pit [2024 5760] is overgrown but shows traces of red, brown and yellow clay and sand. The wall rock exposed at the east side is dolomite, although the country rock beyond is predominantly limestone. A smaller pit [2025 5741], partly flooded, shows sections of banded pebbly clay, ranging in colour from red and purple to yellow, white

and brown, and exposures of brown cherty clay (head) can also be seen. Black mudstone is present at the south end of the pit (*see also* Walsh and others, 1972, p. 533): miospores indicate a Namurian E_2 age (Neves *in* Walsh and others, 1972, p. 534). In one of the pits Howe (1897, pp. 145–146) recorded the presence of fine white, yellow and reddish mottled sand, with a slight indication of bedding and very scarce pebbles; at the east side, a 15 m face showed laminated red and yellow clay, with sand above and below, and at the north end was a bed with lignified wood, probably the Kenslow Member. Jackson and Charlesworth (1920, p. 488) noted chert, dolomite, dolerite and grit in the drift above the Pocket Deposits in one of the pits at Minninglow.

LONGCLIFFE

A group of interconnecting pits in an irregular pocket [228 557] at Longcliffe has now been filled in. The surrounding country rock is dolomite, but a borehole [SK 25 NW/23] near the north side of the pocket encountered limestone beneath 19.2 m of made ground. The limestone was partially decalcified and joint-fissures were filled with grey and yellow clay. Another borehole [SK 25 NW/24] proved made ground and Pocket Deposits to 45.2 m, on limestone to 50.6 m on Pocket Deposits to 55.3 m. Howe (1897, p. 146) recorded the presence of white and yellowish sand with a few small quartz pebbles, and distorted masses of stiff red mottled and black clay: he quotes chemical analyses of two samples of sandy clay (1897, p. 147).

BRASSINGTON

Only the northern end of the sinuous Brassington pocket lies within the Buxton district. A pit [239 551], known to Yorke as the Old Pit, formerly showed dolomite walls (1961, p. 12 and plate 1) and deposits of white sand and drab sandy clay (Howe, 1920, p. 173), but the exposures are now much degraded. To the south, just beyond the district boundary, lie the Green Clay Pit and the Bees Nest Pit, the latter chosen by Boulter and others (1971) as the type section of the Brassington Formation (for details of the exposures, *see* Yorke, 1961; Walsh and others, 1972; 1980). JIC

CHAPTER 11

Structure

The structure of the exposed rocks is essentially that which was imposed by the stresses of the Hercynian earth-movements in late Carboniferous (Stephanian) times. The Derbyshire area then lay on a continental foreland well to the north of the Hercynian orogenic belt and its accompanying ocean closure (Anderton and others, 1979, p. 170), and the stresses were of a relatively gentle nature. Most of the structures are attributed to a phase of east–west compression followed by a period of tension. They are superimposed on a larger-scale structural feature, the 'Derbyshire Dome', which is responsible for the broad distribution of outcrops, and which is probably partly, or mainly, of post-Hercynian age (George, 1963). The radiometric ages of the Derbyshire olivine-dolerite intrusions (287 to 311 Ma; see p. 96) imply that these were emplaced before the deformation began. Dating of the lead-zinc mineralisation (Fitch, Miller and Williams, 1967, p. 784; Ineson and Mitchell, 1973) indicates that hydrothermal activity began at about the same time as the main earth-movements, but continued much longer.

The structural response of the Carboniferous rocks in the western part of the district was quite different from that in the eastern part (Figure 39). The western area was relatively mobile and is characterised by strong linear folds with vertical axes and mainly north–south trends, whereas the eastern area was relatively stable and now shows gentle folds of less regular shape and more random orientation. The line between the two tectonic styles lies close to that of the Asbian apron-reef belt, and follows a slightly sinuous course from west of Buxton through Hartington to Parwich.

Differences in the structure of the basement beneath the two areas are believed to be responsible, directly or indirectly, for the differences in structural style. Thus a relatively stable basement 'block' beneath the eastern area appears to have protected the cover-rocks from the effects of the earth movements, and also to have exerted an indirect control on the structure through the greater rigidity of the predominantly limestone sequence that was deposited there in Dinantian times. In the western area the equivalent sequence is more argillaceous and has been more liable to folding. The structures in the Namurian rocks of the two areas are broadly similar to those in the Dinantian.

The district was tilted and uplifted during the Mesozoic and Cainozoic eras but there was little folding or faulting. The post-Hercynian (Triassic) sediments that rest unconformably on the Carboniferous strata are only slightly deformed.

PRE-CARBONIFEROUS BASEMENT

The basement in the Eyam Borehole, just outside the district boundary, consists of cleaved Ordovician mudstone dipping at 45° to 60° (Dunham, 1973, p. 84). Altered volcanic rocks with a rough parting dipping at about 40° are believed to represent the basement in the Woo Dale Borehole (Cope, 1979). The composition and structure of the basement are otherwise known from only indirect evidence. Thus, the existence of a relatively stable block beneath the eastern part of the district is deduced from the nature of the Hercynian structures (Figure 40) and from variations of thickness and facies of the late-Dinantian and Namurian sediments (Kent, 1966, fig. 3; Aitkenhead and Chisholm, 1983, fig. 8). Both lines of evidence suggest that to the west the rigid block terminates abruptly, perhaps at a fault. From here the basement descends towards the north-east, as proved in the Woo Dale and Eyam boreholes, but no sharp north-eastern margin to the block can be inferred from the facies or structure of the cover rocks.

Positive gravity anomalies (Institute of Geological Sciences, 1977a) are associated with the block, and were interpreted by Maroof (1976) in terms of an upstanding basement mass. He suggested that the north-eastern side slopes more steeply than the western side, a conclusion which apparently conflicts with that based on the structural and sedimentary evidence outlined above. However, apron reefs do not necessarily overlie block margins (Miller and Grayson, 1982, fig. 3), and an alternative interpretation of the basement structure more consistent with the geophysical evidence has been proposed by Smith and others (in press). According to this interpretation two basement blocks, tilted to the south-west, are present beneath the late Dinantian shelf limestones, rather than a single block sloping to the north-east.

A basement fracture probably lies beneath the Cronkston–Bonsall Fault Zone (see below). It had no obvious effect on Dinantian sedimentation, but moved in Hercynian and post-Permian times.

INTRA-CARBONIFEROUS MOVEMENTS

Examples of folds and faults that moved during Dinantian or Namurian times are few. All lie close to the supposed margin of the stable basement block. A regional movement along this line, with downthrow to the south-west, can be inferred in mid-Dinantian times, for it was then that the basin of deposition became clearly differentiated into the shelf and off-shelf provinces of limestone deposition (p. 6). Smaller movements are indicated along the same line in later Dinantian times: at Banktop (p. 37) there is evidence that can be interpreted in terms of pre-Brigantian movement, and farther north some late Brigantian fault movements are inferred from outcrop patterns around Chrome Hill and High Edge (p. 23). Examples of such small movements in the district to the north have been described by Stevenson and Gaunt (1971, fig. 7 and plate 9A).

Regional disconformities within the Carboniferous sequence were formerly attributed to earth movements

Figure 39
Main structural elements of the district.

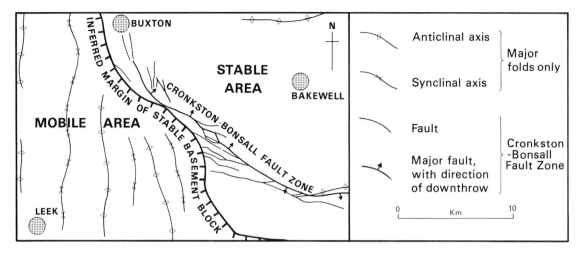

(Stevenson and Gaunt, 1971, p. 322) but are now open to alternative interpretation in terms of eustatic (world-wide) changes of sea-level (Ramsbottom, 1973, pp. 588–589). Dip discordances observed in the reef belt (pp. 36-39) between Monsal Dale and Bee Low limestones, for example, can be explained by the banking up of later sediments against steep slopes cut into earlier beds during a period of low sea-level, and do not necessarily imply a period of folding, or uplift, of the Bee Low Limestones before the Monsal Dale Limestones were laid down. Similar arguments apply to the disconformity recorded in places at the base of the Eyam Limestones (p. 42; Stevenson and Gaunt, 1971, p. 32) and that widely mapped at the base of the Longstone Mudstones and Namurian mudstones (p. 70; Stevenson and Gaunt, 1971, pp. 34, 162). The existence of a mid-Asbian (mid-D_1) unconformity, postulated by Parkinson (1950; 1973) in the Wolfscote Dale area, has not been confirmed by the present survey.

Numerous smaller gaps in the Dinantian shelf limestones, recognised by the presence of palaeokarst features and palaeosols (pp. 17, 29), can be interpreted in terms either of eustatic changes of sea-level or of broad earth movements; the cyclicity observed in Namurian and Westphalian sediments (pp. 69, 93) is of a similar nature.

Thickness variations in sediments of shallow-water origin are normally interpreted in terms of differential subsidence within the basin of deposition, and vertical movement of this nature was presumably taking place throughout the period during which the Carboniferous rocks were accumulating.

Synsedimentary movements of a quite different type, not directly related to tectonic forces, have affected the Ashover Grit in the east of the district (Chisholm, 1977). Large masses of deltaic sediment foundered along curved fault planes (listric growth faults) that developed during deposition. The geometry of the resulting structures is similar to that of rotational landslips, and the faults level off in mudstones below the displaced masses of sediment. The underlying limestones were not affected. A similar structure has been recognised in the Roaches Grit in the west of the district (Jones, 1980, p. 51).

HERCYNIAN STRUCTURES

Mobile (western) area

The main folds trend in a roughly north–south direction (Figure 40). Where they affect late Namurian or Westphalian rocks they are simple open structures in which dips greater than 40° are rare. Where early Namurian and Dinantian rocks are involved, however, the main folds contain many subsidiary ones in which the beds are locally steeply inclined or vertical. Axial planes are generally vertical or nearly so. Faults that cut through the folds include normal dip-slip faults and small oblique and wrench faults.

The **Todd Brook Anticline** lies mainly in the ground to the north (Stevenson and Gaunt, 1971, p. 326). It extends into the present district for about 2 km but then dies out, the folding being continued en échelon by the smaller **Macclesfield Forest Anticline** about 1 km to the west. The latter dies out southwards near Wildboarclough. Dips up to 85°, probably associated with an axial fault, have been recorded in the core of the Todd Brook Anticline.

The **Gun Hill Anticline** has the form of an elongate dome. It extends northwards for about 4 km from Gun Hill; to the south its continuation is largely concealed by Triassic rocks at Leek. An anticline near South Hillswood Farm [991 583] probably represents a subsidiary fold rather than the main structure. In the core of the main anticline, late Pendleian and Arnsbergian beds are faulted in an irregular pattern and thrown into subsidiary folds with steep dips.

The **Goyt Syncline**, first recognised (as the 'Goyte Trough') by Farey (1811), can be traced from near Leek northwards for a total distance of about 30 km, to Rowarth in the Chapel en le Frith district (Stevenson and Gaunt, 1971, p. 324). Dips lie mainly in the range 10 to 40°. Thick Namurian sandstones crop out in well-marked craggy features along both flanks, and the structure has a clear expression in the landscape, especially at the south end, around Hen Cloud [009 616], Ramshaw Rocks and The Roaches.

Small outliers of gently dipping Westphalian strata (p. 93) are present in the axial area at Goldsitch Moss [010 649] and Goyt's Moss [020 718]. The Blackclough, Cut-thorn and Little Hillend faults cut across the axis obliquely suggesting that they were initiated as wrench faults during the folding.

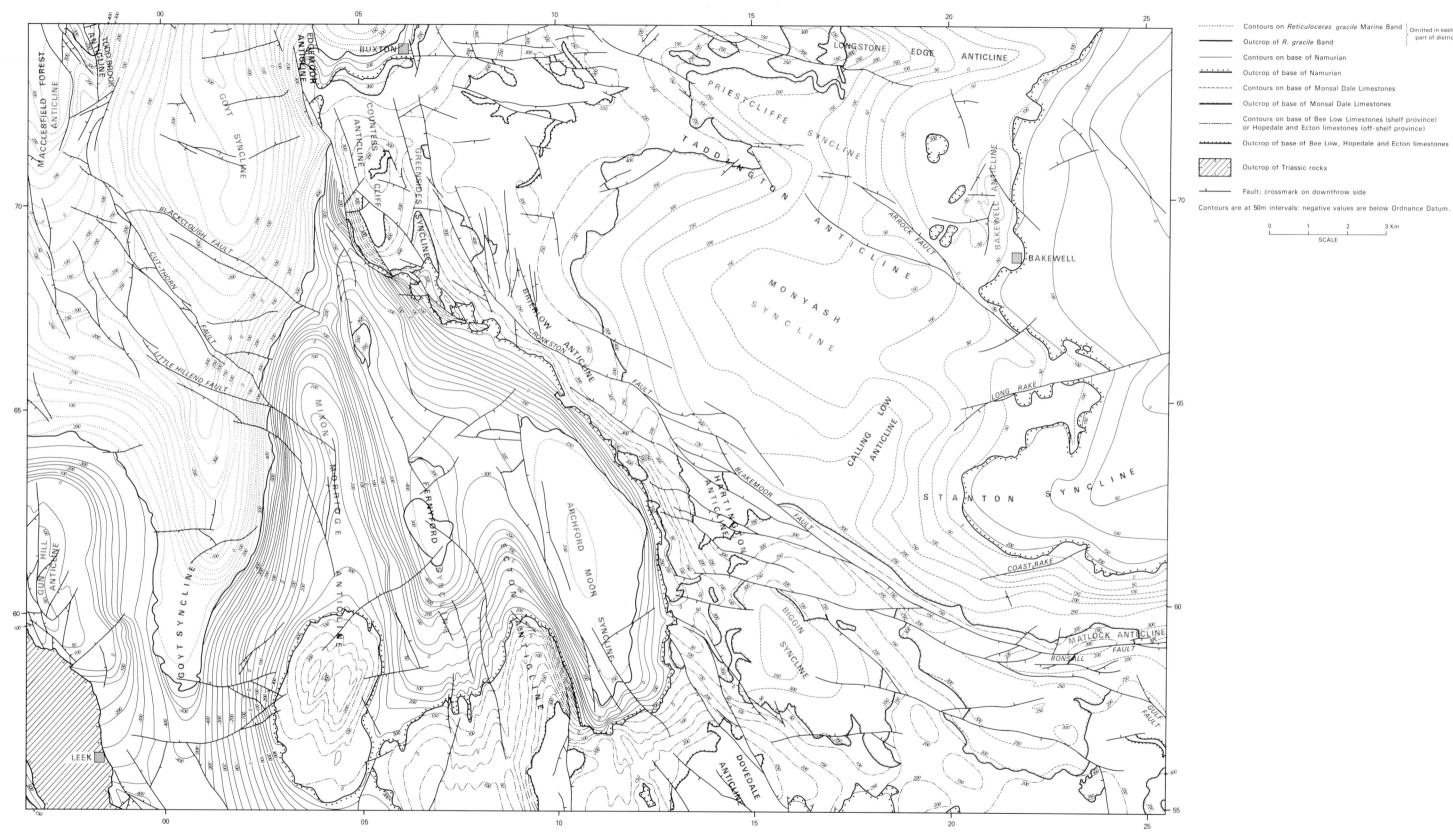

Figure 40   Structure map of the Carboniferous rocks.

They displace the outcrops in the manner characteristic of normal dip-slip faults, however, and the present throws across them are undoubtedly of this type.

'Contorted beds' (Cope, 1946), are small-scale structural features which occur at certain relatively constant horizons in mudstone sequences in the syncline. The beds are usually from 0.15 to 0.76 m thick, contain numerous closely-spaced shear planes and drag folds, and commonly stand out as ribs in weathered outcrops. A well-exposed section in Yeadonian shales near Orchard Farm [SK 06 NW/5] contains several fine examples. Cope concluded that these structures resulted from bedding-plane slip in the least competent beds during folding. Francis (1970) thought that a 'close laminar fissility' in such beds was the main factor responsible for the weakness.

The **Mixon–Morridge Anticline** extends 15 km from Brand Side [044 686] in the north to the southern boundary of the district just west of Onecote. The core of the fold broadens in the south so that the Dinantian outcrop around Mixon is in the form of a pericline with strong subsidiary folds. The axes of these latter diverge slightly (N 15° E) from the main axial trace. However, in the extreme south near Waterhouse [038 558], subsidiary folds trend north-west, parallel to the strike direction of the adjacent lower Namurian sandstones. Subsidiary folding subparallel to the axial trend is also present in the lower Namurian rocks as far north as Smallshaw Farm where, in a stream section [0366 6532 to 0431 6536], they are inferred to have a wavelength of about 200 m. The folds are indicated on the geological map by the sinuous nature of the outcrops. Although these folds tend to die out in the more competent sandstones at a higher stratigraphical level to the north, steep and even overturned dips are still present on the east limb of the major fold near Dun Cow's Grove [0430 6707].

The **Fernyford Syncline** is a broad structure lying east of the Mixon–Morridge Anticline and is complementary to it. The eastern limb is poorly defined and dips are generally low. To the south the syncline loses its identity in a broad area of smaller but more intense folds between the major anticlines. The strata involved in the folds are Onecote Sandstones and Mixon Limestone-Shales, together with mudstones of the early Namurian. The folds have a marked topographical expression in places, notably in the valley of the Warslow Brook around Brownlow [070 577 to 077 576]. Here, erosion has stripped the shale cover off the top of the Onecote Sandstones to leave the northward-plunging anticlines standing out as promontories separated by embayments marking the synclines.

The **Ecton Anticline** extends for about 9 km, from beyond the southern boundary of the district, south of Wettonmill, to around Fawfieldhead [075 636]. It has a marked northward plunge, and the intensity of subsidiary folding decreases northwards in the higher stratigraphical horizons. The structure is also strongly asymmetric with a steep eastern limb showing near-vertical and locally overturned beds, especially in the sector between Haysgate [098 597] and Manor House [104 566]. This asymmetry is probably due to the presence, not far to the east, of the more rigid limestones of the late-Dinantian shelf margin. In the axial part of the structure the subsidiary folding, which affects the Ecton Limestones, is the most intense in the district. These folds are well exposed in several roadside quarries west of Apes Tor [0983 5868 to 1005 5865] where the Manifold valley cuts across the structure (Plate 8; sketch-section by Prentice, 1951, fig. 5). They are upright, with axial planes around vertical, and plunge at angles up to 13° in directions ranging from 318° to 343° (*see also* Critchley, 1979). The axial trends are subparallel to the NNW trend of the major structure. The average wavelength is about 20 m. The folds are of concentric type and minor thrust faulting, veining and brecciation are common around the hinges, where accommodation stresses were at their greatest during folding. The underlying Milldale Limestones, exposed on the side of Ecton Hill [0985 5821] and examined underground in Clayton Level (p. 47), have also been involved in this folding.

The axial part of the Ecton Anticline south of Ecton Hill includes large poorly bedded knoll-reefs in the Milldale Limestones. These have acted as rigid masses (Prentice, 1951), mainly affected by faulting and tilting rather than folding. The reef mass extending from immediately south-east of Wettonmill to Thor's Cave just outside the district [097 561 to 098 550] has probably been tilted towards the west. However, the relative proportions of the depositional and tectonic dip components have not been determined. The thinly bedded limestones around Wettonmill are less strongly folded than the beds farther north, probably due to a degree of protection afforded by the reef masses in the vicinity. In the western part of the outcrop, steep westward dips predominate, while the central strip [099 550 to 100 565] between the reef masses contains a single tight anticline whose inferred axis has been traced for 1.2 km in a rather sinuous but roughly north–south line. The fault bounding this strip west of Manor House and its northwards extension [0989 5593 to 1024 5802] is probably a high-angle reversed fault.

In the north, the **Archford Moor Syncline** is a relatively simple structure affecting Namurian strata, with gentle dips at the centre and steeper dips, up to 60°, on the limbs. The eastern limb lies close to the margin of the stable basement block, in a region of rapid eastward attenuation of the sequence. The higher beds thus dip less steeply there than the lower beds, a fact which gives rise in the structure contour map (Figure 40) to a spurious impression of disconformity within the sequence. To the south, the Namurian outcrop ends at a line of Dinantian knoll-reefs extending from Wetton Hill [105 562] to Gratton Hill [132 571]; beyond here the structure becomes a synclinorium between the Ecton and Dovedale anticlines. Several folds, with wavelengths between 200 and 800 m, can be inferred from scattered dip observations in the Hopedale Limestones. The axes generally trend north–south but appear to be deflected in places by the knoll-reefs. Again the present topography to some extent reflects this folding (see this page), the clearest example being the synclinal hollow containing an outlier of Namurian mudstone at Hope Marsh [122 554].

The **Dovedale Anticline** is a broad structure extending into the district from the south. The main feature is a steady axial plunge to the NNW. Its termination in this direction, before reaching the Namurian outcrop, is probably due to the more massive nature of the Bee Low Limestones and the rigidity of the apron-reef limestones which cross the axial area obliquely at the north-west end of Wolfscote Dale. The

'discordant strike' hereabouts, used by Parkinson (1950, p. 283) as evidence of intra-D_1 (Asbian) unconformity, more likely represents merely the change in strike over this axis. No subsidiary folds have been detected in this northern part of the anticline, although in Hopedale, just across the southern boundary of the district, large folds with a wavelength of 600 to 800 m are present within the major structure.

Stable (eastern) area

This area consists of a broad asymmetrical upfold with its culmination and steep south-western limb over the western edge of the stable basement block, and a gentle eastward or north-eastward dip elsewhere. Superimposed on this arch are gentle folds (Figure 40), with dips only locally above 15°. Most are short periclines or basins with indefinite axial trends between east and south-east; a subordinate set runs between north-east and north, and northerly trends are also found over the block margin. Within the stable area, unusually steep dips are present on the south limb of the asymmetrical **Longstone Edge Anticline** (p. 31), where they range up to 45°, and on the north limb of the **Matlock Anticline**, where they locally reach 30°.

At the south-western margin of the stable area the fold structures are transitional to those of the mobile belt, as are the directions of jointing (Weaver, 1974, figs. 4, 5). Anticlines show approximate north-south axial trends, with the western limbs dipping more steeply than the eastern. Thus, the limestones on the west limb of the **Countess Cliff Anticline** commonly dip at angles up to 45°, whereas in the ill-defined eastern limb they rarely exceed 15°. In the **Hartington Anticline**, likewise, dips on the west limb reach 30°, but those on the east only locally exceed 15°. The **Greenside Syncline** terminates southwards against the rigid apron-reef limestones at the **Chrome Hill Fault** (p. 25), the maximum throw of which is located at the synclinal axis. The zone of steep south-westward dips coincides with the Asbian reef-belt, and there is little doubt that it overlies a basement facture which moved in early Asbian times to give rise to the shelf and off-shelf depositional provinces. A minor movement along the same line probably took place in the late Brigantian, causing the small depositional basin at Banktop (p. 37) to assume a more marked synclinal form. During the main Hercynian movements, the zone of steep dips developed above the edge of the stable block, with associated strike-faulting in places.

Within the stable area prominent and widespread groups of mineralised vertical fractures are now believed to have been initiated by Hercynian wrench-faulting (Ford, 1969, p. 76; Firman, 1977). Most of the fractures can be classified into complementary sets trending broadly north-east and north-west, though both sets show a tendency to curve round to east (Figure 41) which suggests that they were produced by the phase of east–west compression that gave rise to the north–south folds of the mobile area. The two dominant sets of joints in the limestones of the stable area, which also trend north-west and north-east (Weaver, 1974), are believed to have formed in response to the same stress field.

Displacements across the mineralised fractures are generally small, and only those that have a significant effect on the outcrop pattern are included in the structure contour map (Figure 40). Such displacements apparently resulted from dip-slip movements unrelated to the wrench-fault stress pattern, and may be partly or mainly of post-Hercynian date. A good example of such a structure is Long Rake, a major vein trending between ENE and east. Its walls are essentially vertical, though with deviations of up to 15° in either direction, as recorded at Long Rake Spar Mine [1873 6422]. Horizontal slickensides on the walls (Ineson and Al-Kufaishi, 1970, p. 340) indicate a wrench-fault movement. A downthrow to the south of between 20 and 25 m proved in workings and boreholes near Youlgreave probably post-dates some or all of the mineralisation, for at Raper Mine [215 652] the vein material is itself fractured and slickensided (Shirley, 1959, p. 423; Ineson and Al-Kufaishi, 1970, p. 341). Similarly Dirtlow Rake (Fieldgrove Vein) is a north-westerly fracture which, near Sheldon [175 688], is inclined at 80° to the north, with a downthrow of about 25 m in the same direction (Butcher, 1975).

Tunstead Quarry (p. 17) lies in a zone of general east–west faulting. Several dip-slip faults of this trend intersect the 1.5 km-long quarry face, and one of these [097 739] has been observed by Mr P. F. Dagger (personal communication) to have reversed both its throw and the dip direction of the fault plane, as the face has moved back over a number of years.

A belt of faults, probably overlying a basement fracture, crosses the stable area in a WNW direction from Winstermoor [241 594], near which it enters the district, to near Harpur Hill [064 711]; it has been termed the **Cronkston–Bonsall Fault Zone** (Figure 39). The main fault splits repeatedly until, at the western margin of the stable area, its branches can be recognised from Hartington to the southern outskirts of Buxton. All the faults in the Zone appear to be dip-slip faults, some reversed and others with a normal throw; they die out at the western margin of the stable area. The faults are regarded as essentially Hercynian structures, in view of the presence of minor lead–zinc mineralisation along some parts of them, but they are also reputed (Shirley, 1959, p. 423) to cut some of the veins and are, therefore, probably in part of later date. Evidence of later movement has recently been revealed by workings at Balleye Quarry [288 574], not far to the east of the district, where the **Bonsall Fault** brings dolomitised limestone against unaltered limestone, suggesting a phase of movement post-dolomitisation and thus post-Hercynian. The fault plane dips at 25° to 40° to the north-east and the throw is reversed. Movement on the underlying basement fracture may still be going on, for two earthquakes were experienced in 1952 with epicentres at Winster, which lies 1.5 km north of the outcrop of the Bonsall Fault (Worley and Nash, 1979). Other faults in the Zone are well exposed in quarries near Hartington. A reversed east–west fault in Heathcote Quarry [149 606] dips at 55° to the north and juxtaposes Bee Low and Monsal Dale limestones; a vein of calcite about 3 m thick occupies the fault plane. In Hartington Station Quarry [152 612] a similar reversed fault dips to the north at 50° and contains a calcite vein up to 1.5 m thick with traces of galena and hematite. Several faults can be seen in the sides of Long Dale. One of these, by Vincent House [137 632], is normal, dipping at 75° to the north and containing a calcite vein;

another by a bend in the road [136 635] nearby is a steep reversed fault with a small anticline on its north side. The **Cronkston Fault** is exposed in an old railway cutting [132 645] near Cotesfield. It dips north at 60° to 70° and the throw is also to the north; the fault is mineralised, with a vein of calcite about 5 m wide, and minor baryte. The northward continuation of this fault is probably represented by a vertical sheared and veined zone 18 m wide, in the face at Hillhead Quarry [074 692]. This is the westernmost of a set of associated faults, with trends between NNW and north, that are exposed in quarries between Harpur Hill and Earl Sterndale. The fault planes are vertical or steeply dipping (greater than 80°), and are marked by zones of close jointing, brecciation and minor mineralisation. A NW-trending fault [066 685] at High Edge appears to have moved in late Dinantian times, for it has no detectable effect on the Namurian mudstones that lie across the line of fracture. A continuation to the south-east has displaced the mudstones, however, as proved in the Glutton Bridge Borehole [SK 06 NE/17].

POST-HERCYNIAN STRUCTURES

Mesozoic and Tertiary sediments are very limited in outcrop in the Buxton district, and the history of post-Hercynian earth-movements is necessarily based on evidence from a wider area. A regional eastward tilt of the southern Pennine region is shown by the structure of the Permian and Triassic rocks in, for example, the Chesterfield and Derby districts (Smith and others, 1967, p. 220; Frost and Smart, 1979, p. 99). It probably originated during the development of the large north-eastern basin of Mesozoic deposition (Kent, 1949; 1957), and later became accentuated during Tertiary warping and uplift of the 'Pennine Axis' (George, 1963). The evidence for Teriary uplift is largely geomorphological and has been summarised by Walsh and others (1972, pp. 541–542).

In the Buxton district the only undoubted Triassic rocks that have survived to the present time are those of the Leek outlier. The strata here are thought to have been affected by gentle warping about an east–west axis (p. 99), and they are cut by a single fault of roughly east–west trend just west of the district (Figure 36). The movements were clearly of post-Triassic age but closer dating is not possible.

There are a few faults in the Carboniferous rocks for which a post-Hercynian phase of movement can be demonstrated. A reversed throw that developed on the Bonsall Fault after the Permo-Triassic dolomitisation of the limestone (*see above*) is of this type, and Shirley (1959, pp. 423–424) believed that several of the faults to the south of it, around Aldwark, had also moved in post-Hercynian times. Brecciation and slickensiding within the larger mineral veins also provide evidence of small post-Hercynian movements, as recorded, for example, in Long Rake at Raper Mine (*see above*).

Such minor movements can be dated as 'post-Triassic', 'post-dolomitisation' or 'post-mineralisation', as appropriate, but their relation to the broader Mesozoic and Tertiary movements are at present uncertain. JIC, NA, IPS

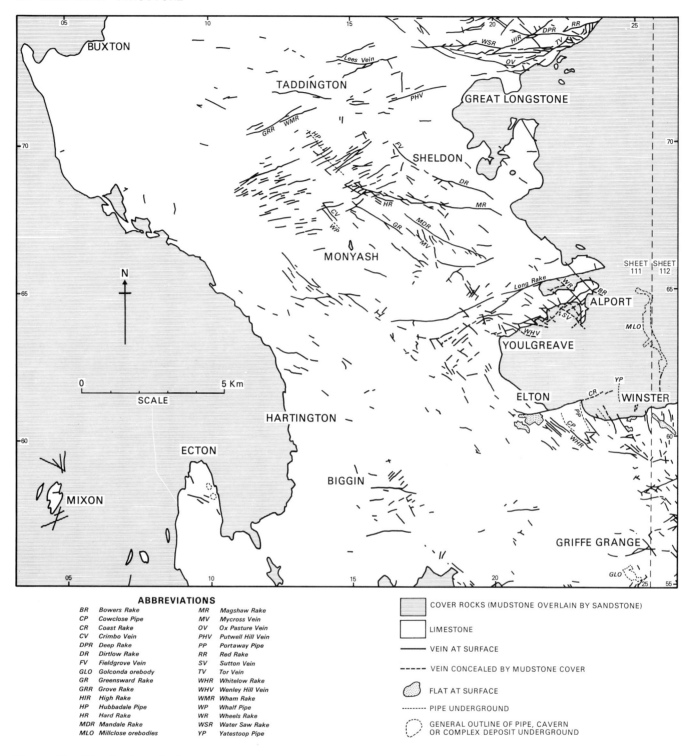

ABBREVIATIONS

BR	Bowers Rake	MR	Magshaw Rake
CP	Cowclose Pipe	MV	Mycross Vein
CR	Coast Rake	OV	Ox Pasture Vein
CV	Crimbo Vein	PHV	Putwell Hill Vein
DPR	Deep Rake	PP	Portaway Pipe
DR	Dirtlow Rake	RR	Red Rake
FV	Fieldgrove Vein	SV	Sutton Vein
GLO	Golconda orebody	TV	Tor Vein
GR	Greensward Rake	WHR	Whitelow Rake
GRR	Grove Rake	WHV	Wenley Hill Vein
HIR	High Rake	WMR	Wham Rake
HP	Hubbadale Pipe	WP	Whalf Pipe
HR	Hard Rake	WR	Wheels Rake
MDR	Mandale Rake	WSR	Water Saw Rake
MLO	Millclose orebodies	YP	Yatestoop Pipe

COVER ROCKS (MUDSTONE OVERLAIN BY SANDSTONE)

LIMESTONE

VEIN AT SURFACE

VEIN CONCEALED BY MUDSTONE COVER

FLAT AT SURFACE

PIPE UNDERGROUND

GENERAL OUTLINE OF PIPE, CAVERN OR COMPLEX DEPOSIT UNDERGROUND

Figure 41 Distribution of mineral deposits in the district.

CHAPTER 12

Mineralisation

The Dinantian limestones of the Peak District contain a suite of hydrothermal mineral deposits which have been worked, mainly for lead, since Roman times (Figure 41). The early history of lead and zinc mining was summarised by Green and others (1887, pp. 118–121) and is the subject of continuing research by the Peak District Mines Historical Society. Large parts of the orefield have been drained by levels ('soughs') constructed in the eighteenth and early nineteenth centuries and many of the productive veins have been worked down to these levels or to even greater depths with the help of pumps. At present the deposits and the waste tips are being worked again for fluorite and baryte.

MINERALS

Except at Ecton the main primary sulphide minerals are galena and sphalerite; the gangue minerals are normally calcite, baryte and fluorite. Minor amounts of other sulphides, notably of copper, iron and nickel, have also been recorded. Oxidation products, mainly cerussite (lead carbonate) and smithsonite (zinc carbonate), are widespread, though seldom abundant. Details of the distribution of these, and of numerous less common mineral species, have been listed by Ford and Sarjeant (1964). The lead ore, galena, was the mineral normally worked, the zinc ores being less common (Green and others, 1887, pp. 125–127). Zinc sulphide (sphalerite) has been noted in the present district at Sheldon, Coast Rake and Millclose, and smithsonite was worked at Hard Rake (see below). Cerussite was worked among the Pocket Deposits at Washmere Pit (p. 107) near Friden (Green and others, 1887, p. 126; Howe, 1920, p. 169). Wad, an earthy oxide of iron and manganese, was worked in the Elton area (Green and others, 1887, p. 159). At Ecton ores (sulphides and oxidation products) predominated, but were accompanied by galena (Green and others, 1887, p. 158).

Mining operations now concentrate on the gangue minerals (Ford and Ineson, 1971), with the sulphide ores as by-products. Fluorite and baryte are the most valuable, but calcite is also worked to a limited extent. There is a generalised zoning of these minerals, with fluorite commonest in the east and calcite in the west (Wedd and Drabble, 1908; Dunham, 1952b; Mueller, 1954). Firman and Bagshaw (1974) have summarised the available information on the distribution of gangue minerals and have shown that the zonal concept, though broadly valid, cannot be refined to the degree proposed by Mueller (1954). In detail the proportions of the three minerals do not follow regular or predictable patterns, and it has been suggested that the anomalies in zoning may be due to the overlapping effects of successive phases of mineralisation, a conclusion supported by detailed studies of the mineral paragenesis (Ineson and Al-Kufaishi, 1970).

SOURCE OF MINERALISING FLUIDS

The zoning apparently reflects a temperature gradient, the minerals having crystallised at higher temperatures in the east than in the west, this implying that the source of mineralising fluids lay to the east. This source was formerly supposed to be a deep-seated granitic body located to the east of the limestone outcrop (Traill, 1939; Ford, 1961), though as was pointed out by King (in Sylvester-Bradley and Ford, 1968, p. 123) there is no other evidence for the existence of such a body. More recently it has become accepted that lead–zinc deposits such as those in Derbyshire have in general arisen not from magmatic sources but by lateral migration of deeply buried formation waters into basin-margin areas where, under suitably reducing conditions, sulphides have been deposited. The composition of the transporting brines and the origin of the sulphur in the radicals with which the metals eventually precipitate are still matters of dispute; opinions have been reviewed by Dunham (1970) and, for the Pennine orefields, by Worley and Ford (1977) and Emblin (1978). In the case of the Derbyshire mineral field, the Lower Carboniferous sediments to the east are believed to have provided the mineralising fluids (Worley and Ford, 1977, p. 205), although a study of the isotopic composition of the minerals (Robinson and Ineson, 1979) suggests that local materials have become mixed with the elements derived from more distant sources. The possible role of Tournaisian evaporites as a contributory source for transporting brines has been emphasised by Dunham (1973). The copper ores at Ecton and Mixon were probably generated by the same processes, but were clearly derived from a different source lying to the west.

AGE OF DEPOSITS

The majority of the fissures occupied by veins were initiated as joints and faults during the Hercynian earth-movements (p. 114), and the mineralisation is thus younger than late Carboniferous. Some deposits post-date the dolomitisation of the limestone and are probably of post-Permian age. Moorbath (1962) obtained from lead-isotope studies a mean model age of 180 ± 40 Ma for Derbyshire galenas, and suggested a mid-Triassic to late Jurassic age for the mineralisation, though the values of the constants used in the dating calculations have been questioned by Mitchell and Krouse (1971). Ineson and Mitchell (1973) obtained radiometric ages between 289 and 186 Ma from altered igneous rocks adjacent to mineral veins in Derbyshire, and concluded that the alteration was produced by the mineralising fluids. Peaks at around 270 and 235 Ma were taken to indicate that the main mineralisation took place in the Permian, and the younger dates were attributed tentatively to continued flow, at a reduced rate, into Mesozoic times.

FORM OF DEPOSITS

Most of the minerals have been deposited in the joint fissures and other cavities through which the mineralising fluids circulated, but there has also been some replacement of the country rocks, especially by fluorite. The deposits are classified, on the basis of their shape, into veins, pipes and flats.

Veins In the Derbyshire orefield the larger veins are known as rakes, the smaller as scrins. They occupy vertical or near-vertical fissures, most of which were probably initiated as wrench faults (p. 114; Firman, 1977). Most veins are simple fissure-fills, with some metasomatic alteration of the wall rocks (Ineson, 1970). The larger veins have had a complex history that includes several phases of mineral deposition and slight earth-movement, as has been demonstrated at Long Rake by Ineson and Al-Kufaishi (1970). Veins normally become thin or disappear where the fissures pass through volcanic rocks (Green and others, 1887, pp. 123–124), although there have been some notable finds of ore minerals in such situations, as where Whitelow Rake passes through the Lower Matlock Lava (op. cit. p. 145). A list of similar occurrences has been compiled by Walters and Ineson (1980).

Pipes are linear deposits that generally parallel the stratification of the gently-dipping country rocks. Exceptionally, as in the strongly folded beds at Ecton (p. 113), they have a near-vertical inclination. They contain a mixture of replacement and cavity-fill deposits, and appear to occupy pre-existing solution-cavities, the position of which has been determined by the relative permeability of different beds in the sequence. Pipes near Elton, for example, have been deposited in cavities in dolomite above the impervious Lower Matlock Lava (Worley, 1977).

Flats also conform roughly to the bedding of the country rocks; in plan, flats differ from pipes by being irregular rather than linear. They have formed partly by replacement and partly by the filling of cavities, and range from small wing-deposits located where veins have intersected impervious beds, to large complex ore-bodies like that of Golconda (see below), where a system of caverns along the dolomite–limestone contact has been mineralised.

Mineralisation in the Derbyshire orefield is of 'manto' or blanket type, for the distribution of mineral deposits within the limestones has clearly been controlled by the presence of impervious beds, the most important of which is the thick mudstone cover that formerly extended over the whole limestone area. Although it had probably been breached locally by Permian times (p. 103), it was still far more extensive during the period of mineralisation than it is at present, and its existence is generally believed to have prevented the mineralising fluids from rising farther, and to account for the fact that the highest parts (stratigraphically) of the limestone contained the richest deposits (Green and others, 1887, p. 122). Veins in the top of the limestone normally pinch out rapidly as they pass up into the mudstone, as was illustrated by Parsons (1897, plate 4, fig. 2) at Millclose Mine. An exceptional occurrence of a vein of galena well above the top of the limestone has been recorded by Lacey (1862) near Axe Edge [035 700]. Small veins of baryte have also been record-ed in the Chatsworth Grit in a stream [004 689] in the Dane valley.

Lavas and clay wayboards are also relatively impervious, and local enlargements of veins into flats or pipes beneath these beds have been recorded, as at Millclose Mine (Traill, 1939; 1940). A similar enrichment has been noted at Magpie Mine (Butcher, 1971, p. 404), where a vein was worked below a clay wayboard, but not above. The situation is complicated by the effect of the same impervious beds on downward percolation of groundwaters, for this can lead to a concentration of solution cavities above lavas and wayboards, and these may in turn become mineralised. The matter is discussed further by Ford (1969). Another control on the development of ore deposits is the nature of the limestone itself, pale coarse-grained lithologies having proved more favourable to mineralisation than dark fine-grained rocks (Traill, 1939, p. 879).

The broad structure of the region has determined the location of the mineralisation at the east side of the Derbyshire Dome, but control by smaller-scale structures is not everywhere evident. An association between major veins and anticlines was suggested by Shirley and Horsfield (1945) and can be demonstrated in the present district by the Longstone Edge Anticline, which is mineralised both along its axis and flanks. South of here, however, the veins are concentrated rather in the broad flanks of folds, as at the north and south sides of the Stanton Syncline. The distribution of ore deposits in the district is shown in Figure 41.   JIC

DETAILS

Longstone Edge and Taddington

Mineralisation at Longstone Edge shows a close control by structure. The major vein system lies near to the axis of the east-trending Longstone Edge Anticline. In addition, vein mineralisation is developed on both limbs of the fold. The eastern (and east-trending) part of the major vein system is known as Deep Rake. To the west [213 736] of Bleaklow the vein swings WSW for 1 km and is called High Rake. Watersaw Rake makes up the east-trending western end of the system; it continues as far as Cressbrook Dale, where it bears only calcite as gangue.

Mr J. D. Hedges (written communication) states that the Deep Rake – High Rake – Watersaw Rake system is a vertical or near-vertical fissure filling which has been worked opencast for fluorite and is now worked underground from Sallet Mine [2193 7409]. The vein dips at about 80° south, though shallower dips occur locally, particularly at intersections with volcanic beds; vein widths range from 3 m to 15 m. Vertical displacement on the vein rarely exceeds 5–10 m, and there is a downthrow of some 4 m to the south near the Sallet Hole crosscut. The mineral assemblage within the vein-fill is commonly complex, indicating repeated phases of movement and polyphase mineralisation; it typically comprises a marginal zone of crystalline fluorite and galena succeeded by a wall zone of colloform baryte studded with galena (aligned parallel to the vein-banding) and a central core of randomly orientated baryte blocks in a coarsely-crystalline fluorite matrix. A general increase in the baryte content is evident in a westerly direction. Fluoritisation of the wall rocks, notably in areas where there is a pronounced structural and/or lithostratigraphical control, has resulted locally in the formation of extensive replacement deposits. At Sallet Hole Mine, for example, such a deposit, about 100 m in strike length, 25 m wide, and 10 to 15 m in height, is associated with the Bow Rake vein plexus, immediately overlying the Cressbrook Dale Lava.

Red Rake is a vein diverging from the main system at the eastern end. The average width is about 2 m, though there is local variation from 6.10 m [2356 7403] to as little as 0.31 m near Red Rake Mine [2401 7416]. The vein bears fluorite and some galena. Ford and Ineson (1971, p. B200) have suggested that the structure continues east beneath the Namurian cover. On the south side, Red Rake is joined by a group of roughly north–south veins including Catlow Rake [2342 7401] and Dog Rake [2352 7404].

Tributary veins farther west include Bow Rake in the angle between High Rake and Deep Rake: Strawberrylees Vein, Brandy Bottle Vein, Unwin Vein and Wager's Flat radiate from the same general position (Ford and Rieuwerts, 1983, p. 64).

On the steeply dipping southern limb of the Longstone Edge Anticline a plexus of veins extends from a point [201 730] near the eastern end of Watersaw Rake eastwards to the limestone margin at Brightside Mine [2290 7326]. The veins bear much fluorite and lesser amounts of baryte, calcite and galena. The plexus is intersected by Tor Vein, seen in a 3.66 m open working [2186 7296]; the vein extends westwards to join Ox Pasture Vein.

Putwell Hill Vein forms an important westward extension of the Longstone Edge structures. It extends for over 2 km from Monsal Dale to Taddington Dale; from the latter it continues westwards as a partly mineralised fault. At outcrop mineralisation is interrupted by the Upper Miller's Dale Lava; beyond this the vein continues to near Calton Hill. Putwell Hill Vein is 3.0 m wide in an open working [1734 7170] where it hades to the north, although there is no throw. Shaw (1980, pp. 342–343) gives the width underground as 2 to 4.5 m, and notes the presence of a substantial vein-fill of calcite with traces of galena and baryte. Carruthers and Strahan (1923, p. 61) state that the vein was 'formerly worked for galena and calamine' (smithsonite). There is little information on the western continuation of the vein system in the Taddington Anticline, though tips in the vicinity of Glory Mine [1328 7176] show abundant calcite and some baryte; a northerly hade was also present here (Carruthers and Strahan, 1923, p. 61). Lees Vein lies north of the Taddington Anticline near Priestcliffe Lees.

On the south side of the Taddington Anticline a WSW-trending group of veins has been worked in the lower part of the Monsal Dale Limestones. Grove Rake is the most important; there is, however, little information available regarding it except that the tips show relatively abundant baryte in addition to calcite. Wham Rake, to the south, is seen in an open working [1305 7097], 9 m wide, west of Taddington; nearby, tips show abundant baryte, some fluorite and limonite, and traces of malachite.   IPS

SHELDON AND MONYASH

The major veins are arcuate, trending between north-west and west. A set of smaller veins east of Monyash follows the same trend and another set west of Sheldon trends between north-east and east. The majority occupy vertical or near-vertical joint fissures.

Fieldgrove Vein occupies a fault plane with a downthrow to the north-east. In Deep Dale [1657 6989] the vein exposed at surface is of calcite with sparse baryte, and varies from 1.0 to 2.0 m in width. It dips steeply to the south-west. The Engine Shaft [1695 6949] is 115 m deep, and surrounding tips contain minor amounts of baryte, fluorite and galena. In an open working [181 688] near Kirk Dale the vein is about 10 m wide, consisting of an irregular intergrowth of baryte and fluorite with calcite veinlets and blocks of limestone. The walls dip at 70° to the SSW but at depth, where the vein is cut by Magpie Sough, the dip is in the opposite direction (Butcher, 1975). The downthrow, as estimated from measurements in the sough, is 25 m to the NNE. Details of mining records and recent exploration of the workings on Fieldgrove Vein are given by Robey (1966). East of Kirk Dale the vein is known as Dirtlow Rake and

has been worked opencast for fluorite to the vicinity of Dirtlow Farm [188 686]. Where exposed near Kirk Dale, the vein was vertical and up to 10 m wide, consisting mainly of calcite and baryte; the working has now been backfilled.

A group of major veins trends almost east–west through Magpie Mine [1725 6816], near Sheldon. At the west end, around High Low [156 683], tips contain baryte, fluorite and galena, and at Hard Rake an old openwork 3.0 m wide is visible [1603 6816]. Calamine (smithsonite) was among the minerals worked here (Green and others, 1887, p. 141). Around Magpie Mine a plexus of veins is recorded (Carruthers and Strahan, 1923, pp. 62–64); for details of workings and the drainage sough see Butcher (1975), who also cites earlier literature. The sough runs north to the River Wye and drains to about 170 m below the surface. Tips near the mine contain baryte, fluorite and galena, and a minor vein not far to the north contains sphalerite; details are given by Worley (1975). East of Kirk Dale a single vein, Magshaw (Mogshawe) Rake, is present. It has been worked opencast in recent years and backfilled, the width at the west end [182 679] being 10 m, and the walls vertical. A branch vein on Bole Hill [182 676] is 3 m wide, with fluorite and baryte.

Hubbadale Pipe trends south-east from Taddington Moor [135 705] and intersects a group of north-easterly veins. The pipe contained a rich deposit of galena (Green and others, 1887, p. 140; Kirkham, 1964; Robey, 1965, pp. 2–4; Willies, 1976). According to old records it is about 137 m wide and lies between about 70 and 90 m below the surface, with a gentle but irregular dip towards the south or south-east. It seems to run roughly parallel to the bedding in the limestone not far above the Upper Miller's Dale Lava, but the precise control is not known; it may be related to the presence of an overlying pair of thin clay wayboards recorded on one of the plans. The area was drained by the Wheal (Whale) Sough, which emerges in Deep Dale [1613 6952] at about 250 m above OD. Baryte and goethite are recorded on tips [1384 6995] above the northern end of the pipe, and traces of fluorite, baryte and galena have been seen in tips [1536 6975] on the associated north-easterly veins. Recent exploration (Worley and others, 1978) has confirmed the stratigraphical position of the pipe, and has shown that the richer galena deposits were probably concentrated by solution of the associated calcite. An unusual feature is the presence of marcasite.

North-easterly trending veins have been mapped west of Flagg and south to Monyash; old records (summarised by Robey, 1965, pp. 5–7; 1973) show that pipes are present below. Traces of fluorite and galena have been noted on tips [1414 6785] on Crimbo Vein, and baryte occurs in an openwork at Hillocks Mine [1450 6724] on Whalf Pipe.

Several veins trending between south-east and east crop out between Monyash and Over Haddon. Mandale Rake was drained by a sough into Lathkill Dale [197 661]. Tips along its outcrop are rich in baryte and contain some fluorite; a small open working [1834 6665], seen in 1971, showed a loose brown breccia about 20 m wide consisting of baryte, calcite and fluorite, with traces of galena. The workings can be entered via the sough, and according to Rieuwerts (1963) the vein is slickensided by wrench-fault movements, consisting in places of breccia of limestone and gangue minerals in a matrix of brown clay and limonite. Old records (see Rieuwerts, 1963; 1966a; 1973) date back to 1278, and show that the galena was found mainly in widenings of the vein where it cuts the dark limestones. The deposits are probably best regarded as pipes (Ford, 1969, p. 80). The closely parallel Pasture Vein has been described by Worley and Ford (1976) at the Forefield Shaft [1904 6639]. Horizontal slickensides are present on wall rocks and cut through all the phases of mineral fill, which in order of appearance are: i, calcite; ii, fluorite; iii, baryte and galena. Breccias similar to those noted in Mandale Rake are also present. Displacement of a clay wayboard across the vein indicates a vertical downthrow of 0.3 m to the south-west.

Lesser veins to the south of Mandale Rake were drained in part by Lathkill Dale Sough and Small Penny Sough (Rieuwerts, 1963; 1966b; Robey, 1965; Pickin, 1975). Workings in some of the veins are visible in the sides of Lathkill Dale; at [1866 6581], there are traces of calcite and fluorite on the walls of an open working 0.6 m wide. The largest vein, 4.0 m wide, consists mainly of calcite [1798 6614]. Baryte is common on tips along Mycross Vein [1690 6719].

Tips along Greensward Rake and minor veins to the south contain mainly calcite and baryte [e.g. 1549 6774], but a pipe about 3 m deep by 8 m wide at Ricklow Quarry [1655 6614] contains laminated calcite, dolomite and fluorite. The deposit is described in more detail by Ford (1969, p. 80).

A set of veins trends south-east from Sparklow [127 660]. Trial pits on an east-west vein [1494 6486], 4 m wide, showed abundant calcite, some baryte, and traces of fluorite and galena. The westernmost record of fluorite is of the purple form, in tips [1432 6482]; beyond here only calcite and baryte have been noted. The veins die out southwards at the Cronkston – Bonsall Fault Zone.   JIC, NA

ALPORT AND YOULGREAVE

A group of veins lies south of Long Rake, along the north side of the Stanton Syncline. North-east and north-west trends predominate. The field was a rich one, and some of the veins were followed to considerable depth beneath the mudstone cover. Local drainage soughs into the rivers Bradford and Lathkill were eventually replaced by the long Hillcarr Sough, which discharges into the River Derwent [2585 6372] at about 97 m above OD. Some details are given by Green and others (1887, pp. 141 – 144), and the history of mining operations has been summarised by Willies (1976) and Rieuwerts (1981a).

Long Rake is the largest of the group and extends for 8 km, mainly with ENE trend. In the western part [167 639 to 199 647] galena is scarce and the gangue is mainly calcite; it was formerly worked from shafts. At Long Rake Spar Mine [1873 6422] the shaft record shows that down to 76 m the vein dips south at about 85°, but below this the dip is to the north at 75°. Similar variations take place along the length of the vein, as can be seen in the openwork; at [1925 6443] the open fissure is 3 m wide and the walls show a broad horizontal fluting suggestive of transcurrent movement. The vein varies from 2 to 6 m in width and is locally in two parts, separated by limestone, (Carruthers and Strahan, 1923, p. 65). Further details are given by Ford (1967, p. 63). East of the calcite workings there is a gap in the outcrop [201 647] where the vein intersects a basalt, probably the Conksbury Bridge Lava. This body is not shown on the 1:50 000 map, as its existence was only recently discovered by Dr N. J. D. Butcher.

East of the basalt outcrop, the vein is rich in fluorite and baryte, which have been worked on both sides of Conksbury Lane [210 650]. The vein is faulted here, with a downthrow to the south of between 5 and 15 m. A branch vein, which may take some of the throw, trends south-west into a complex of small veins north and west of Middleton; baryte has been noted on tips around Crossflat Plantation [192 637].   JIC

In the Lathkill valley the vein splits in a bow-like manner. Raper Mine [217 652], now filled in, was a large open working in a mineralised shatter-zone at the eastern end of the split. The main throw is on the south branch and this lets down Longstone Mudstones and Namurian mudstones on the south side. The mineralised ground was a confused mass of largely fluoritised limestone and veins, the latter containing fluorite together with some galena, baryte and calcite; cerussite and smithsonite have been noted in small quantities (Ford, 1967, p. 64). The maximum width of the working was about 24 m and the greatest depth about 30 m. A detailed investigation by Ineson and Al-Kufaishi (1970) has shown that the fissure-fill vein is flanked on the south side by a

replacement deposit probably related to the presence of the mudstone cover. Mineralisation by baryte and fluorite took place in several phases, with periods of brecciation between. Most of the fluorite is pale brown or yellow, but the final phase is purple.

From Raper Mine eastwards across Haddon Fields to the edge of the limestone outcrop, the vein can be traced beneath boulder clay by the presence of minor workings and trials. Boreholes show a downthrow to the south between 20 and 30 m but the vein dips steeply to the north (Green and others, 1887, p. 142). An inferred continuation of the fault on the same trend is shown on the map, and Ford and Ineson (1971, p. B202) quote geochemical evidence showing that this is mineralised up to about 1 km beyond the limestone outcrop. Geophysical work by J. D. Cornwell (unpublished), however, failed to locate any continuation of Long Rake beneath the mudstone cover and suggests the presence of a parallel fault [241 660] some 300 m to the north.

Bowers Rake crosses Long Rake with a north-west trend. At the intersection [2290 6551] it is said to be shifted laterally by 10 to 15 m (Green and others, 1887, p. 142), probably by a dextral wrench movement on Long Rake. Near Bowers Hall [235 649] it passes beneath the mudstone cover; a section showing the structure is given by Rieuwerts (1981a, p. 6).

Wheels Rake is a north-west vein with a fault-displacement of 0.6 m down to the north-east (Green and others, 1887, p. 143). Wheels Rake Shaft [2280 6485], 54.86 m deep, penetrated the limestone beneath the Conksbury Bridge Lava, and a borehole was made into the underlying Lathkill Lodge Lava for a further 17.07 m (Walters and Ineson, 1981, p. 95), not 102.41 m as stated by Green and others (1887, p. 143). A section along the vein is illustrated by Walters and Ineson (1981, p. 96).   IPS, JIC

Sutton Vein runs south-east from Broad Meadow [224 643], entirely beneath the mudstone cover. A section along part of it is shown by Rieuwerts (1981a, p. 7).

In Bradford Dale, Wenley Hill Vein dips steeply to the north, with an open-work 1.2 m wide [2040 6387]. From here it runs SSE, and was worked beneath the mudstone cover at Mawstone Mine [2119 6339] until a gas explosion, presumably of methane, brought operations to an end in 1932 (Ford and Rieuwerts, 1983, p. 100). The vein is stated to be from 0.3 to 1.2 m wide, with strings of galena in baryte (Carruthers and Strahan, 1923, p. 69).

ELTON AND WINSTER

A set of north-west veins, with associated pipes and flats, lies along the southern flank of the Stanton Syncline. The precise nature of many of these deposits is not known, and on the published maps all are shown as veins; in Figure 41, however, an attempt has been made to differentiate between veins, pipes and flats. The limestones are dolomitised to a variable depth below the surface, and the sequence contains several impervious clays and lavas. Most of the workings lay in the beds above the Lower Matlock Lava. The richest deposits were pipes which, according to Worley (1977), formed preferentially beneath clay bands in dolomitised limestone above the Lower Matlock Lava. Flats have probably formed, by analogy with the example at Golconda (see below), at the base of the dolomite zone (Ford, 1969, p. 83). Some details of the deposits and workings are given by Green and others (1887, pp. 144 – 146), and a comprehensive account of the history of mining operations has been published by Rieuwerts (1981b).

Coast Rake can be traced for 3.5 km, from the region of Gratton Dale [211 607] to near Ivy House [242 616] north of Winster. Effectively it forms the limit of the Elton – Winster mining field, for few deposits have been found to the north of it. Along part of its length the vein is faulted, with a downthrow to the north, but recent boreholes through the mudstone cover around Whiteholmes [236 614] have proved little displacement across it. Details of the vein are few, but near its intersection with Portaway Pipe [230 612] it is said to have contained sphalerite and 'steel ore', a variety of galena (Green and others, 1887, p. 144). Sphalerite was also proved, with galena, in the boreholes near Whiteholmes.

Small north-west veins south and south-east of Elton have been worked mainly in dolomite above the Lower Matlock Lava, but Whitelow Rake has been worked also in the beds below (Green and others, 1887, p. 145). Calcite and baryte have been noted on tips [2339 5953]. To the north-west, the veins run into areas of irregular surface workings, perhaps several flats, near Hungerhill Farm [212 604]. Cowclose Pipe (Leadmines Vein), a richer deposit south of Elton, is marked by tips, some with baryte, near Leadmines Farm [225 604]. Old records from the Elton area are few, but an account of recent mine explorations is given by Bird (1970). Portaway Pipe contained rich deposits between about 10 to 90 m below the surface (Green and others, 1887, p. 145), in beds above the Lower Matlock Lava. At Portaway Mine [230 611] it lay in limestone beneath a dolomite roof, not far above the lava: it consisted of an outer lining of fluorite round a central mass of baryte and galena (Dunham, 1952b, p. 103).

A set of veins and pipes in dolomite above the Lower Matlock Lava at Winster has been worked, mainly beneath the mudstone cover (Green and others, 1887, p. 145; Rieuwerts, 1981b, figs. 1, 12, 17). Tips contain mainly calcite and baryte. Worley (1977) has described details of a pipe in dolomite, about 12 m above the top of the lava, in Wills Founder Shaft [2359 6074]. The pipe is a solution cavity 7.5 m deep, containing in the lower part a loose fill of laminated brown clay, 'wad', dolomitic limestone and fragments of baryte, galena and fluorite. The wall rocks have locally been replaced by intergrown baryte and fluorite. In the upper part, a breccia of mammillated baryte cemented by calcite was present. Yatestoop Pipe runs north beneath the mudstone cover from near Painter's Way Farm [245 610]. It contained rich deposits of lead ore, and was followed to depths exceeding 180 m below the surface at the north end, where it was intersected by Yatestoop Sough (Green and others, 1887, p. 145; Rieuwerts, 1981b, pp. 128–142). The veins, which are of north-west trend, continue as a group to the south-east and have been worked on Bonsall Moor [250 595] in limestone and dolomite beneath the lava. Old surface workings on one of them [2494 5943] showed remnants of calcite and fluorite with strings of galena.

MILLCLOSE MINE

Millclose Mine [259 626], situated just east of the district boundary, worked the richest lead deposit in Derbyshire until its closure in 1940. The workings extended into the Buxton district beneath Pilhough [250 650], and reached depths of about 150 m below sea level. An account of the oldest workings is given by Rieuwerts (1981b, pp. 112–124), and details of the main vein at the south end, where it was worked below the mudstone cover, are described by Parsons (1897). The vein runs almost north–south. As it was worked northwards, rich finds were made in lateral deposits at progressively deeper levels where mineralising solutions had been trapped below impervious lavas and clay bands. The deposits were described in detail by Traill (1939; 1940); the main vein ('Main Joint') acted as a feeder for mineralisation in fissures, flats, pipes and caverns on the up-dip side. The primary minerals were galena, sphalerite, pyrite, calcite, fluorite and baryte, with bravoite, a copper-nickel sulphide, in the deepest levels (Shirley, 1950). Further comments on the nature of the deposits and their extraction have been provided by Varvill (1937; 1959). JIC

EARL STERNDALE, HARTINGTON AND GRIFFE GRANGE

The majority of fault planes within the Cronkston–Bonsall Fault Zone contain veins of sparry calcite. Trial workings on these are common, but only to the north of Hartington, where tips

[1415 6147; 1271 6236; 1259 6297] contain traces of galena, is there any sign of sustained exploitation. Pockets of copper ore were recorded by Green and others (1887, p. 158) in one of the veins, near Ludwell Mill [124 623]. The metalliferous parts of the veins in this area are located at the highest topographical levels, perhaps a result of control by the former presence of the mudstone cover. Baryte has been noted in minor quantities farther north, in veins between Parsley Hay station [146 635] and Cotesfield [134 645]. No fluorite has been seen here, but there are minor occurrences of a blue variety of this mineral near the summit of Chrome Hill [069 674] and in a vein associated with the Chrome Hill Fault (p. 114). Traces of baryte and galena are also present; the latter has been mined at Hitter Hill [086 668] in an eastward extension of the vein.

South of the fault zone, few areas of significant mineralisation are known. In the Biggin Syncline, between Biggin [155 593] and Hawks Low [170 567], small veins trending north-east, with a few of east and north-west trend, have been mapped. Calcite is accompanied by traces of galena [1685 5682] and hematite [1675 5717]. Near Griffe Grange, at Golconda Mine [246 553], a major flat was worked beneath a NW-trending vein and pipe, at the junction of the dolomite with unaltered limestone. It contained mainly baryte with some galena (Carruthers and Strahan, 1923, p. 81; Dunham and Dines, 1945, p. 91), and lies between 80 and 130 m below the surface. An examination of deposits remaining in the old workings (Ford and King, 1965) has shown a great variety of mineral associations, including layered sedimentary accumulations, cavity linings, collapse-breccias, metasomatic replacements and placer deposits. Additional details have been provided by King (1966). The caverns in which many of the deposits were found are believed to have developed by preferential solution above the dolomite–limestone interface. NA, JIC

ECTON AND MIXON

This area lies on the west side of the Peak District, where the facies and structure of the host rocks are quite different from those of the main orefield to the east. The mineralisation is more isolated and is restricted to the central parts of two major structures, the Ecton and Mixon–Morridge anticlines, where limestones of Dinantian age are flanked by younger mudstones. The main sulphide ores are those of copper, in contrast to the lead–zinc ores that predominate elsewhere.

According to Robey and Porter (1972) the Ecton Copper Mines produced ore intermittently from early in the 17th century until their abandonment in 1891, with a peak production of over 4000 tons in 1786. The workings underlie an area of about one square kilometre on Ecton Hill (Plate 11). The main orebodies were irregular near-vertical structures, Deep Ecton Pipe [0991 5840] and Clayton Pipe [1007 5809], both of which are now flooded and inaccessible. Old plans (Kirkham and Ford, 1967; Robey and Porter, 1972) show the pipes to have been worked to depths about 300 m below the level of the nearby River Manifold. The largest cavity in the Deep Ecton Pipe was said to extend 40 yards (36.58 m) east–west and 200 yards (182.88 m) north–south (Green and others, 1887). Ore was produced from infillings, linings and wall-rock replacements in these and several other smaller cavities and veins, some of the latter being fault-controlled.

The chief ore mineral extracted was chalcopyrite, with lesser amounts of chalcocite and bornite, galena and sphalerite. Calcite, together with some baryte and fluorite, were the main gangue minerals. These can still be seen underground and in dumps together with much dolomitised limestone. Many other minerals are noted in the literature, the latest list being that of Critchley (1979). Large-scale zoning of the sulphide minerals recorded by Green and others (1887) is shown by the predominance in several orebodies of galena at higher levels and copper ores below. This

Plate 11   Manifold valley, Ecton; waste tips of the old Ecton Copper Mines.

At top left are some of the old buildings of Ecton Mine itself and the tip halfway up the hill is that of Dutchman Mine. In the second half of the 18th century, the mines were among the richest in Europe (L 1204).

probably reflects fractionation during the progressive cooling of the rising hydrothemal mineral-bearing fluids (Ford *in* Robey and Porter, 1972). The fluids may have dissolved the limestone to produce the pipe cavities prior to, or during mineralisation (Critchley, 1979).

Observations in Clayton Level during the present survey, and more extensive underground exploration by Critchley (1979), have shown that the Ecton Hill orebodies appear to lie just to the east of the broad axial culmination of the anticline, largely in the thicker bioclastic and peloidal beds of the Ecton Limestones. The dark thickly bedded argillaceous facies of the Milldale Limestones forms an inlier at Ecton and is present underground in subsidiary anticlines in Clayton Level. One such fold forms a small inlier on the western slope of Ecton Hill, though this is disputed by Critchley.

The location of these deposits links them with stratabound occurrences of copper ores at or near the edge of the Triassic basins of Cheshire and Staffordshire, with the possibility of a common source (Allen, 1980, fig. 133). The fluids may have migrated via pathways within the Dinantian limestones or in the sub-limestone strata; in the Caldon Low Borehole (Institute of Geological Sciences, 1978), drilled in an anticline some 10 km to the south of Ecton, chalcopyrite was found in sandstones below the limestones.

Other small mines in the Ecton area, worked mainly for galena, include Dale Mine [0929 5882] and Hayesbrook Mine [0895 5987] (Green and others, 1887, p. 152; Aitkenhead *in* Porter and Robey, 1972a). The former lies in strongly folded Ecton Limestones and was said to have exploited a gently dipping pipe deposit; the latter worked a vein mainly in the Mixon Limestone-Shales.

The Mixon Copper Mine [046 573] was worked at least as early as 1730 and was abandoned in 1858. It lies on the outcrop of the Mixon Limestone-Shales just south of the main inlier of Ecton Limestones that forms the core of the southern part of the Mixon – Morridge Anticline. Information from old documents and from an 1853 plan reproduced by Robey and Porter (1970) indicates that the orebodies were probably cavity-fills rather than replacements, highly irregular in form, and mainly associated with an east – west 'lode' or 'lum' crossed by a north-east to south-west 'counter lode', both intersecting a closely-spaced group of folds or 'saddles' (Watson, 1860). These 'lodes' are shown as veins on the 1:50 000 map; their orientation and position, taken from the Old Series one-inch map of 1867, differ somewhat from the 1853 plan, which shows the workings extending to a depth of 85 fathoms (155.45 m) below the surface, though Green and others (1887, p. 158) give a figure of 100 fathoms (182.88 m). The main host rocks at depth are probably the Ecton Limestones, the more argillaceous Mixon Limestone-Shales tending to have acted as a 'cap rock' to the rising mineralising fluids. The ore extracted consisted largely of chalcopyrite, with calcite as the main gangue mineral. There were also several other minerals including galena, sphalerite and fluorite (*see* summary by Robey and Porter, 1970).

Several other small copper mines are known to have operated in the Mixon Inlier (Porter and Robey, 1972b), including the Royledge Mine [0453 5911, probable site of adit] and the New York Mine [0484 5910]. They worked a set of intersecting veins or 'lodes' whose position is known only from the Old Series one-inch sheet. Porter and Robey (1972b) reproduced a plan and section dated 1850 showing the shafts and levels, but little geological information is given.   NA

CHAPTER 13

Geophysics

Geophysical studies in the district consist of regional gravity and aeromagnetic surveys, and more detailed local surveys using a variety of methods. The regional surveys have particular relevance to an understanding of large-scale problems, including the structure of the Derbyshire Dome and the nature of the underlying rocks. Detailed surveys have been made to investigate localised features such as mineralised structures.

The area covered by this description extends beyond the margins of the district (Figure 42) to provide a regional background for the geophysical account.

PHYSICAL PROPERTIES OF ROCKS

The densities and velocities of the main rock types (Table 3) are based on results obtained by the British Geological Survey (*formerly* Institute of Geological Sciences) and from published sources. For the interpretation of Bouguer anomalies related to near-surface sources, an important density change occurs at the top of the Dinantian limestone sequence where it is succeeded by lower-density mudstones and sandstones. In the eastern part of the district this change occurs close to the Dinantian–Namurian boundary but in the west it occurs within the Dinantian, around the base of the Mixon Limestone-Shales. Although the density of 2.70 g/cm^3 is typical for limestones of Dinantian age, an overall value of 2.65 g/cm^3 is preferred, perhaps with local increases due to the presence of dolomites, with a maximum density of 2.87 g/cm^3. Bouguer anomalies of deeper origin could arise from lateral density changes in the pre-Carboniferous basement rocks, as indicated by the variable densities ($2.60 – 2.75 \text{ g/cm}^3$) from the three deep boreholes in the region (Table 3). However, for general interpretation purposes, a density similar to that of the Dinantian limestones has been assumed.

The magnetisation values of Carboniferous sediments are low, but the igneous rocks in the Derbyshire Dome contain quantities of magnetite. The susceptibilities of the igneous rocks are typified by the figures from the Cressbrook Dale Lava in the Eyam Borehole (Figure 43); this shows three groups of values of about 5×10^{-4}, 5×10^{-3} and 5×10^{-2} SI units. The maximum values recorded for the lavas are about 3×10^{-2} SI units, but dolerite samples gave values up to 5×10^{-2} SI units. Titman (1971) reports higher values (Table 4).

The magnetic properties of some of the Derbyshire lavas and intrusions have been examined in palaeomagnetic studies by Everitt and Belshé (1960). More recently Titman (1971) has described the magnetic properties (*see* Table 4) of some lavas and dolerite intrusions; these results indicate that the remanence is generally weak, with values less than 20% of those for the induced magnetisation. One exception is the suite of samples from the Upper Miller's Dale Lava in which

the Curie temperatures suggest that the weak magnetisation is probably due to the presence of hematite, rather than magnetite. The mean directions of magnetisation (Table 4) are considered to represent (A) the effect of remagnetisation during Permo-Triassic times, (B) and (C) reversed and normal palaeofields for middle Carboniferous times. The last two are comparable with the results of Everitt and Belshé and indicate a palaeolatitude of about 20°S for northern Britain during the Lower Carboniferous and also that the igneous activity in Derbyshire occurred both before and after a geomagnetic field reversal.

Although none have been proved in the Derbyshire Dome, Evans and Maroof (1976) have argued that magnetic Caledonian granodiorite intrusions, similar to that at Mountsorrel in Leicestershire, might exist at depth in the region; density and velocity values for this rock type have therefore been included in Table 3.

The electrical resistivities of the main rock types, as recorded in various detailed surveys, are summarised as follows:

Overburden	10–100	ohm metres
Mudstone (Namurian and Dinantian)	20–80	ohm metres
Limestone (Dinantian)	100–3000	ohm metres

Table 3 Densities and velocities of the main rock types in and near the Buxton district

	Density g/cm³	Velocity km/s
Carboniferous		
Westphalian (average of all lithologies)	2.55[2]	3.6[1]
Namurian (average of all lithologies)	2.55[2]	4.0[1]
Dinantian (limestone)	2.70[1 2 4]	5.2[1]
Igneous		
Basic sills (3 sites)	2.85[3]	
Basic dyke	2.70[3]	
Basic lava	2.65[3]	
	2.61[1]	
Pre-Carboniferous rocks		
?Devonian sandstones (Caldon Low Borehole)	2.64[1]	3.8[1] (dry samples)
Ordovician slates (Eyam Borehole)	2.75[4]	
Volcanic rocks (Woo Dale Borehole)	2.60[4]	
Charnian (average of all lithologies)	2.80[4]	5.4–5.65[6] 6.4 at depth[6]
Mountsorrel granodiorite	2.66[5]	6.2[6]

1 BGS data	4 Maroof (1976)
2 Whetton and others (1961)	5 Evans and Maroof (1976)
3 Harrison *in* Stevenson and Gaunt (1971)	6 Whitcombe and Maguire (1980)

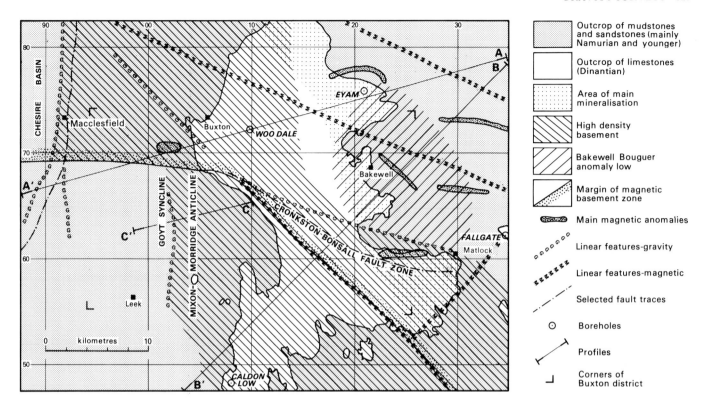

Figure 42   Compilation of the main geophysical and geological features of the Buxton district and surrounding areas.

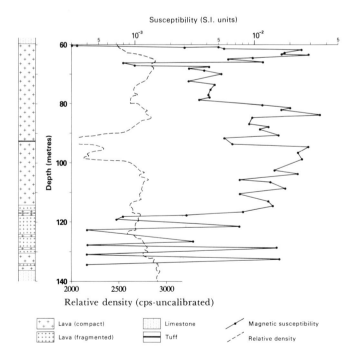

Figure 43   Relative density log and magnetic susceptibility value for the Cressbrook Dale Lava in the Eyam Borehole.

Ponsford (1955) has reported the occurrence of radioactive shale horizons in both Dinantian and Namurian rocks of the Castleton and Matlock areas, and Cosgrove (*in* Ramsbottom and others, 1962) reports comparable shale horizons in the upper part of the E_{1a} Zone in boreholes in the Ashover area. Radioactive mudstones at the same horizon (the *Cravenoceras leion* Band) have also been proved at Wardlow Mires in the Chapel en le Frith district (Stevenson and Gaunt, 1971, p. 164) and in boreholes on Long Rake and Coast Rake by Mr M. J. Brown (unpublished). Dr T. K. Ball (personal communication) reports that the *C. leion* Band is particularly radioactive and carries considerable quantities of the radioelements U and Th; the concentrations vary up to a limit of 160 ppm U and 140 ppm Th in the Wardlow Mires boreholes. The band is at its most radioactive where there is a marked transition between the underlying limestone and shale and is least where there is a gradual transition through a turbiditic limestone – calcareous shale sequence.

GRAVITY SURVEYS

The results of a regional gravity survey including the district have been presented on 1:250 000 maps (Institute of Geological Sciences, 1977a; 1977c). Additional gravity surveys have been carried out, mainly in the vicinity of the Dinantian – Namurian boundary, and comprise in-fill of the regional gravity coverage and detailed traverses (Cornwell *in* Frost and Smart, 1979).

Table 4   Magnetic properties of some igneous rocks (from Titman, 1971)

Igneous unit	Grid reference SK	Number of samples	Magnetic susceptibility $\times10^{-3}$ SI units	Remanence intensity $\times10^{-3}$ SI units	Q value‡	Curie temperature °C	Palaeo-magnetic field direction†
Bonsall Sill	275 593	2	119	9	0.16	520–575	A
Ible Sill	253 570	1	75	2	0.06	560–575	A
Waterswallows Sill	085 750	7	155	9	0.11	530–570	B
'Ashford Sill'*	180 698	3	436	36	0.16	610–650	C
Tideswell Dale Sill	154 744	1	163	2	0.02	530	A
Upper Miller's Dale Lava	155 743	3	81	24	0.58	610–640	B

* Now regarded as part of the Shacklow Wood Lava

‡ Q = Remanent magnetisation/induced magnetisation

† Mean directions of sites in group (*see text for explanation*)

	Declination	Inclination
A	230°	– 28°
B	199°	+ 40°
C	23°	– 61°

The Bouguer anomaly map (Figure 44) shows a high in the central part of the area elongated north–south with an easterly extension south of Matlock. It coincides in a general way with the Derbyshire Dome. The Bouguer anomaly values decrease gradually westwards away from the high, as far as Macclesfield beyond which there is a rapid decrease coinciding with the edge of the thick, low-density Permo-Triassic sediments of the Cheshire Basin. The decrease of values to the east of the central high occurs mainly towards the north-east and is due partly to the increasing thickness of Upper Carboniferous and younger sediments. A local closure around Bakewell is marked on its southern side by a well-defined gradient zone passing through Matlock.

The high Bouguer anomaly values over the Derbyshire Dome probably represent the combined effect of several structures at different levels in the crust. One structure, which occurs deep in the crust, is responsible for a regional increase of values over a large part of central England and culminates in this area. Superimposed on the gravity effect of this very deep structure is the anomaly, due to dense pre-Carboniferous basement rocks (Maroof, 1976), which is shown by the broad highs in Figure 44. The density contrast of about 0.15 g/cm^3 between the Dinantian and the Namurian and younger sediments also gives rise to anomalies but detailed gravity surveys across this interface are sometimes necessary to reveal these.

Bouguer anomalies of near-surface origin

Bouguer gravity anomalies associated with the density change from the Dinantian limestone sequence to the overlying mudstones are seen to the west of the Derbyshire Dome where the Namurian sediments thicken greatly (Figure 44). The pronounced north–south trending folds are reflected on the Bouguer anomaly map by the small anomalies with a similar trend [around 100 600]. Along the profile CC′ (Figure 45 iii), the steep Bouguer anomaly gradient over the eastern flank of the Goyt Syncline coincides with, and accentuates, the gradient due to the more deep-seated high density basement. In Figure 45iii the model shown for the Namurian

rocks is comparable with that derived from the geological evidence and also produces a gravity effect which, when added to the observed profile, yields a smooth curve considered to reflect the deeper-seated, high density body. This deeper body must underlie much of the Derbyshire Dome and is postulated (Maroof, 1976) to explain the main Bouguer anomaly high in the area.

The main Bouguer anomaly high

This high extends through the district from Grid line 13E, at the southern margin of Figure 44, northwards to cross the northern margin near 08E. The high represents the effect of a block of dense pre-Carboniferous basement rocks which narrows near the southern edge of the area shown in the figure but can be extended southwards along Grid line 28E through the western margin of the Derby district, where it forms the core of a structural high crossing the thick Namurian sediments of the Widmerpool Gulf (Cornwell *in* Frost and Smart, 1979) to underlie the Ashby Anticline in Leicestershire. Along this length the high density basement rocks coincide with structural highs in Carboniferous sediments, but the anomaly cannot be explained simply in terms of a corresponding rise in the pre-Carboniferous basement (contrast with Figure 45 i). The nature of the high-density basement is also unknown and the evidence of the few deep boreholes in the area of the high is conflicting, with pre-Carboniferous rocks consisting of volcanics (Woo Dale Borehole; Cope, 1979), Ordovician mudstones (Eyam Borehole; Dunham, 1973) and ?Devonian sediments (Caldon Low Borehole; Institute of Geological Sciences, 1978a). It is clear for example, that the low-density volcanic rocks at Woo Dale should give rise to a Bouguer anomaly low rather than the high observed (Figure 45 i), the conclusion being, therefore, that they must be part of a body of restricted size.

Maroof (1976) presented interpretations in which the Bouguer anomaly high is considered to be due to a topographic high in basement rocks, with a postulated density (2.80 g/cm^3) similar to the average for Charnian rocks and

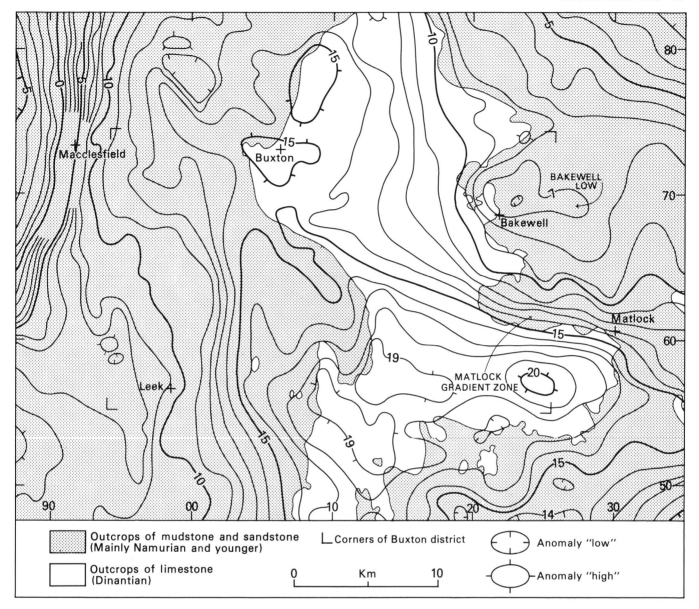

Figure 44 Bouguer gravity anomaly map of the Buxton district and adjacent areas.

extending from depths of − 0.1 km to − 2.0 km with steeper slopes on the eastern side. Whitcombe and Maguire (1981), however, found a seismic refractor with a velocity similar to that of Charnian rocks at a depth of about 1.7 km near Ballidon [204 546] on the southern edge of the Derbyshire Dome. These authors suggested that the upper part of the high density zone was made up of Old Red Sandstone and Lower Palaeozoic sediments, but this seems unlikely in the light of the density data in Table 3.

The Bouguer anomaly data do not necessarily require the high density basement to be as near the surface as indicated by Maroof (1976) and it might be more realistic to correlate them with a northern extension of the 5.7 km/s refractor reported by Whitcombe and Maguire (1981). In another paper (Whitcombe and Maguire, 1980) these authors report the existence of a 5.8–6.0 km/s refractor extending from a depth of 2 km down to 10 km beneath Buxton, using the results of the LISPB refraction survey (Bamford and others, 1978). Examples of the numerous alternative models for the high density basement, based on the interpretation of smooth profiles corrected for the effect of Namurian and later sediments, are shown in Figure 45 i and ii. In the southern part of the area the basement slopes downwards both to the east and to the west (Figure 45 ii) but in the north, around Buxton, the high density rocks appear to extend westward beneath the Cheshire Basin (Figure 45 i).

The Dinantian shelf area (Figure 42) transgresses the Bouguer anomaly high, and the shelf margin area is not clearly defined by the Bouguer anomaly data, although it might coincide with the edge of a deeper magnetic unit (Figure 42). Beyond the margins of Figure 44 the southern margin of the shelf is, however, reflected by the Bouguer anomaly gradient zone along the northern margin of the Widmerpool Gulf (Frost and Smart, 1979).

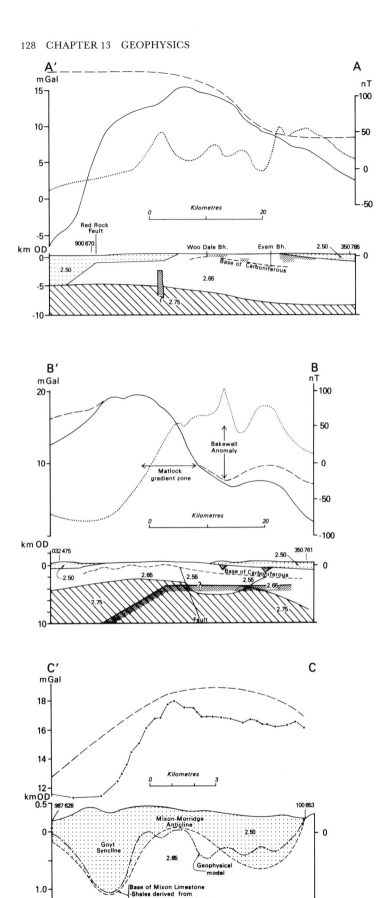

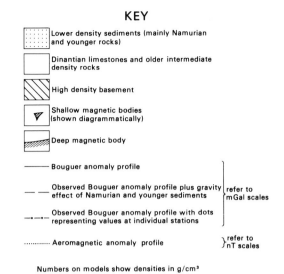

Figure 45 Bouguer gravity and aeromagnetic anomaly profiles (i) AA′, (ii) BB′, and (iii) CC′ (gravity anomaly only) (see Figure 42 for location), together with models for these profiles.

Matlock Bouguer anomaly gradient zone

A westward displacement of the high-density basement occurs at about Grid line 70N (Figure 44), approximately coincident with a north-westward bulge of the Dinantian outcrop [080 670]. The line of the southern margin of this displacement can be followed to the ESE across the Dinantian limestone outcrop into the zone of steep Bouguer anomaly gradient near Matlock. This zone is particularly distinctive because of its linearity, its coincidence in places with the Dinantian–Namurian junction and the steepness of the gradient. The latter is not due to the thickening Namurian sediments (Figure 45 ii) but the top of the structure responsible cannot lie deeper than about 2.0 km below OD. The form of the anomaly zone strongly suggests the presence of an ESE-trending fault in the pre-Carboniferous basement. Part of this anomaly is probably due to the presence of a low-density body in the basement in the Bakewell area (*see below*), but the deeper, high-density basement must also be affected. The apparent westward displacement of the latter near Buxton (*see above*) suggests the possibility that the proposed fault also had a transcurrent component (Figure 42). It is also suggested that the basement fault could have been re-activated in post-Dinantian times, giving rise to the Cronkston–Bonsall Fault Zone (p. 114 and Figure 42), and perhaps also determining the position of the two ESE-trending segments of the Dinantian–Namurian boundary on opposite sides of the Derbyshire Dome. This ESE trend is shown also by several magnetic structures (Figure 46).

Bakewell Bouguer anomaly low

North of the Matlock Bouguer anomaly gradient zone, the contours in Figure 44 form a negative closure centred on Bakewell, near the Dinantian–Namurian boundary. The maximum depth to the source of this anomaly, assuming a cylindrical model, is about 2.0 km for a density contrast of 0.1 g/cm³ and, as the low-density Namurian sediments cannot be responsible (Figure 45 ii), this may be due to the presence of either a basin of low-density sediments of Lower Carboniferous or greater age, or a granite intrusion. Examination of the second derivatives of profiles through this anomaly fails to produce convincing evidence on the direction of slope of the density interface, and the aeromagnetic map shows that the low-density body must be intersected or overlain by the igneous material responsible for anomaly 4 in Figure 46.

The possible presence of a granite to the east of the Derbyshire Dome has been considered on several occasions (as by King and Ford, 1968) to account for the concentration of mineralisation in the eastern part of the Dinantian outcrop (Figure 42) and the existence of such an intrusion at Bakewell cannot be ruled out on the basis of geophysical evidence. However, the mineralisation is no longer thought to be related to any igneous source (p. 117) and so provides no supporting evidence for the presence of granite here. The considerable thickness of Dinantian sediments in the Eyam Borehole (Dunham, 1973), at the edge of the Bakewell anomaly, seems to support the alternative interpretation of the Bouguer anomaly low as a sedimentary basin. However, the total thickness at the centre of the structure would need to be several times the 1.7 km recorded at Eyam to account for the full amplitude of the anomaly and the observed density contrast with the underlying rocks is not great (Table 3). A basin of sediments of possible Devonian age, similar to those discovered in the Caldon·Low Borehole, is an alternative explanation.

The model chosen (Figure 45 ii) assumes that the Bakewell low is due to a basin at the top of the pre-Carboniferous basement, but the observed profile could be interpreted equally well by a granite model. The deep high-density basement is assumed to produce only the smooth background change in Bouguer anomaly values but is penetrated by the lower part of the calculated basin model, which consequently has to be given a larger density contrast at depth (Figure 45 ii).

MAGNETIC SURVEYS

The Buxton district was included in an aeromagnetic survey carried out on contract in 1955. Total magnetic field measurements were made at a flying height of 1000 feet (305 m) along east–west flight lines 1 mile (1.61 km) apart and north–south tie lines 6 miles (9.65 km) apart. The map shown in Figure 46 is based on the original compilation and includes minor anomalies omitted for the sake of clarity in the published 1:250 000 scale maps (Institute of Geological Sciences, 1977b; 1978b).

The aeromagnetic map indicates the presence of magnetic bodies at several levels in the Carboniferous and pre-Carboniferous rocks of the district. These range from the lavas and intrusions at outcrop, which form small magnetic closures, [such as 115 795; 112 750], to the belt of magnetic rocks at a depth of about 4 km below OD that is responsible for the broad zone of high magnetic values across the northern and eastern parts of the region (Figure 46).

The anomalies due to the surface or near-surface Dinantian igneous rocks tend to be small in area and amplitude and are restricted to the area where these rocks have been mapped. However, ground magnetic surveys show that the individual lavas can cause anomalies, often of several hundred nanotesla amplitude, which accurately define the margins of the igneous rocks. All the lavas and intrusive bodies examined appear to be magnetic but the tuff beds are non-magnetic or only weakly magnetised.

In addition to the sharply defined anomalies due to lavas and intrusions at outcrop, several anomalies on the ground magnetic profiles indicate magnetic sources at depths of some tens of metres. These are probably due to faulting or the flow edges of the down-dip extensions of lavas and offer a means of mapping deeper structures. An anomaly [160 720] on the aeromagnetic map occurs in an area devoid of surface igneous rocks and possibly represents the effect of a large concealed igneous mass, perhaps a westward extension of the structure indicated by anomalies 4 and 5 (Figure 46).

An example of the response of a ground magnetic survey to the lavas is shown in Figure 47 for the area around the eastern part of Long Rake. The Conksbury Bridge Lava underlies the area and it is likely that the magnetic anomalies are due to its displacement by Long Rake and Bowers Rake. Extensions of these rakes can be postulated on the evidence of anomaly alignment, including a short eastward extension of Long Rake beneath the alluvium of the River Wye.

Over the lavas there are no anomalies on the aeromagnetic map (Figure 46) with amplitudes greater than a few tens of nanotesla: their absence is consistent with the observed size and magnetisation of the flows. It can be shown for example that the amplitude of the anomaly due to a 70 m flow with susceptibilities similar to those of the Cressbrook Dale Lava (Figure 45) would be only about 18 nanotesla at a height of 300 m, compared with several hundred nanotesla on the ground. Similarly, the lack of larger anomalies suggests that there are no large downward extensions of the dolerite intrusions.

Several of the more pronounced magnetic anomalies in the area occur over Dinantian and Namurian sediments near the eastern margin of the dome. These anomalies (2 to 7 in Figure 46) are characterised by comparatively large amplitudes, ESE to WNW elongations and interpreted depths to the magnetic sources of several hundred metres.

Several features suggest that these anomalies indicate thick accumulations of Dinantian igneous rocks which were probably the sources of the lavas and perhaps the intrusions that are known to be present in the eastern part of the dome. The known distribution of the lavas (Walters and Ineson, 1981) suggests the existence of four separate volcanic centres (p. 59): magnetic anomaly 4 (Figure 46) may correlate with a source inferred near Alport; anomaly 2 may represent the centre from which the volcanic rocks of Longstone Edge, among others, were derived; anomaly 7 may represent the Bonsall centre, the source of the Matlock lavas. The source of the Miller's Dale lavas, inferred near Tunstead (p. 61), is not represented by any pronounced magnetic anomaly, however.

The exceptionally thick (over 293 m) assemblage of volcanic rocks proved in the Fallgate Borehole in the Ashover inlier (Ramsbottom and others, 1962, fig. 4) coincides with a local magnetic high (6 Figure 46). The presence of at least 177 m of basalt in the bottom of the Fallgate

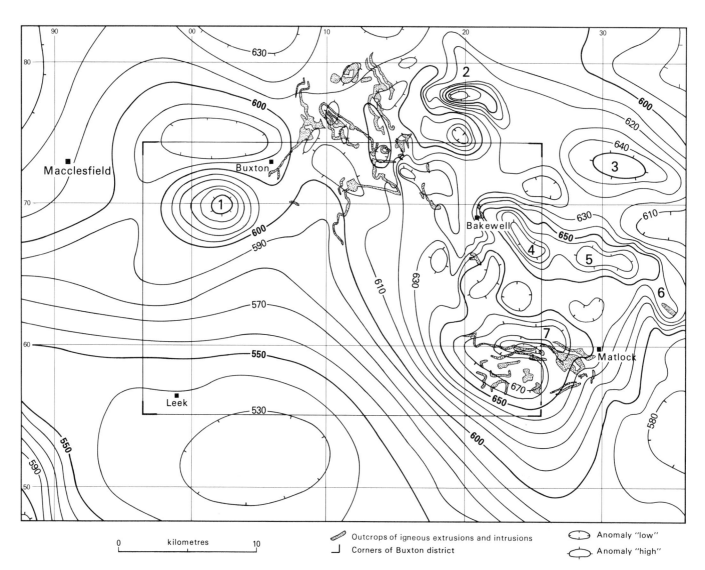

Figure 46 Aeromagnetic map of the Buxton district and surrounding areas with contours at 10 nT intervals.

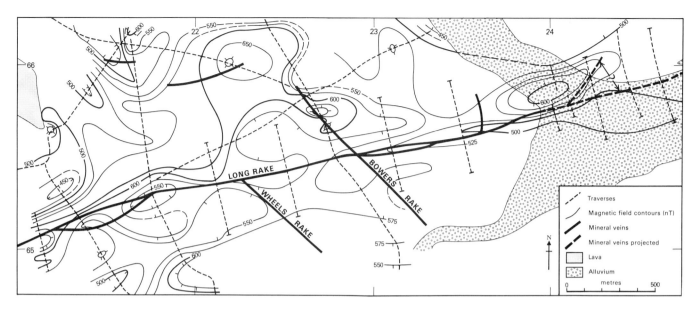

Figure 47   Magnetic map of part of Long Rake based on ground measurements along the traverses shown.

Borehole led these authors to suggest that the flank of an ancient volcano had been intersected. The Ashover magnetic anomaly however is a local feature superimposed on an elongated belt of anomalies (4 and 5 in Figure 46), which extends to the ESE for a further 30 km and suggests the existence of a large, linear, volcanic accumulation, perhaps fissure-controlled. There is geophysical support for the correlation of the magnetic anomalies with Dinantian igneous rocks in the presence of several magnetic horizons in the Fallgate Borehole (they include lavas and 'lava breccias' but not the tuff) and in the existence of ground magnetic anomalies around Ashover.

The interpreted depths to the sources of the magnetic anomalies 2 to 5 are consistent with positions beneath Namurian sediments but near the top of the Dinantian sequence and their amplitudes require a thicker sequence of igneous material than that exposed at the surface. The nature of the ground magnetic profiles suggest a variety of forms for the magnetic bodies.

The anomaly to the north of Eyam (2 Figure 46) occurs over Namurian sediments adjacent to an area of Dinantian sediments with few surface indications of igneous rocks, but underground evidence indicates a considerable development of igneous rock and corresponds with the thickest known development (Stevenson and Gaunt, 1971, pp. 28 and 68) of the Cressbrook Dale Lava (Figure 45i).

The dominant elongation of the magnetic anomalies 2 to 5, which lie just outside the outcrop of the Dinantian limestones, suggests control by ESE to WNW faulting, and faults with this trend occur in places in the Dinantian (Figure 40).

On the western side of the dome an isolated circular anomaly is situated just south-west of Buxton (1 of Figure 46; Figure 47). It is attributed to a magnetic body, at an estimated depth of 1.5 to 2.0 km below OD, suggesting that it lies below the top of the Dinantian or in the pre-Carboniferous basement. The deeper origin of this magnetic body, compared with those to the east of the Derbyshire Dome, is reflected in a comparison of the magnetic anomaly amplitudes on the ground with those recorded in the airborne survey. For anomalies 2, 3 and 4 (Figure 46) the amplitudes recorded at 308 m were about 40–50 per cent of those recorded on the ground but for anomaly 1 there was little difference between the two.

The anomalies due to the deepest magnetic bodies make up a broad band of high values extending across the northern part of the region and down to the south-east. The magnetic zone indicated is extensive, continuing northwards for a further 90 km and south-eastwards for 60 km, but terminating in the west just 10 km beyond the margin of Figure 46. The south-western margin of the zone (Figure 45 ii and 44) does not correspond with any significant geological feature at the surface but all the magnetic bodies at shallower depths overlie the deep-seated magnetic zone. While it seems probable from the great depth and lateral extent of the deep-seated zone that it represents a magnetite-bearing assemblage of Precambrian rocks, there is a possibility that its presence exerted some control over the location of later igneous activity. The absence of any clear correlation between the main Bouguer anomaly features and the SE-trending magnetic anomaly zone suggests that the body responsible for the latter has no density contrasts with the adjacent rocks (*compare* Figure 45 ii).

In the north-western corner of the region the margin of the deep magnetic body swings round to a westerly direction and crosses beneath the Red Rock Fault, apparently without any large change in depth. The magnetic zone in this area coincides approximately with the high-density basement rocks (Figure 42).

The group of contours in the extreme south-western corner of Figure 46 is the end of a pronounced NNE-trending anomaly associated with the Red Rock Fault. It is clear that there is no continuity between the igneous rocks probably responsible for this anomaly and those in the Derbyshire

Dome. For the main part of the south-western area the low magnetic gradients suggest the presence of a considerable thickness of non-magnetic rocks, perhaps Lower Palaeozoic sediments; some ?Devonian sediments, similar to those in the Caldon Low Borehole, may be present near the top.

OTHER SURVEYS

Various geophysical methods have been used in Derbyshire to determine the exact locations of known mineral veins and their extensions. There has been little reported success in detecting the conductive metallic ore minerals, probably because these are simply not present in sufficient quantity to provide targets. Recent mining interest has been in the gangue minerals, notably fluorspar, and various electrical and electromagnetic geophysical techniques, as well as the magnetic method (Figure 47), have proved useful, mainly in locating the fault structures which usually provided the locations for the mineralisation. Seaborne and others (1979) have demonstrated the use of Slingram electromagnetic equipment in detecting electrical resistivity contrasts due to changes in the lithology and/or thickness of bedrock and overburden across faults in Derbyshire. Brown and Ogilvy (1982) describe a feasibility study involving the application of a wide variety of geophysical and geochemical techniques to the location of Long Rake.

Resistivity surveys have been applied successfully to the location of concealed pocket deposits (p. 105) (Dr J. R. Brown, unpublished).   JDC

CHAPTER 14

Quaternary deposits

The distribution of the drift deposits of the district is shown in Figure 48. In general much of the ground is drift-free, though a substantial area of boulder clay masks the solid rocks near Leek and extensive spreads of peat are present on the moors west of Buxton. The paucity of glacial deposits is because much of the higher ground was free of ice during the most recent (Devensian) glaciation. Head deposits of probable periglacial origin occur throughout the district, while recent alluvium and some terrace deposits are present in the main valleys. The most extensive recent deposit is peat.

Jowett and Charlesworth (1929), in a classic work, described many of the glacial features of the district, including the limit of the 'Newer Drift' of the Cheshire plain and the marginal drainage features around Leek. They also noted the relatively scattered occurrences of glacial deposits on the higher ground. Reference to other more recent literature will be found in the apropriate parts of the chapter.

The district shows a sharp contrast between the relatively isolated occurrences of glacial deposits in the limestone area and the more continuous western drift, which forms an extension of that in the Cheshire Plain.

In the limestone area the available information has been summarised by Burek (1977) for the whole of the Derbyshire Dome. The greater part of the boulder clay of the Wye valley has been attributed to the over-riding of the area by western ice during the penultimate (Straw and Lewis, 1962, pp. 78–79) or Wolstonian (Mitchell and others, 1973) glaciation. Mappable deposits representing younger Pleistocene stages are not present in this area though evidence for these stages occurs in certain caves. The Ipswichian interglacial is represented by the mammalian warm fauna from Hoe Grange (p. 136), while a cold fauna of proved late-Devensian age occurs at Ossum's Cave in the Manifold valley; a Devensian age is also attributed to the mammalian fauna from Fox Hole Cave, Earl Sterndale.

The glacial deposits of the western part of the district show, in places, fresh constructional forms, are associated with meltwater channels, and form a continuation of similar deposits referred to the Weichselian in the Cheshire area (Evans and others, 1968, pp. 251–253; Worsley, 1970, pp. 102–104); these are now called Devensian (Mitchell and others, 1973). They were the products of a large ice sheet that extended southwards over the west side of Britain during the Late Devensian (Mitchell and others, 1973) and abutted against the western slopes of the Pennines. The limits of this ice at its maximum extent (Figure 49) largely agree with the 'limit of extraneous drift' represented by erratics of Lake District or south-west Scotland origin shown by Jowett and Charlesworth (1929, plate XIX). These occur on the slopes to the west of the main watershed between Shining Tor, Axe Edge, and Morridge at heights up to 427 m. Sparsely distributed erratics elsewhere in the district are thought to be relics of one or more earlier glaciations (Burek, 1977).

The glacial sands and gravels (Figure 49) are of varying origin, but appear to be associated with the retreat of the Devensian ice and, therefore, the extreme ice limit probably lay to the east of these deposits. Only the outcrops at Bottom of the Oven and around Hilly Lees show a delta form (p. 136) and, as Johnson (1965b, p. 98) pointed out, there is little evidence for the existence of most of the extraglacial lakes postulated by Jowett and Charlesworth (1929). However, some of the glacial drainage channels, mapped by these authors and by Cazalet (1969) to the north and south of Thorncliff and to the east of Leek, are confirmed. Probably these were mostly eroded sub-glacially, but the presence of spreads of alluvium (too small to show on Figure 49) and sand and gravel with terrace-like form suggest that the channels contained small ephemeral lakes during the final ice retreat phase. Two channels to the north and south-west of Shutlingsloe may be of sub-glacial origin, the peak of Shutlingsloe standing out at this time as a nunatak, as suggested by Jowett and Charlesworth. NA, IPS

BOULDER CLAY

Boulder clay (till) is most extensive in the south-west and west of the district, the deposits being a continuation of those described in the adjacent Macclesfield district (Evans and others, 1968). This material is in most cases a stiff brown clay with erratics of north-western origin (Stevenson and Gaunt, 1971, p. 331) and a proportion of locally-derived material. Large erratics occur in places. These deposits are mainly preserved on the lower ground but also extend up the western hill slopes to heights of up to 380 m.

DETAILS

WILDBOARCLOUGH AND MACCLESFIELD FOREST

South of Wildboarclough, boulder clay is present on the slopes of Clough Brook valley up to a height of 380 m; these deposits extend westwards into the Macclesfield district. On High Moor [968 702] an area of boulder clay lies at a height of some 400 m, while at Macclesfield Forest an unusually high patch occurs at 457 m near Whitehills.

Isolated occurrences of erratics are known outside the main areas of glacial deposits. In one [983 670], near Allgreave, boulders of Borrowdale Volcanic type occur at a height of about 380 m. IPS

LEEK

Boulder clay forms an extensive fill in the Meerbrook valley and is also present in more restricted areas south and west of Leek.

Around Meerbrook the deposit forms a featureless spread in the valley bottom and gently sloping tracts on the valley sides. Good sections are rare, the best being in a stream bank [9859 6045] near Alder Leigh, where 3.66 m are seen. Boulder clay is also present on

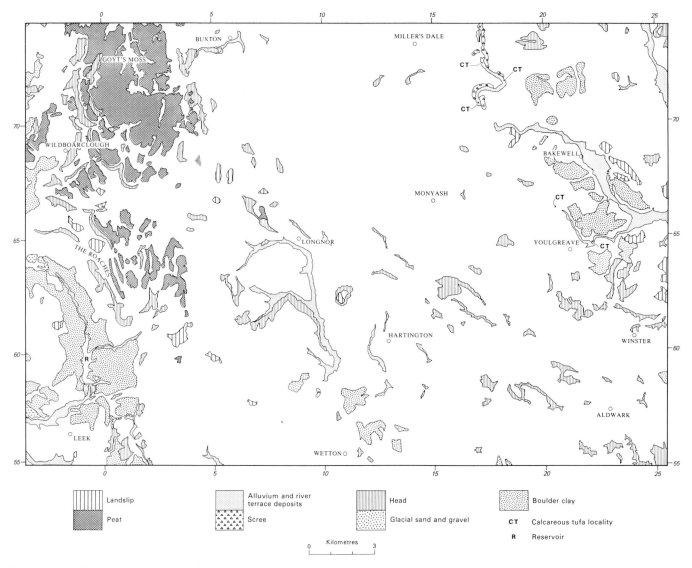

Figure 48 Sketch-map showing Quaternary deposits.

the west bank of Tittesworth Reservoir, and in one place [9903 5959] shows the effects of cryoturbation in the upper part of a section 2.13 m thick (*see also* p. 136). To the west of the reservoir, boreholes have proved the presence of thick drift deposits two examples being:

SJ 95 NE/15b, boulder clay to 15.09 m, sand and gravel to 19.20 m, sand with some silty clay bands to 24.54 m, silt and silty clay to 30.33 m; solid rocks were not reached, rock head being thus below 169.03 m O.D.

SJ 95 NE/15i, boulder clay (coarse below 6.40) to 10.36 m, sand and silt to 15.54 m, boulder clay to 21.34 m, silt to 21.64 m, mudstone to 24.69 m; height of rock head, 197.71 mO.D.

These boreholes prove the presence of a sub-drift valley substantially deeper at its north-eastern end than the present valley between Fould and Tittesworth Reservoir. Barke (1929) suggested that the present valley may have been a former course of the River Churnet, a suggestion disputed by Johnson (1965, p. 102). The new evidence neither supports nor denies Barke's suggestion but, contrary to Johnson's view, there is 'structural advantage' in the present course of the Churnet as it follows the strike of the Namurian mudstones around the southern culmination of a south-easterly plunging anticline, and this favours Barke's hypothesis. NA

WYE VALLEY

A terrace-like area of boulder clay has been recognised at Cressbrook House [170 730], and boulders of limestone (one striated) and some igneous rocks occur at the surface nearby [1698 7298]. The deposit occurs at a height between 220 and 243 m, the valley floor being at about 173 m and the general plateau level around 290 m. There is a similar, but smaller occurrence at Upperdale, and here [1772 7223] a pile of boulders of up to 0.2 m diameter includes pale and dark limestone and some fine-grained sandstone.

LITTLE LONGSTONE AND HASSOP

An extensive, though dissected, area of boulder clay lies in the valley south of Longstone Edge. Its presence is shown by the occurrence of erratics at the surface, for example in Hassop Park [217 714]. Sections are few, though one [2020 7165], near Great

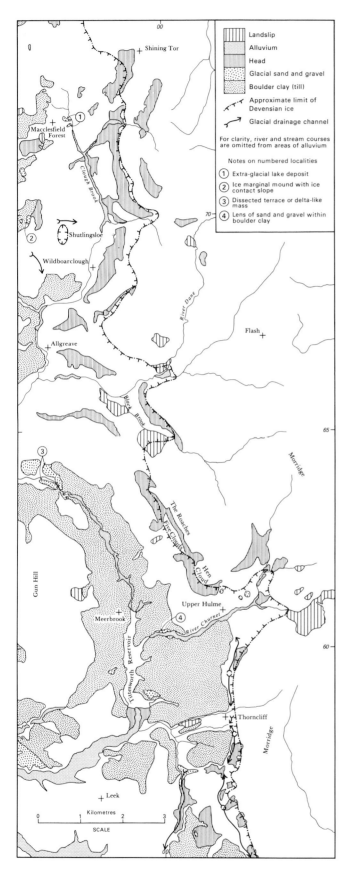

Figure 49   Sketch map showing drift deposits in the western part of the district (peat excluded).

Longstone, shows 0.91 m of boulder clay on mudstone. The Buskey Cottage Borehole [SK 27 SW/26] proved 17.23 m of stiff brown clay, with limestone boulders up to 0.17 m across and some chert, resting on Longstone Mudstones. Temporary sections in the deposit have been noted by Straw and Lewis (1962, pp. 73–75) at Great Longstone [200 715] and Rowland [211 720, 215 726].

BAKEWELL AND STANTON

Extensive spreads of boulder clay occur on both sides of the Wye valley between Bakewell and Rowsley and on the south side of the Lathkill valley.

A section at Bakewell Hospital [2124 6769] proved 1.1 m of brown clay with sandstone, chert and limestone pebbles and cobbles up to 0.15 m in diameter. Farther south the deposit is coarser and limestone boulders are abundant. Boreholes at Haddon Fields have passed through substantial thicknesses of boulder clay, an example being SK 26 NW/16, which proved 8.96 m of it.

The best exposure is in Shining Bank Quarry [SK 26 NW/35] where 9 m of boulder clay contain limestone boulders, up to 0.9 m diameter, and smaller sandstone boulders. A former section on the south side of an open working at Raper Mine [2175 6524] showed 2.3 m of boulder clay with limestone boulders, some striated, up to 1.0 m. On the north side of the working, Burek (1977, pp. 123–130) has recorded a 1.3 by 2 m inclined block of sand and gravel within the boulder clay; the block appears to have been transported while frozen.   IPS

An isolated occurrence of boulder clay [1832 6666] was recorded in the side of an open working at Mandale Rake, to the west of Over Haddon. The section was: brown stratified loam (head) 3 m, on grey clay with rounded limestone boulders 2 m, on rotted vein material.

On the south side of the Lathkill valley, areas of boulder clay occur near Bradford and below Stanton; in the latter area a borehole [SK 26 SW/47] proved 20 m of the deposit on 3 m of sand before entering solid rock.

HARTINGTON TO LONGCLIFFE

Small occurrences of boulder clay have been recorded overlying some of the Pocket Deposits (p. 107–108) in this area. Jackson and Charlesworth (1920, p. 488) recorded boulders of Eskdale granite and Borrowdale volcanic rocks in the drift above Pocket Deposits near Parsley Hay [147 636], and erratics of more local origin at Washmere [168 602], Minninglow [202 576] and Longcliffe [228 557]. Yorke (1961) recorded striated boulders of limestone in boulder clay at Kenslow Top [182 616] and at Heathcote [154 610]. An occurrence at the last locality was noted during the present survey (p. 107).   JIC

ALSTONEFIELD AND WARSLOW

Boulder clay is present in a number of patches between these places. It is, in general, little exposed, though clay with igneous erratics of north-western type is seen in a sink-hole [1168 5630] near Under Wetton.   NA

SAND AND GRAVEL

Deposits of glacial sand and gravel are restricted to the western part of the district. They are much less extensive than the boulder clay, though they extend vertically through the same height range.

The deposits are normally clean sands with pebbles and cobbles of north-western origin; in places they show constructural form.

DETAILS

WILDBOARCLOUGH AND MACCLESFIELD FOREST

Sand and gravel forms a delta-like mass in the valley of Clough Brook at Bottom of the Oven [978 721] at a height of about 380 m (Figure 49, note 1). The maximum thickness of the deposit, which is ill-exposed, is about 10 m. It is interpreted as having been formed by sediment debouching from a drainage channel (since largely modified by erosion) into standing water ponded in the Clough Brook valley (*see also* Johnson, 1965a, p. 78). The ponding probably occurred near Wildboarclough as boulder clay here reaches a height of 380 m.

At Macclesfield Forest [968 710] an ill-exposed patch of sand and gravel occurs at a height of 290–300 m. A little to the south at the Shooting Box [967 698] two patches of sand and gravel occur at about 373 m. The eastern and larger of these shows a marked feature, interpreted as an ice-contact slope, on its western side (Figure 49, note 2). These deposits form the north-eastern end of a discontinuous belt of sand and gravel extending into the Macclesfield district.

Smaller patches of sand and gravel also occur south of the Shooting Box at 358 m [966 694] and near Lower Nabbs at 275 m [972 680].

IPS

ALLGREAVE AND LEEK

Sand and gravel forms a dissected delta-like mass around Hillylees [968 642] (Figure 49, note 3). The deposit, up to about 30 m thick, lies between boulder clay, on the east and south, and solid rock.

Small areas of sand and gravel are present on both sides of Tittesworth Reservoir. One of these [999 604] is probably a lens within boulder clay (Figure 49, note 4), though the others rest on top of it. Small patches also occur [019 612] in the Churnet valley a little beyond the limit of the boulder clay.

At Foker Grange, an area [967 575] of sand and gravel extends westwards into the Macclesfield district near Rudyard and lies on the margin of a glacial drainage channel (Evans and others, 1968, pp. 190–191).

NA

BONE CAVES

Bone Caves are scattered over much of the limestone outcrop of the district. Their detailed description falls outside the scope of this memoir and the subject is summarised by Bramwell (1973; 1977).

The earliest mammalian fauna known was described by Arnold-Bemrose and Newton (1905, pp. 43–63) from a cavern [2219 5602] revealed by quarrying at Hoe Grange, near Longcliffe. The fauna included hyena (*Crocuta crocuta*), cave lion (*Felis leo*), brown bear (*Ursus aretos*), straight-tusked elephant (*Palaeotoxodon antiquus*), steppe rhinoceros (*Dicerorhinus hemitoechus*) and several species of deer, including *Megaloceros giganteus*. These forms are indicative of a warm climate and have been referred by Bramwell (1977, pp. 277–278, 284–289) to the Ipswichian interglacial; they are the oldest fauna in the district. An Ipswichian fauna is also known from Elder Bush Cave, south of the present district. Fox Hole Cave, Earl Sterndale, yielded a Devensian fauna with lion and cave bear (*Ursus spelaeus*): faunas of a similar age are known at several places in the Peak District (Bramwell, 1977, p. 278).

Post-Pleistocene faunas occur at a number of places.

Ossum's Cave [096 557], in the Manifold valley, has yielded a fauna including Przewalski's horse (*Equus przewalskii*), reindeer (*Rangifer tarandus*), red deer (*Cervus elephus* and steppe bison (*Bison priscus*). A radiocarbon dating on reindeer bone gave an age of 8460 ± 70 BC. Wetton Mill Minor has yielded Mesolithic artifacts, dated at 6897 ± 210 BC: at Dowel Cave Neolithic artifacts have been found. For a fuller description of these caves, see Bramwell (1973, 1977).

IPS

HEAD

Head deposits are generally accepted as mainly of periglacial origin. This is largely confirmed by their more extensive development in the present district in those areas which were free of the most recent (Devensian) ice. They are divisible into scarp-slope deposits, valley-fill deposits and deposits on the limestone plateau (*see also* Stevenson and Gaunt, 1971, p. 337).

Scarp-slope head accumulated beneath the steeper exposures of Namurian and Westphalian sandstones; it comprises locally-derived angular detritus up to boulder size in a sandy clay matrix, the coarsest material tending to lie closest to source. Downslope, this passes into finer valley-fill head in which the angularity of the clasts is more evident than in the coarser deposits. Glacial material is locally incorporated in both of these types though it is more common in valley-fill head. Both varieties are extensive on the Namurian outcrop, though they are only locally thick enough to show on the map. On the limestone outcrop the head consists of brown silty loam with varying amounts of chert; in places [195 736] the deposit is a cherty rubble derived from underlying cherty limestones. Pigott (1962, p. 152) found evidence that the finer deposits were largely loessic in origin though some limestone solution residue was also present. More recently both Cazalet (1969, p. 79) and Bryan (1970, p. 269) confirmed the aeolian origin of much of the silt fraction in material here described as head, though they placed greater emphasis than Pigott on the residual component. The coarser, cherty deposits are at least partly local in origin, and have been compared (Stevenson and Gaunt, 1971, p. 334) to the clay-with-flints of the Chalk outcrop.

Head on the Namurian–Westphalian outcrop has been derived mainly from the effects of periglacial weathering on the exposed strata, and the subsequent solifluction of the detritus. Some incorporation of glacial material occurred in places, particularly in the western part of the district. A good example was noted by E. A. Francis (personal communication) on the east shore of Tittesworth Reservoir. Here [9903 5959] about 1 to 2 m of head, containing abundant angular fragments of local sandstone and mudstone and scattered north-western erratics, overlies 0.3 to 1.4 m of purplish-brown till with abundant well-rounded erratics, mainly sandstone. There is some mixing of the two lithologies by solifluction near their interface, which is highly contorted by cryoturbation. The section shows that head formation here post-dated the till, but the two deposits were strongly affected by periglacial or post-glacial processes. On the limestone plateau, the head deposits are less certainly due to solifluction, though this process may have contributed to their formation in places; here the head comprises both the products of residual weathering and aeolian loess, the latter a product of periglacial conditions (Pigott, 1962, p. 154).

IPS, NA

PEAT

Peat is extensive on the high ground of Goyt's Moss and Axe Edge Moor, and in lesser areas in and around the southern part of the Goyt Syncline. Its maximum observed thickness is 4 m [0289 6935] near Axe Edge. Peat is also present at a lower level in parts of the upper valleys of the Dove and Manifold, the largest patch [073 659] being near Moss Carr.

At Goyt's Moss ([003 714], Tallis (1964, pp. 324–331) recognised five horizons within the upland peats. The earliest of these horizons (some 22 cm from the base of the deposit) was allotted an age of 1000–1200 BC, using pollen analysis, while the latest was considered to have been formed about 900–1000 AD. These results conflict with those obtained by Conway (1954, p. 145), in the Kinder Scout and Bleaklow area, where she concluded that peat formation commenced around 6000 BC.

NA, IPS

CALCAREOUS TUFA

The deposition of calcium carbonate from springs has produced deposits of calcareous tufa at a few places in the district. The location of the springs in relation to the topography has determined the form of the tufa deposits.

In Monsal Dale, between Cressbrook Mill and Lees Bottom, calcareous tufa makes up several low terrace-like features at the edge of the flood-plain. Exposures are lacking, though debris can be seen in places.

In Lathkill Dale, calcareous tufa occurs near Lathkill Lodge where it is at least 1 m thick [203 661]. The deposit owes its presence to springs issuing from the valley floor. Farther downstream, at Alport [221 646], calcareous tufa forms a sloping area on the valley side. The maximum thickness seen is some 9 m and the deposit is due to a spring some distance up the valley side.

IPS

RIVER TERRACES

River terrace deposits are significant only around Bakewell in the Wye valley. Here they lie at 1 to 2 m above the level of the flood-plain; sections are in general lacking, though one near Rowsley [2488 6557] shows 0.75 m of rounded limestone and sandstone cobbles up to 20 cm across. This composition indicates a derivation from the local boulder clay.

Small and isolated areas of terrace deposits are also present in the Manifold valley near Hulme End [101 592] and The Holmes [094 619].

The isolated nature of these occurrences and their positions only in the upper reaches of the valleys makes correlation between them impracticable.

ALLUVIUM

Alluvium is generally present in the valleys of the more important rivers and streams. It is usually silt or silty clay, commonly with gravel beneath. The deposit is thickest and most extensive in the broader valleys, such as those of the Wye below Ashford and the Manifold above Hulme End. Where the rivers flow through limestone gorges, as in the upper reaches of the Wye, the deposits are thin and discontinuous.

SCREE

Scree deposits are restricted to some of the deeper valleys on the limestone outcrop, particularly in Cressbrook Dale and Monsal Dale. Prentice and Morris have described (1959, pp. 16–19) the occurrence of cemented screes in the Manifold valley. At Ecton [097 581] these deposits reach 24 m in thickness, and smaller patches are present elsewhere [such as 102 654]. Cementation of screes is not normal under present-day conditions, and it was attributed by the above authors to the effects of freeze–thaw conditions with abundant groundwater presumably in a periglacial environment. Burek, however (1977, p. 95), postulated that cementation occurred during a climatic amelioration following perglacial conditions in the late Devensian.

IPS, NA, JIC

LANDSLIPS

On the Namurian–Westphalian outcrop in the western area, landslipping has occurred at many horizons. Most of the slips are of rotational type, involving either sandstones overlying mudstones or mainly mudstone sequences. Examples of the former occur in the Dane valley near Burntcliff Top [999 671], at Fough [058 677], and on Sheen Hill [108 622]; the latter type occur at the 'Mermaid Pool' northwest of Merryton Low [039 612], near Sprink [123 615], and near Moseley [051 666]. At the last-named locality, cracks in the turf and crescentic depressions at the back of the slip showed that it was still active at the time of survey (1967). Landslips due to bedding-plane slip are common both within major sandstone units and in mudstone-with-sandstone sequences. A spectacular example in sandstone occurs at 'Luds Church' [987 655], where a chasm 2 to 3 m wide and 9 to 14 m deep (Plate 12; Hull and Green, 1866) has formed by down-slope movement, probably along an argillaceous bed within the Roaches Grit. The best example affecting a mudstone-with-sandstone sequence occurs [033 606] southwest of Merryton Low in the $E_{1c}-E_{2a}$ sequence.

In the eastern outcrop, rotational landslipping is common in the mudstones underlying the steep scarps of the Ashover Grit; locally the escarpment has collapsed, as near Stanton Lees [254 631], where masses of sandstone large enough to have been quarried have slipped down to levels between 45 and 75 m below their original position.

On the Dinantian limestone outcrop, landslips are not widespread. However, rotational slips occur due to movement on softer-weathered igneous rocks, for example on the

Plate 12 Lud's Church near Gradbach; a chasm formed at the back of a landslip in Roaches Grit.

The sandstone face on the right, formed along a major joint plane, has moved about 3.5 m from the face on the left (L 1253).

Lower Miller's Dale Lava [129 734, 135 740] and Ravensdale Tuff [171 738]. The Hob's House landslip [175 713] involves the Shacklow Wood Lava.

An uneven area overlying the dark facies of the Monsal Dale Limestones on the side of the Wye valley [187 694], south-west of Ashford, is interpreted as a landslip formed by bedding-plane slip.

Few of the landslips show major activity at the present time though there are instances of minor movement; an example can be seen in the toe of a landslip [2370 6765] at Bakewell. Recent work by Johnson and Walthall (1979, pp. 150–152) on the landslips of the Longdendale area, near Glossop, indicates that these date from early post-glacial to 2000 B.P.; in the present district a similar date may apply.

NA, IPS, JIC

CAMBERING

Two instances of cambering have been noted. The first lies on the escarpment of the Sheen Sandstones near Harris Close [120 622], where an outlier of sandstone is inferred to dip gently towards the Dove valley, contrary to the regional dip. The second occurs where Monsal Dale Limestones overlie the Lower Matlock Lava at Aldwark [228 573]. Here, blocks of limestone have foundered into the clay and now lie at angles of up to 65°.

NA, JIC

CHAPTER 15

Mineral products

LIMESTONE

The quarrying of limestone is an important industry in the district. The main workings are concentrated in the Buxton and Wye valley area; isolated working quarries are also present at Grangemill, Ballidonmoor and Alport. They are mostly situated outside the Peak District National Park, the boundary having been drawn in 1951 to exclude them. Since then, little quarrying has been allowed in the Park showing that amenity is almost as important as geology as a factor in quarry location, although the large Tunstead Quarry of Imperial Chemical Industries is now being extended into the Park at Old Moor [108 737].

The main formation worked is the Bee Low Limestones, on account of its very high purity (mainly 98.5% $CaCo_3$; see for instance, Harrison, 1981, p. 19). However, the Woo Dale Limestones are also worked at Topley Pike Quarry (Tarmac Roadstone) and at Tunstead Quarry. Of the seven major quarries in the Bee Low Limestones south of Buxton, four are currently in production: Buxton Quarry [082 691] (Tarmac Roadstone); Brier Low [089 687] (Peakstone); Hindlow [095 677] (ICI); and Dow Low [099 674] (Steetley Minerals). Three quarries near Grangemill and one at Ballidonmoor also work the Bee Low Limestones: Longcliffe Quarry [239 570] worked by Longcliffe Quarries; Grangemill Quarry [241 573] (Ben Bennett Junior); Prospect Quarry [245 573] (Tarmac Roadstone); and Ballidon Quarry [202 554] (Tilcon). Ivonbrook Quarry [233 584] near Grangemill (Ogden Group) and Shining Bank Quarry [229 650] near Alport (North Lonsdale Tarmacadam) are in the Monsal Dale Limestones; the overlying Eyam Limestones are also worked in the latter quarry.

Uses for limestones from the district include chemical, iron and steel, and agricultural lime manufacture, in addition to roadstone and aggregate production (for fuller information see Harris, 1982). Detailed accounts of the limestone resources have been published by the British Geological Survey (Cox and Bridge, 1977; Cox and Harrison, 1980; Bridge and Gozzard, 1981; Harrison, 1981; Gatliff, 1982; Bridge and Kneebone, 1983) and the data in all of these has recently been summarised by Harrison and Adlam (1985) that will complete the assessment of the district.

Some limestones have been worked for ornamental purposes. Examples include the Ashford Black Marble and the Rosewood Marble from the Monsal Dale Limestones in the Ashford area (Ford, 1964) and polished crinoidal limestones particularly from the flank facies of knoll-reefs in the Eyam Limestones around Monyash.

SANDSTONE

Many of the Namurian sandstones have been worked in places for building stone, the most important examples being the Rough Rock at Danebower and Reeve Edge quarries in the west. Sandstone has been mined on a small scale near Longnor (p. 84). In the east, the Ashover Grit has been worked extensively for freestone and millstones at Stanton up to the present time.

SILICA SAND AND SILICA ROCK

The Pocket Deposits (p. 105) yield a silica sand used in the production of siliceous refractories. The most extensive workings are in the Friden area.

The protoquartzitic sandstones ('crowstones') of the Namurian have in the past been worked in a number of places as a source of refractories. None of the workings was extensive, however. These rocks also have the properties of a good road aggregate, but the common presence of thinly interbedded mudstone or siltstone reduces their potential for this use.

CHERT

This mineral was formerly worked extensively in the Bakewell area. The deposits are stratiform in appearance, being due to the replacement (p. 34) of preferred horizons high in the Monsal Dale Limestones. Working was by adit, and the chert was used in the manufacture of pottery at Stoke on Trent (Brown, 1976).

COAL

Although coal is no longer worked in the district, it was formerly got in a substantial area in the Goyt Syncline between Axe Edge Moor and Goyt's Moss. The workings were in both the Ringinglow and Yard coals. There was also minor working of these and higher seams (p. 95) around Goldsitch Moss.

FLUORSPAR

This term in here used for fluorite as a commercial product. A high proportion of the production of this mineral in the South Pennine Orefield comes from the Buxton district. The principal current sources (1982) are underground at Sallet Hole Mine (Laporte Industries) in the Longstone Edge area, and opencast near Youlgreave (Dresser Minerals). Detailed description of earlier working has been given by Dunham (1952b) while exploitation and uses of the mineral have been summarised by Notholt (1971). Ford and Ineson (1971) have discussed the fluorspar mining potential of the orefield as a whole.

The two above-named companies use flotation to produce acid-grade fluorspar for the chemical industry, and also supply metallurgical-grade fluorspar to the steel industry. The flotation process also produces useful by-products (*see below*).

BARYTES

This term is here used for baryte as a commercial product. In the past, production of this mineral has been closely associated with that of lead, though in places it has been worked separately. In recent years, Bleaklow Mining Co. has worked a group of veins on the south side of the Longstone Edge Anticline. Beneficiation is by washing. The mineral is also produced as a by-product of fluorite flotation.

In addition to such long established uses as the manufacture of paint, and as a filler in the production of paper (Stevenson and Gaunt, 1971, p. 356), barytes is now in demand as a component of heavy muds used in drilling for oil and gas (Collins, 1972).

CALCITE

The Long Rake Spar Mine, Youlgreave, recently ceased production of calcite, and now processes the mineral from other local sources. Calcite has also been mined in Horseshoe Dale, Chelmorton. The chief use is for decorative concrete aggregates.

LEAD

Working of lead in the South Pennine Orefield dates from Roman times and continued until the closure of Millclose Mine in 1940 (Varvill, 1959, p. 189). Since the latter date, mining interest has turned mainly to the production of fluorspar, though lead-zinc concentrates are produced as a by-product of the flotation process.

The lead-mining and mines of the South Pennine Orefield have been well summarised by Ford and Rieuwerts (1983). The mineralisation is described separately in Chapter 12. IPS

COPPER

Copper was formerly worked in the Ecton and Mixon areas (p. 121); in the late 18th century, the mine at Ecton was said to be amongst the most productive in Britain (Robey and Porter, 1972).

REFERENCES

ADAMS, A. E. 1980. Calcrete profiles in the Eyam Limestone (Carboniferous) of Derbyshire: petrology and regional significance. *Sedimentology*, Vol. 27, No. 6, 651–660.

— and COSSEY, P. J. 1978. Geological history and significance of a laminated and slumped unit in the Carboniferous Limestone of the Monsal Dale region, Derbyshire. *Geol. J.*, Vol. 13, Pt. 1, 47–60.

AITKENHEAD, N. 1977. The Institute of Geological Sciences Borehole at Duffield, Derbyshire. *Bull. Geol. Surv. G.B.*, No. 59, 1–38.

— and CHISHOLM, J. I. 1979. Excursion report: Lathkill Dale and Pilsbury, Derbyshire. *Mercian Geol.*, Vol. 7, No. 1, 75–77.

— — 1982. A standard nomenclature for the Dinantian formations of the Peak District of Derbyshire and Staffordshire. *Rep. Inst. Geol. Sci.*, No. 82/8.

— and HOLDSWORTH, B. K. 1974. Report on field meeting to North Staffordshire–Derbyshire Borders. *Proc. Yorkshire Geol. Soc.*, Vol. 40, Pt. 1, 58–62.

ALLEN, P. M. 1980. Copper mineralisation in Great Britain. Pp. 266–276 *in* JANKOVIC, S. and SILLITOE, R. H. (Eds.) *Spec. Publ.* UNESCO–IGCP Proj. 169 and 63, No. 1. (Belgrade: Society for Geology Applied to Mineral Deposits; *and* [Djusina 7, 1100 Belgrade]: Dept. of Economic Geology, Faculty of Mining and Geology.)

ANDERTON, R., BRIDGES, P. H., LEEDER, M. R. and SELLWOOD, B. W. 1979. *A dynamic stratigraphy of the British Isles.* (London: Allen and Unwin.)

ARNOLD-BEMROSE, H. H. 1894. On the microscopical structure of the Carboniferous dolerites and tuffs of Derbyshire. *Q.J. Geol. Soc. London*, Vol. 50, Pt. 4, 603–644.

— 1899. On a sill and faulted inlier in Tideswell Dale (Derbyshire). *Q.J. Geol. Soc. London*, Vol. 55, Pt. 2, 239–250.

— 1907. The toadstones of Derbyshire: their field relations and petrography. *Q.J. Geol. Soc. London*, Vol. 63, Pt. 3, 241–281.

— and NEWTON, E. T. 1905. On an ossiferous cavern of Pleistocene age at Hoe-Grange Quarry, Longcliff, near Brassington (Derbyshire). *Q.J. Geol. Soc. London*, Vol. 61, Pt. 1, 43–63.

ASHTON, C. A. 1974. Palaeontology, stratigraphy and sedimentology of the Kinderscoutian and lower Marsdenian (Namurian) of North Staffordshire and adjacent areas. Unpublished PhD thesis, University of Keele.

BAMFORD, D., NUNN, K., PRODEHL, C. and JACOB, B. 1978. LISPB – IV Crustal structure of northern Britain. *Geophys. J. R. Astr. Soc.*, Vol. 54, 43–60.

BARKE, F. 1929. The old course of the River Churnet. *Trans. North Staffordshire Field Club*, Vol. 63, 90–97.

BATHURST, R. G. C. 1959. The cavernous structure of some Mississippian *Stromatactis* reefs in Lancashire, England. *J. Geol.*, Vol. 67, No. 5, 506–521.

— 1982. Genesis of stromatactis cavities between submarine crust in Palaeozoic carbonate mud buildups. *J. Geol. Soc. London*, Vol. 139, Pt. 2, 165–181.

BIRD, R. H. 1970. Notes on the western Elton Mining Field. *Bull. Peak Dist. Mines Hist. Soc.*, Vol. 4, Pt. 4, 306–310.

BISAT, W. S. 1957. Upper Viséan goniatites from the Manifold valley, North Staffordshire. *Palaeontology*, Vol. 1, 16–21.

— and HUDSON, R. G. S. 1943. The Lower *Reticuloceras* (R_1) goniatite succession in the Namurian of the North of England. *Proc. Yorkshire Geol. Soc.*, Vol. 24, Pt. 6, 383–438.

BOLTON, T. 1978. The palaeontology, sedimentology and stratigraphy of the upper Arnsbergian, Chokerian and Alportian of the North Staffordshire Basin. Unpublished PhD thesis, University of Keele.

BOSWELL, P. H. G. 1918. A memoir of the British resources of refractory sands, Pt. 1. (London: Taylor and Frances.)

BOULTON, M. C. 1971. A palynological study of two of the Neogene plant beds in Derbyshire. *Bull. Br. Mus. (Nat. Hist.) Geol.*, Vol. 19, No. 7, 359–410.

— and CHALONER, W. G. 1970. Neogene fossil plants from Derbyshire (England). *Rev. Palaeobot. Palynology.*, Vol. 10, 61–78.

— FORD, T. D., IJTABA, M. and WALSH, P. T. 1971. Brassington Formation, a newly recognised Tertiary formation in the southern Pennines. *Nature, London*, Vol. 231, 134–136.

BOUMA, A. H. 1962. *Sedimentology of some flysch deposits: a graphic approach to facies interpretation.* (Amsterdam: Elsevier.)

BRAMWELL, D. 1973. *Archaeology in the Peak District.* (Buxton: Moorland Publishing.)

— 1977. Archaeology and palaeontology. Pp. 263–291 *in* FORD, T. D. (Ed.). *Limestones and caves of the Peak District.* (Norwich: Geo Abstracts.)

BRIDGE, D. McC. and GOZZARD, J. R. 1981. The limestone and dolomite resources of the country around Bakewell, Derbyshire. Description of 1:25 000 sheet SK 26 and part of SK 27. *Miner. Assess. Rep. Inst. Geol. Sci.*, No. 79.

— and KNEEBONE, D. S. 1983. The limestone and dolomite resources of the country north and west of Ashbourne, Derbyshire. Description of 1:25 000 sheet SK 15 and parts of SK 04, 05 and 14. *Miner. Assess. Rep. Inst. Geol. Sci.*, No. 129.

BROADHURST, F. M. and SIMPSON, I. M. 1973. Bathymetry on a Carboniferous reef. *Lethaia*, Vol. 6, 367–381.

BROWN, I. J. 1976. The Pretoria and Greenfield Chert Mine, Bakewell, Derbyshire. *Bull. Peak Dist. Mines Hist. Soc.*, Vol. 6, No. 3, 169–172.

BROWN, M. J. and OGILVY, R. D. 1982. Geophysical and geochemical investigations over the Long Rake, Haddon Fields, Derbyshire. *Miner. Reconnaissance Programme Rep. Inst. Geol. Sci.*, No. 56.

BRUNTON, C. H. C. and CHAMPION, C. 1974. A Lower Carboniferous brachiopod fauna from the Manifold Valley, Staffordshire. *Palaeontology*, Vol. 17, Pt. 4, 811–840.

BRYAN, R. B. 1970. Parent materials and texture of Peak District soils. *Z. Geomorph.*, Vol. 14, 262–274.

BUREK, C. 1977. The Pleistocene ice age and after. Pp. 87–128 *in* FORD, T. D. (Ed.). *Limestones and caves of the Peak District.* (Norwich: Geo Abstracts.)

BUTCHER, N. J. D. 1971. Some recent surface and under-ground observations at Magpie Mine. *Bull. Peak Dist. Mines Hist. Soc.*, Vol. 4, Pt. 6, 403–412.

— 1975. The geology of Magpie sough and mine. *Bull. Peak Dist. Mines Hist. Soc.*, Vol. 6, No. 2, 65–70.

— and FORD, T. D. 1973. The Carboniferous Limestone of Monsal Dale, Derbyshire. *Mercian Geol.*, Vol. 4, No. 3, 179–196.

CARRUTHERS, R. G. and STRAHAN, A. 1923. Special reports on the mineral resources of Great Britain. Vol. 26. Lead and zinc ores of Durham, Yorkshire and Derbyshire. *Mem. Geol. Surv. G.B.*

CAZALET, P. C. D. 1969. Correlation of Cheshire Plain and Derbyshire Dome glacial deposits. *Mercian Geol.*, Vol. 3, No. 1, 71–84.

CHALLINOR, J. 1921. Notes on the geology of the Roches district. *Trans. North Staffordshire Field Club*, Vol. 55, 76–87.

— 1928. Notes on the geology of the Mixon district. *Trans. North Staffordshire Field Club*, Vol. 62, 96–109.

— 1929. Notes on the geology of the North Staffordshire moorlands. *Trans. North Staffordshire Field Club*, Vol. 63, 111–113.

CHARSLEY, T. J. 1982. A standard nomenclature for the Triassic formations of the Ashbourne district. *Rep. Inst. Geol. Sci.*, No. 81/14.

CHISHOLM, J. I. 1977. Growth faulting and sandstone deposition in the Namurian of the Stanton Syncline, Derbyshire. *Proc. Yorkshire Geol. Soc.*, Vol. 41, Pt. 3, 305–323.

— and BUTCHER, N. J. D. 1981. A borehole proving dolomite beneath the Dinantian limestones near Matlock, Derbyshire. *Mercian Geol.*, Vol. 8, No. 3, 225–228.

— MITCHELL, M., STRANK, A. R. E. and HARRISON, D. J. 1983. A revision of the stratigraphy of the Asbian and Brigantian limestones of the area west of Matlock, Derbyshire. *Rep. Inst. Geol. Sci.*, No. 83/10, 17–24.

COCKS, L. R. M. and FORTEY, R. A. 1982. Faunal evidence for oceanic separations in the Palaeozoic of Britain. *J. Geol. Soc. London*, Vol. 139, Pt. 4, 465–478.

COFFEY, R. 1969. The geology of the Mixon and Ecton areas, North Staffordshire. Unpublished PhD thesis, University of Sheffield.

COLLINS, R. S. 1972. Barium Minerals. *Miner. Dossier Miner. Resour. Consult. Comm.*, No. 2. (London: Her Majesty's Stationery Office.)

COLLINSON, J. D. 1968. Deltaic sedimentation units in the Upper Carboniferous of Northern England. *Sedimentology*, Vol. 10, 233–254.

— 1970. Deep channels, massive beds and turbidity current genesis in the central Pennine basin. *Proc. Yorkshire Geol. Soc.*, Vol. 37, Pt. 4, 495–519.

CONWAY, V. M. 1954. Stratigraphy and pollen analysis of southern Pennine blanket peats. *J. Ecol.*, Vol. 42, 117–147.

COPE, F. W. 1933. The Lower Carboniferous succession in the Wye Valley region of North Derbyshire. *J. Manchester Geol. Assoc.*, Vol. 1, 125–145.

— 1936. The *Cyrtina septosa* Band in the Lower Carboniferous succession of North Derbyshire. *Summ. Prog. Geol. Surv. G.B.*, for 1934, Pt. 2, 48–51.

— 1937. Some features in the D$_1$–D$_2$ limestones of the Miller's Dale region, Derbyshire. *Proc. Yorkshire Geol. Soc.*, Vol. 23, Pt. 3, 178–195.

— 1939. The mid-Viséan (S$_2$–D$_1$) succession in north Derbyshire and north-west England. *Proc. Yorkshire Geol. Soc.*, Vol. 24, Pt. 1, 60–66.

— 1946. Intraformational contorted rocks in the Upper Carboniferous of the southern Pennines. *Q.J. Geol. Soc. London*, Vol. 101, Pts. 3 and 4, 139–176.

— 1949a. Woo Dale Borehole near Buxton, Derbyshire. *Q.J. Geol. Soc. London*, Vol. 105, Pt. 1, iv.

— 1949b. Correlation of the Coal Measures of Macclesfield and the Goyt Trough. *Trans. Inst. Min. Eng.*, Vol. 108, Pt. 10, 466–483.

— 1958. The Peak District, Derbyshire. *Geol. Assoc. Guide*, No. 26.

— 1972. Some stratigraphical breaks in the Dinantian massif facies in North Derbyshire. *Mercian Geol.*, Vol. 4, No. 2, 143–148.

— 1973. Woo Dale Borehole near Buxton, Derbyshire. *Nature, London*, Vol. 243, 29–30.

— 1979. The age of the volcanic rocks in the Woo Dale Borehole, Derbyshire. *Geol. Mag.*, Vol. 116, No. 4, 319–320.

COX, F. C. and BRIDGE, D. McC. 1977. The limestone and dolomite resources of the country around Monyash, Derbyshire: Description of 1:25 000 resource sheet SK 16. *Miner. Assess. Rep. Geol. Sci.*, No. 26.

— and HARRISON, D. J. 1980. The limestone and dolomite resources of the country around Wirksworth, Derbyshire: Description of parts of sheets SK 25 and 35. *Miner. Assess. Rep. Inst. Geol. Sci.*, No. 47.

CRITCHLEY, M. F. 1979. A geological outline of the Ecton Copper Mines, Staffordshire. *Bull. Peak Dist. Mines Hist. Soc.*, Vol. 7, No. 4, 177–191.

DONALDSON, C. H. 1978. Petrology of the uppermost upper mantle deduced from spinel-lherzolite and harzburgite nodules at Calton Hill, Derbyshire. *Contrib. Mineral. Petrol.*, Vol. 65, 363–377.

DUNHAM, K. C. 1952a. Age-relations of the epigenetic mineral deposits of Britain. *Trans. Geol. Soc. Glasgow*, Vol. 21, Pt. 3, 395–429.

— 1952b. Special reports on the mineral resources of Great Britain. Vol. 4. Fluorspar (4th Edn). *Mem. Geol. Surv. G.B.*

— 1970. Mineralisation by deep formation waters: a review. *Trans. Inst. Min. Metall.*, Vol. 79, B127–136.

— 1973. A recent deep borehole near Eyam, Derbyshire. *Nature, Phys. Sci., London*, Vol. 241, 84–85.

— and DINES, H. G. 1945. Barium minerals in England and Wales. *Wartime Pam. Geol. Surv. G.B.*, No. 46.

EAGAR, R. M. C. 1947. A study of a non-marine lamellibranch succession in the *Anthraconaia lenisulcata* Zone of the Yorkshire Coal Measures.

EARP, J. R., MAGRAW, D., POOLE, E. G., LAND, D. H. and WHITEMAN, A. J. 1961. Geology of the country around Clitheroe and Nelson. *Mem. Geol. Surv. G.B.*, Sheet 68.

EDEN, R. A., ORME, G. R., MITCHELL, M. and SHIRLEY, J. 1964. A study of part of the margin of the Carboniferous Limestone 'massif' in the Pin Dale area, Derbyshire. *Bull. Geol. Surv. G.B.*, No. 21, 73–118.

EMBLIN, R. 1978. A Pennine model for the diagenetic origin of base-metal ore deposits in Britain. *Bull. Peak Dist. Mines Hist. Soc.*, Vol. 7, No. 1, 5–20.

EVANS, A. M. and MAROOF, S. I. 1976. Basement controls on mineralisation in the British Isles. *Mining Mag.*, Vol. 134, 401–411.

EVANS, W. B., WILSON, A. A., TAYLOR, B. J. and PRICE, D. 1968. Geology of the country around Macclesfield, Congleton, Crewe and Middlewich. *Mem. Geol. Surv. G.B.*, Sheet 110.

EVERITT, C. W. F. and BELSHÉ, J. C. 1960. Palaeomagnetism of the British Carboniferous System. *Phil. Mag.*, Vol. 5, 675–685.

FAREY, J. 1811. *General view of the agriculture and minerals of Derbyshire*, Vol. 1. (London.)

FIRMAN, R. J. 1977. Derbyshire wrenches and ores – a study of the rakes' progress by secondary faulting. *Mercian Geol.*, Vol. 6, No. 2, 81–96.

— and BAGSHAW, C. 1974. A re-appraisal of the controls of non-metallic gangue mineral distribution in Derbyshire. *Mercian Geol.*, Vol. 5, No. 2, 145–161.

FITCH, F. J. and MILLER, J. A. 1964. The age of the paroxysmal Variscan orogeny in England. *In* 'The Phanerozoic timescale'. *Q.J. Geol. Soc. London*, Vol. 120S, 159–175.

— and WILLIAMS, S. C. 1970. Isotopic ages of British Carboniferous rocks. *C.R. 6e Congr. Int. Stratigr. Geol. Carbonif., Sheffield 1967*, Vol. 2, 771–789.

FOLK, R. L. 1959. Practical petrographic classification of limestones. *Bull. Am. Assoc. Petrol. Geol.*, Vol. 43, No. 1, 1–38.

FORD, T. D. 1960. White Watson (1760–1835) and his geolological sections. *Proc. Geol. Assoc.*, Vol. 71, Pt. 4, 349–363.

— 1961. Recent studies of mineral distribution in Derbyshire and their significance. *Bull. Peak Dist. Mines Hist. Soc.*, Vol. 1, No. 5, 3–9.

— 1963. The dolomite tors of Derbyshire. *East Midlands Geogr.*, Vol. 3, 148–154.

— 1964. The black marble mines of Ashford-in-the-Water. *Bull. Peak Dist. Mines Hist. Soc.*, Vol. 2, Pt. 4, 179–188.

— 1967. Some mineral deposits of the Carboniferous Limestone of Derbyshire. Pp. 53–75 in NEVES, R. and DOWNIE, C. (Eds.) *Geological excursions in the Sheffield region.* (Sheffield: J. W. Northend.)

— 1969. The stratiform ore-deposits of Derbyshire. *In* JAMES, C. H. (Ed.). Sedimentary ores ancient and modern (revised). *Proc. 15th Inter-Univ. Geol. Congr., 1967, Leicester, England.*

— 1972. Field meeting in the Peak District: report of the Director. *Proc. Geol. Assoc.*, Vol. 83, Pt. 2, 231–236.

— (Ed.). 1977. *Limestones and caves of the Peak District.* (Norwich: Geo Abstracts.)

— and INESON, P. R. 1971. The fluorspar mining potential of the Derbyshire ore field. *Trans. Inst. Min. Metall.*, Vol. 80, B186–210.

— and KING, R. J. 1965. Layered epigenetic galena–barite deposits in the Golconda Mine, Brassington, Derbyshire, England. *Econ. Geol.*, Vol. 60, No. 8, 1686–1701.

— — 1969. The origin of the silica sand deposits in the Derbyshire limestone. *Mercian Geol.*, Vol. 3, No. 1, 51–69.

— and RIEUWERTS, J. H. (Eds.) 1983. *Lead Mining in the Peak District.* (3rd edn.). (Bakewell: Peak Park Joint Planning Board.)

— and SARJEANT, W. A. S. 1964. The Peak District mineral index. *Bull. Peak Dist. Mines Hist. Soc.*, Vol. 2, Pt. 3, 122–150.

FRANCIS, E. A. 1970. Excursion report: the geology of the southern part of the Goyt Syncline, north Staffordshire. *Mercian Geol.*, Vol. 3, No. 4, 415–418.

FRANCIS, E. H. 1968. Effect of sedimentation on volcanic processes, including neck-sill relationships, in the British Carboniferous. *23rd Sess. Int. Geol. Congr., Czechoslovakia, 1968*, Sect. 2, 163–174.

FROST, D. V. and SMART, J. G. O. 1979. Geology of the country north of Derby. *Mem. Geol. Surv. G.B.*, Sheet 125.

GATLIFF, R. W. 1982. The limestone and dolomite resources of the country around Tideswell, Derbyshire. Description of 1:25 000 sheets SK 17 and parts SK 18 and 27. *Miner. Assess. Rep. Inst. Geol. Sci.*, No. 98.

GEIKIE, A. 1897. *The ancient volcanoes of Great Britain*, Vol. 2. (London: Macmillan.)

GEORGE, T. N. 1963. Tectonics and palaeogeography in northern England. *Sci. Prog.*, Vol. 51, 32–59.

—, JOHNSON, G. A. L., MITCHELL, M., PRENTICE, J. E., RAMSBOTTOM, W. H. C., SEVASTOPULO, G. D. and WILSON, R. B. 1976. A correlation of Dinantian rocks in the British Isles. *Spec. Rep. Geol. Soc. London*, No. 7.

GIBSOB, W. and WEDD, C. B. 1913. The geology of the northern part of the Derbyshire Coalfield and bordering tracts. *Mem. Geol. Surv. G.B.*

GILLIGAN, A. 1920. The petrography of the Millstone Grit of Yorkshire. *Q.J. Geol. Soc. London*, Vol. 75, Pt. 4, 251–294.

GREEN, A. H., FOSTER, C. LE NEVE and DAKYNS, J. R. 1869. The geology of the Carboniferous Limestone, Yoredale rocks and Millstone Grit of North Derbyshire. *Mem. Geol. Surv. England and Wales.*

— — — 1887. The geology of the Carboniferous Limestone, Yoredale Rocks and Millstone Grit of North Derbyshire. 2nd edition with additions by A. H. Green and A. Strahan. *Mem. Geol. Surv. England and Wales.*

HAMAD, S. EL D. 1963. The chemistry and mineralogy of the olivine nodules of Calton Hill, Derbyshire. *Mineral Mag.*, Vol. 33, No. 261, 483–497.

HARRIS, P. M. (compiler). 1982. Limestone and dolomite. *Miner. Dossier Miner. Resour. Consult. Comm.*, No. 23. (London: Her Majesty's Stationery Office).

HARRISON, D. J. 1981. The limestone and dolomite resources of the country around Buxton, Derbyshire. Description of 1:25 000 sheet SK 07 and parts of SK 06 and 08. *Miner. Assess. Rep. Inst. Geol. Sci.*, No. 77.

— and ADLAM, K. A. McL. 1985. The Limestone and dolomite resources of the Peak District of Derbyshire and Staffordshire. Description of 1:50 000 geological sheets 99, 111, 112, 124 and 125. *Miner. Assess. Rep. Br. Geol. Surv.*, No. 144.

HEATH, C. W. 1973. A sedimentological and palaeogeographical study of the Namurian Rough Rock in the southern Pennines. Unpublished PhD thesis, University of Keele.

HECKEL, P. H. 1975. Carbonate buildups in the geologic record: a review. Pp. 90–154 *In* LAPORTE, L. F. (Ed.). *Reef in time and space. Spec. Publ., Soc. Econ. Paleontol. and Mineral.*, No. 18. (Tulsa: Society of Economic Paleontologists and Mineralogists.)

HESTER, S. W. 1932. The Millstone Grit succession in north Staffordshire. *Summ. Prog. Geol. Surv. G.B.* (for 1931), Pt 2, 34–48.

HIND, W. 1910. Staffordshire. In *Geology in the field*, 564–591. MONCKTON, H. W. and HERRIES, R. S.(Eds.). Jubilee Vol. Geol. Assoc. (1858–1908). (London: Stanford).

HOLDSWORTH, B. K. 1963a. The palaeontology, stratigraphy and sedimentology of Namurian rocks in the Longnor–Hollinsclough–Morridge region, north-east Staffordshire and south-west Derbyshire. Unpublished PhD thesis, University of Manchester.

— 1963b. Pre-fluvial, autogeosynclinal sedimentation in the Namurian of the southern Central Province. *Nature, London*, Vol. 199, 133–135.

— 1964. The 'crowstones' of Staffordshire, Derbyshire and Cheshire. *North Staffordshire J. of Field Studies*, Vol. 4, 89–102.

— 1966. A preliminary study of the palaeontology and palaeoenvironment of some Namurian limestone 'bullions'. *Mercian Geol.*, Vol. 1, No. 4, 315–337.

—, EDWARDS, C. A. and TREWIN, N. H. 1970. Written discussion of MORRIS, P. G., 1969. *Proc. Geol. Assoc.*, Vol. 81, Pt. 1, 175–177.

— and TREWIN, N. H. 1968. Excursion to the Namurian outcrop in south-west Derbyshire and north-east Staffordshire. *Mercian Geol.*, Vol. 2, No. 4, 435–437.

HOWE, J. A. 1897. Notes on the pockets of sand and clay in the limestone of Derbyshire and Staffordshire. *Trans. North Staffordshire Field Club*, Vol. 31, 143–149.

— (Ed.) 1920. Special reports on the mineral resources of Great Britain. Vol. 6, Refractory materials. Resources and geology. (2nd Edn.) *Mem. Geol. Surv. G.B.*

HUDSON, R. G. S. 1931. The pre-Namurian knoll topography of Derbyshire and Yorkshire. *Trans. Leeds Geol. Assoc.*, Vol. 5, 49–64.

— and COTTON, G. 1945.   The Lower Carboniferous in a boring at Alport, Derbyshire.   *Proc. Yorkshire Geol. Soc.*, Vol. 25, Pt. 4, 254–330.

HULL, E. 1869.   The Triassic and Permian rocks of the Midland Counties of England.   *Mem. Geol. Surv. G.B.*

— and GREEN, A. H. 1866.   The geology of the country around Stockport, Macclesfield, Congleton and Leek.   *Mem. Geol. Surv. G.B.*

INESON, P. R. 1970.   Trace-element auroles in limestone wallrocks adjacent to fissure veins in the Eyam area of the Derbyshire ore field.   *Trans. Inst. Min. Metall.*, Vol. 79, B238–245.

— and AL-KUFAISHI, F. A. M. 1970.   The mineralogy and paragenetic sequence of Long Rake Vein at Raper Mine, Derbyshire.   *Mercian Geol.*, Vol. 3, No. 4, 337–351.

— and MITCHELL, J. G. 1973.   Isotopic age determinations on clay minerals from lavas and tuffs of the Derbyshire orefield.   *Geol. Mag.*, Vol. 109, No. 6, 501–512.

INSTITUTE OF GEOLOGICAL SCIENCES. 1977a.   1:250 000 Series, Bouguer Gravity Anomaly map (Provisional Edition), Humber–Trent sheet, 53°N–02°W. (London: Institute of Geological Sciences.)

— 1977b.   1:250 000 Series, Aeromagnetic anomaly map, Humber–Trent Sheet, 53°N–02°W. (London: Institute of Geological Sciences.)

— 1977c.   1:250 000 Series, Bouguer Gravity anomaly map (Provisional Edition), Liverpool Bay sheet, 53°N–04°W. (London: Institute of Geological Sciences.)

— 1978a.   IGS boreholes 1977.   *Rep. Inst. Geol. Sci.*, No. 78/21.

— 1978b.   1:250 000 Series, Aeromagnetic anomaly map Liverpool Bay sheet, 53°N–04°. (London: Institute of Geological Sciences.)

JACKSON, J. W. and CHARLESWORTH, J. K. 1920.   The quartzose conglomerate at Caldon Low, Staffordshire.   *Geol. Mag.*, Vol. 57, 487–492.

JOHNSON, G. A. L. 1981.   Geographical evolution from Laurasia to Pangaea.   *Proc. Yorkshire Geol. Soc.*, Vol. 43, Pt. 3, 221–252.

JOHNSON, R. H. 1965a.   The glacial geomorphology of the west Pennine slopes from Cliviger to Congleton. Pp. 58–93 *in* WHITTOW, J. B. and WOOD, P. D. (Eds.). *Essays in geography for Austin Miller.* (Reading, UK: Univeristy of Reading.)

— 1965b.   The origin of the Churnet and Rudyard valleys. *North Staffordshire J. of Field Studies*, Vol. 5, 95–105.

— and WALTHALL, S. 1979.   The Longdendale landslides. *Geol. J.*, Vol. 14, Pt. 2, 135–158.

JONES, C. M. 1980.   Deltaic sedimentation in the Roaches Grit and associated sediments (Namurian R_{2b}) in the south-west Pennines.   *Proc. Yorkshire Geol. Soc.*, Vol. 43, Pt. 1, 39–67.

— and McCABE. 1980.   Erosion surfaces within giant fluvial crossbeds of the Carboniferous in northern England.   *J. Sediment. Petrol.*, Vol. 50, No. 2, 613–620.

JOWETT, A and CHARLESWORTH, J. K. 1929.   The glacial geology of the Derbyshire dome and the western slopes of the southern Pennines.   *Q. J. Geol. Soc. London*, Vol. 85, Pt. 3, 307–334.

KENT, P. E. 1949.   A structure contour map of the surface of the buried pre-Permian rocks of England and Wales.   *Proc. Geol. Assoc.*, Vol. 60, Pt. 2, 87–104.

— 1957.   Triassic relics and the 1000-foot surface in the southern Pennines.   *East Midland Geographer*, Vol. 1, No. 8, 3–10.

— 1966.   The structure of the concealed Carboniferous rocks of north-eastern England.   *Proc. Yorkshire Geol. Soc.*, Vol. 35, Pt. 3, 323–352.

KEREY, I. E. 1978.   Sedimentology of the Chatsworth Grit sandstone in the Goyt – Chapel en le Frith area. Unpublished MSc thesis, University of Keele.

KING, R. J. 1966.   Epi-syngenetic mineralisation in the English Midlands.   *Mercian Geol.*, Vol. 1, No. 4, 291–301.

— and FORD, T. D. 1968.   Mineralisation. Pp. 112–137 *in* SYLVESTER-BRADLEY, P. C. and FORD, T. D. (Eds.). *The geology of the East Midlands.* (Leicester: Leicester University Press.)

KIRKHAM, N. 1964.   Whale Sough and Hubberdale Mine. *Bull. Peak Dist. Mines. Hist. Soc.*, Vol. 2, Pt. 4, 206–229.

— and FORD, T. D. 1967.   The Ecton Copper Mines.   *Peak Dist. Mine Hist. Soc.*, Spec. Publ., No. 1 (2nd Ed.).

LACEY, E. 1862.   A note on the occurrence of galena at Axe Edge.   *Trans. Manchester Geol. Soc.*, Vol. 3, 135.

LEEDER, M. R. 1982.   Upper Palaeozoic basins of the British Isles – Caledonide inheritance versus Hercynian plate margin processes.   *J. Geol. Soc. London*, Vol. 139, Pt. 4, 479–491.

LEES, A. 1964.   The structure and origin of the Waulsortian (Lower Carboniferous) 'reefs' of west-central Eire.   *Phil. Trans. R. Soc.*, Series B, Vol. 247, 483–531.

— , NOEL, B. and BOUW, P. 1977.   The Waulsortian 'reefs' of Belgium: a progress report.   *Mém. Inst. Géol. Univ. Louvain*, Vol. 29, 289–315.

LUDFORD, A. 1951.   The stratigraphy of the Carboniferous rocks of the Weaver Hills district, North Staffordshire.   *Q. J. Geol. Soc. London*, Vol. 106, Pt. 2, 211–230.

— 1970.   The stratigraphy and palaeontology of the Lower Carboniferous rocks of north-east Staffordshire and adjacent parts of Derbyshire. Unpublished PhD thesis, University of London.

— MADGETT, P. and SADLER, H. E. 1973.   The Carboniferous Limestone margin between Crowdecote and Hartington, Derbyshire.   *Mercian Geol.*, Vol. 4, No. 3, 213–222.

MACDONALD, R., GASS, K. N., THORPE, R. S. and GASS, I. G. 1984.   Geochemistry and petrogenesis of the Derbyshire Carboniferous basalts.   *J. Geol. Soc. London*, Vol. 141, Pt. 1, 147–159.

McKERROW, W. S. (Ed.). 1978.   *The ecology of fossils.* (London: Duckworth.)

MAROOF, S. I. 1976.   The structure of the concealed pre-Carboniferous basement of the Derbyshire Dome from gravity data.   *Proc. Yorkshire Geol. Soc.*, Vol. 41, Pt. 1, 59–69.

MAYHEW, R. W. 1966.   A sedimentological investigation of the Marsdenian grits and associated measures in north-east Derbyshire. Unpublished PhD thesis, University of Sheffield.

— 1967.   The Ashover and Chatsworth Grits in north-east Derbyshire. Pp. 94–103 *in* NEVES, R. and DOWNIE, C. (Eds.). *Geological excursions in the Sheffield region.* (Sheffield: J. W. Northend.)

MILLER, J. and GRAYSON, R. F. 1982.   The regional context of Waulsortian facies in northern England. Pp. 17–33 *in* BOLTON, K., LANE, H. R. and LeMOME, D. V. (Eds.). *Symposium on the paleoenvironmental setting and distribution of the Waulsortian facies.* (El Paso: El Paso geological Society and the University of Texas.

MITCHELL, G. F., PENNY, L. F., SHOTTON, F. W. and WEST, R. G. 1973.   A correlation of the Quaternary deposits in the British Isles.   *Spec. Rep. Geol. Soc. London*, No. 4.

MITCHELL, R. H. and KROUSE, H. R. 1971.   Isotopic composition of sulfur and lead in galena from the Greenhow–Skyreholme area, Yorkshire, England.   *Econ. Geol.*, Vol. 66, No. 2, 243–251.

MOORBATH, S. 1962.   Lead isotope abundance studies on mineral occurrences in the British Isles and their geological significance.   *Phil. Trans. Roy. Soc. London*, Ser. A, Vol. 254, 295–360.

MORGAN, N. 1980.   Palaeoecology and sedimentology of Waulsortian 'reefs' (Lower Carboniferous). Unpublished PhD thesis, University of Oxford.

MORRIS, P. G. 1969.   Dinantian and Namurian stratigraphy, east and south-east of Leek, north Staffordshire.   *Proc. Geol. Assoc.*, Vol. 80, Pt. 2, 145–175.

— 1970.   Carboniferous conodonts in the south-western Pennines.   *Geol. Mag.*, Vol. 106, No. 5, 497–499.

MUELLER, G. 1954.   The distribution of coloured varieties of fluorites within the thermal zones of Derbyshire mineral deposits.   *19th Sess. Int. Geol. Congr., Algiers 1952*, Sect. 13, Fasc. 15, 523–539.

MUNN, D. and JACKSON, D. E. 1980.   Dedolomitisation of Lower Carboniferous dolostone in the Wirksworth area, Derbyshire, England.   *Geol. Mag.*, Vol. 117, No. 6, 607–612.

NOCKOLDS, S. R., KNOX, R. W. O'B. and CHINNER, G. A. 1978.   *Petrology for students.* (Cambridge: Cambridge University Press.)

NOTHOLT, A. J. G. (Compiler). 1971.   Fluorspar.   *Miner. Dossier Miner. Resour. Consult. Comm.*, No. 1. (London: Her Majesty's Stationery Office.)

ORME, G. R. 1970.   The D₂-P₁ 'Reefs' and associated limestones of the Pin Dale – Bradwell Moor area of Derbyshire.   *C. R. 6e Congr. Int. Stratigr. Geol. Carbonif., Sheffield 1967*, Vol. 3, 1249–1262.

PARKINSON, D. 1950.   The stratigraphy of the Dovedale area, Derbyshire and Staffordshire.   *Q. J. Geol. Soc. London*, Vol. 105, Pt. 2, 265–294.

— 1957.   Lower Carboniferous reefs of northern England.   *Bull. Am. Assoc. Pet. Geol.*, Vol. 41, No. 3, 511–537.

— and LUDFORD, A. 1964.   The Carboniferous Limestone of the Blore-with-Swinscoe district, north-east Staffordshire, with revisions to the stratigraphy of neighbouring areas.   *Geol. J.*, Vol. 4, Pt. 1, 167–176.

— 1973.   The mid-D₁ unconformity between Hartington and Alsop, Derbyshire.   *Mercian Geol.*, Vol. 4, No. 3, 205–207.

PARSONS, C. E. 1897.   The deposit at the Mill Close lead-mine, Darley Dale, Matlock.   *Trans. Fed. Inst. Min. Eng.*, Vol. 12, 115–121.

PARSONS, L. M. 1922.   Dolomitisation in the Carboniferous Limestone of the Midlands.   *Geol. Mag.*, Vol. 59, 51–63 and 104–117.

PETTIJOHN, F. J. 1957.   *Sedimentary rocks.* (New York: Harper.)

PICKIN, J. 1975.   Smallpenny Sough, Lathkill Dale.   *Bull. Peak Dist. Mines Hist. Soc.*, Vol. 6, No. 2, 100–101.

PIGOTT, C. D. 1962.   Soil formation and development on the Carboniferous Limestone of Derbyshire. 1 Parent materials.   *J. Ecol.*, Vol. 50, 145–156.

PONSFORD, D. R. A. 1955.   Radioactivity studies of some British sedimentary rocks.   *Bull. Geol. Surv. G.B.*, No. 10, 24–44.

PORTER, L. and ROBEY, J. A. 1972a.   The Dale Mine, Manifold Valley, North Staffordshire.   *Bull. Peak Dist. Mines Hist. Soc.*, Vol. 5, No. 2, 93–106.

— — 1972b.   The Royledge and New York copper mines, Upper Elkstones, ner Leek, Staffordshire.   *Bull. Peak Dist. Mines Hist. Soc.*, Vol. 5, No. 1, 1–9.

PRENTICE, J. E. 1951.   The Carboniferous Limestone of the Manifold valley region, North Staffordshire.   *Q. J. Geol. Soc. London*, Vol. 106, Pt. 2, 171–209.

— 1952.   Note on previous paper (Prentice, 1951).   *Q. J. Geol. Soc. London*, Vol. 107, Pt. 3, 335.

— and MORRIS, P. G. 1959.   Cemented screes in the Manifold Valley.   *East Midlands Geogr.*, Vol. 2, No. 11, 16–19.

RAMSBOTTOM, W. H. C. 1969.   The Namurian of Britain.   *C. R. 6e Congr. Int. Stratigr. Geol. Carbonif., Sheffield 1967*, Vol. 1, 219–232.

— 1973.   Transgressions and regressions in the Dinantian: a new synthesis of British Dinantian stratigraphy.   *Proc. Yorkshire Geol. Soc.*, Vol. 39, Pt. 4, 567–607.

— 1977.   Major cycles of transgression and regression (mesothems) in the Namurian.   *Proc. Yorkshire Geol. Soc.*, Vol. 41.

— and MITCHELL, M. 1980.   The recognition and divisions of the Tournaisian Series in Britian.   *J. Geol. Soc. London*, Vol. 137, Pt. 1, 61–63.

— CALVER, M. A., EAGER, R. M. C., HODSON, F., HOLLIDAY, D. W., STUBBLEFIELD, C. J. and WILSON, R. B. 1978.   A correlation of Silesian rocks in the British Isles.   *Spec. Rep. Geol. Soc. London*, No. 10.

— RHYS, G. H. and SMITH, E. G. 1962.   Boreholes in the Carboniferous rocks of the Ashover district, Derbyshire.   *Bull. Geol. Surv. G.B.*, No. 19, 75–168.

RIEUWERTS, J. H. 1963.   Lathkilldale: its mines and miners.   *Bull Peak Dist. Mines Hist. Soc.*, Vol. 2, No. 1, 9–30.

— 1966a.   Lathkill Dale: its mines and miners (supplementary notes).   *Bull. Peak Dist. Mines Hist. Soc.*, Vol. 3, Pt. 1, 71–74.

— 1966b.   A list of the soughs of the Derbyshire lead mines.   *Bull. Peak Dist. Mines Hist. Soc.*, Vol. 3, Pt. 1, 1–42.

— 1973.   Lathkill Dale: its mines and miners. (Hartington: Moorland Press.)

— 1981a.   The drainage of the Alport mining field.   *Bull. Peak Dist. Mines Hist. Soc.*, Vol. 8, No. 1, 1–28.

— 1981b.   The development of mining and drainage in the Wensley, Winster and Elton areas.   *Bull. Peak Dist. Mines Hist. Soc.*, Vol. 8, No. 2, 109–150.

RITTMANN, A. 1962.   *Volcanoes and their activity.* (New York: John Wiley and Sons.) [tr. E. A. Vincent].

ROBEY, J. A. 1965.   The drainage of the area between the River Wye and the River Lathkill, Derbyshire.   *Proc. Br. Speleol. Assoc.*, No. 3, 1–10.

— 1966.   Field Grove Mine, Sheldon, Derbyshire.   *Bull. Peak Dist. Mines Hist. Soc.*, Vol. 3, Pt. 2, 93–101.

— 1973.   Supplementary notes on the lead mines in the Monyash – Flagg area of Derbyshire.   *Bull. Peak Dist. Mines Hist. Soc.*, Vol. 5.

— and PORTER, L. 1970.   The copper and lead mines of the Mixon area, Staffordshire.   *Bull. Peak Dist. Mines Hist. Soc.*, Vol. 4, No. 4, 256–280.

— — 1972.   *The copper and lead mines of Ecton Hill, Staffordshire.* (Leek: Moorland Publishing Co.; Bakewell: Peak District Mines Historical Society.)

ROBINSON, B. W. and INESON, P. R. 1979.   Sulphur, oxygen and carbon isotope investigations of lead – zinc – baryte – fluorite – calcite mineralisation, Derbyshire, England.   *Trans. Inst. Min. Metall.*, Vol. 88, B107–117.

SADLER, H. E. 1964.   Conditions of sedimentation of the *Cyrtina septosa* Band in the Lower Carboniferous of Derbyshire.   *Mercian Geol.*, Vol. 1, No. 1, 15–22.

— 1966.   A detailed study of microfacies in the mid-Viséan (S₂–D₁) limestones near Hartington, Derbyshire, England.   *J. Sediment. Petrol.*, Vol. 36, No. 4, 864–879.

— and WYATT, R. J. 1966.   A lower Carboniferous S₂ inlier near Hartington, Derbyshire.   *Proc. Geol. Assoc.*, Vol. 77, Pt. 1, 55–64.

SEABOURNE, T. R., SIRIKCI, D. H. and SOWERBUTTS, W. T. C. 1979.   Examples of horizontal loop electromagnetic anomalies controlled by geological faulting.   *Geoexploration*, Vol. 17, 77–87.

SHACKLETON, J. S. 1962. Cross-strata of the Rough Rock (Millstone Grit Series) in the Pennines. *Liverpool and Manchester Geol. J.*, Vol. 3, Pt. 1, 109–118.

SHAW, R. P. 1980. A survey and the geology of Putwell Hill Mine, Monsal Dale. *Bull. Peak Dist. Mines Hist. Soc.*, Vol. 7, No. 6, 342–344.

SHINN, E. A. 1968. Practical significance of birdseye structures in carbonate rocks. *J. Sediment. Petrol.*, Vol. 38, No. 1, 215–223.

SHIRLEY, J. 1950. The stratigraphical distribution of the lead-zinc ores of Mill Close Mine, Derbyshire and the future prospects of the area. *18th Int. Geol. Congr. London*, Pt. 7, 353–361.

— 1959. The Carboniferous Limestone of the Monyash–Wirksworth area, Derbyshire. *Q. J. Geol. Soc. London*, Vol. 114, Pt. 3, 411–429.

— and HORSFIELD, E. L. 1945. The structure and ore deposits of the Carboniferous Limestone of the Eyam district, Derbyshire. *Q. J. Geol. Soc. London*, Vol. 100, Pts. 3 and 4, 289–308.

SIBLY, T. F. 1908. The faunal succession in the Carboniferous Limestone (Upper Avonian) of the midland area (north Derbyshire and north Staffordshire). *Q. J. Geol. Soc. London*, Vol. 64, Pt. 1, 34–82.

SIMPSON, I. M. and BROADHURST, F. M. 1969. A boulder bed at Treak Cliff, north Derbyshire. *Proc. Yorkshire Geol. Soc.*, Vol. 37, Pt. 2, 141–151.

SLEIGH, J. 1862. *A history of the ancient parish of Leek in Staffordshire*. J. Russell Smith; Leek: Robert Nall. (London)

SMITH, E. G., RHYS, G. H. and EDEN, R. A. 1967. Geology of the country around Chesterfield, Matlock and Mansfield. *Mem. Geol. Surv. G.B.*, Sheet 112.

SMITH, K., SMITH, N. J. P. and HOLLIDAY, D. W. *In press*. The deep structure of Derbyshire.

SOMERVILLE, I. D. 1979. Minor sedimentary cyclicity in late Asbian (upper D₁) limestones in the Llangollen district of North Wales. *Proc. Yorkshire Geol. Soc.*, Vol. 42, Pt. 3, 317–341.

STEVENSON, I. P. and GAUNT, G. D. 1971. Geology of the country around Chapel en le Frith. *Mem. Geol. Surv. G.B.*, Sheet 99.

— HARRISON, R. K. and SNELLING, N. J. 1970. Potassium-argon age determination of the Waterswallows Sill, Buxton, Derbyshire. *Proc. Yorkshire Geol. Soc.*, Vol. 37, Pt. 4, 445–447.

STRANK, A. R. E. *In press*.

STRAW, A. and LEWIS, G. M. 1962. Glacial drift in the area around Bakewell, Derbyshire. *East Midland Geographer*, Vol. 3, 72–80.

SYLVESTER-BRADLEY, P. C. and FORD, T. D. (Eds.). 1968. *The geology of the East Midlands*. (Leicester: Leicester University Press.)

TALLIS, J. H. 1964. Studies on southern Pennine peats. 1. The general pollen record. *J. Ecol.*, Vol. 52, No. 2, 323–331.

TEALL, J. J. H. 1888. *British petrography*. (London: Dulau and Co.)

THACH, T. K. 1964. Sedimentology of Lower Carboniferous limestone (Viséan) in north Staffordshire and south-west Derbyshire. Unpublished PhD thesis, University of Reading.

THOMPSON, D. B. 1970. Sedimentation of the Triassic (Scythian) red pebbly sandstones in the Cheshire Basin and its margins. *Geol. J.*, Vol. 7, Pt. 1, 183–216.

TIMMS, A. E. 1978. Aspects of the palaeoecology of productoid and associated brachiopods in the middle to upper Viséan 'reef' limestones of Derbyshire. Unpublished PhD thesis, University of Manchester.

TITMAN, D. J. 1971. A palaeomagnetic study of some Upper Palaeozoic rocks. Unpublished MSc thesis, University of Newcastle upon Tyne.

TOMKEIEFF, S. I. 1928. The volcanic complex of Calton Hill, Derbyshire. *Q. J. Geol. Soc. London*, Vol. 84, Pt. 4, 703–718.

TRAILL, J. G. 1939. The geology and development of Mill Close Mine, Derbyshire. *Econ. Geol.*, Vol. 34, No. 8, 851–889.

— 1940. Notes on the Lower Carboniferous Limestones and toadstones at Mill Close Mine, Derbyshire. *Trans. Inst. Min. Metall.*, Vol. 49, 191–229.

TREWIN, N. H. 1968. Potassium bentonites in the Namurian of Staffordshire and Derbyshire. *Proc. Yorkshire Geol. Soc.*, Vol. 37, Pt. 1, 73–91.

— and HOLDSWORTH, B. K. 1972. Further K-bentonites from the Namurian of Staffordshire. *Proc. Yorkshire Geol. Soc.*, Vol. 39, Pt. 1, 87–89.

— — 1973. Sedimentation in the Lower Namurian rocks of the north Staffordshire basin. *Proc. Yorkshire Geol. Soc.*, Vol. 39, Pt. 3, 371–408.

VARVILL, W. W. 1937. A study of the shapes and distribution of the lead deposits in the Pennine limestones in relation to economic mining. *Trans. Inst. Min. Metall.*, Vol. 46, 463–559.

— 1959. The future of lead-zinc and fluorspar mining in Derbyshire. *In* The future of non-ferrous mining in Great Britain and Ireland; a symposium. *Inst. Min. Metall.*, 175–232.

VAUGHAN, A. 1905. The palaeontological sequence in the Carboniferous Limestone of the Bristol area. *Q. J. Geol. Soc. London*, Vol. 61, Pt. 2, 181–307.

WALKDEN, G. M. 1972. The mineralogy and origin of inter-bedded clay wayboards in the Lower Carboniferous of the Derbyshire Dome. *Geol. J.*, Vol. 8, Pt. 1, 143–160.

— 1974. Palaeokarstic surfaces in the Upper Viséan (Carboniferous) limestones of the Derbyshire Block, England. *J. Sediment. Petrol.*, Vol. 44, No. 4, 1232–1247.

— 1977. Volcanic and erosive events on the upper Viséan carbonate platform, north Derbyshire. *Proc. Yorkshire Geol. Soc.*, Vol. 41, Pt. 3, 347–366.

WALKER, R. G. 1966. *Shale Grit and Grindslow Shales*: transition from turbidite to shallow water sediments in the Upper Carboniferous of northern England. *J. Sediment. Petrol.*, Vol. 36, No. 1, 90–114.

WALKER, P. T., BOULTER, M. C., IJTABA, M. and URBANI, D. M. 1972. The preservation of the Neogene Brassington Formation of the southern Pennines and its bearing on the evolution of upland Britain. *J. Geol. Soc. London*, Vol. 128, Pt. 6, 519–559.

— COLLINS, P., IJTABA, M., NEWTON, J. P., SCOTT, N. H. and TURNER, P. R. 1980. Palaeocurrent directions and their bearing on the origin of the Brassington Formation (Miocene-Pliocene) of the southern Pennines, Derbyshire, England. *Mercian Geol.*, Vol. 8, No. 1, 47–62.

WALTERS, S. G. and INESON, P. R. 1980. Mineralisation within the igneous rocks of the South Pennine Orefield. *Bull. Peak Dist. Mines Hist. Soc.*, Vol. 7, No. 6, 315–325.

— — 1981. A review of the distribution and correlation of igneous rocks in Derbyshire, England. *Mercian Geol.*, Vol. 8, No. 2, 81–132.

— — 1983. Clay minerals in the Basalts of the South Pennines. *Miner. Mag.*, Vol. 47, pp. 21–26.

WARDLE, T. 1862. On the geology of the neighbourhood of Leek. *In* SLEIGH, J. 1862. *A history of the ancient parish of Leek in Staffordshire. (Reprinted separately, 1863)*. (London: J. Russell Smith; Leek: Robert Nall.)

WARRINGTON, G., AUDLEY-CHARLES, M. G., ELLIOTT, R. E.
EVANS, W. B., IVIMEY-COOK, H. C., KENT, P. E., ROBINSON,
P. L., SHOTTON, F. W. and TAYLOR, F. M. 1980. A
correlation of the Triassic rocks in the British Isles. *Spec. Rep.
Geol. Soc. London*, No. 13.

WATSON, J. J. W. 1860. Notes on the metalliferous saddles, or
ore-bearing beds in the contorted strata of the Lower
Carboniferous rocks of certain parts of Derbyshire and north
Staffordshire. *The Geologist*, Vol. 3, 357–369.

WATSON, W. 1811. A delineation of the strata of Derbyshire.
(Sheffield)

WEAVER, J. D. 1974. Systematic jointing in south Derbyshire.
Mercian Geol., Vol. 5, No. 2, 115–132.

WEDD, C. B. and DRABBLE, G. C. 1908. The fluorspar
deposits of Derbyshire. *Trans. Inst. Min. Eng.*, Vol. 35,
501–535.

WELSH, A. and OWENS, B. 1983. Early Dinantian miospore
assemblages from the Caldon Low Borehole, Staffordshire,
England. *Pollen Spores*, Vol. 25, No. 2, 253–264.

WHETTON, J. T., MAYERS, J. O. and BURKE, K. B. S. 1961.
Tracing the boundary of the concealed coalfield of Yorkshire
using the gravity method. *Min. Engineer*, Vol. 120, 657–674.

WHITCOMBE, D. N. and MAGUIRE, P. K. H. 1980. An
analysis of the velocity structure of the Precambrian rocks of
Charnwood Forest. *Geophys. J. R. Astr. Soc.*, Vol. 63,
405–416.

— — 1981. Seismic refraction evidence for a basement ridge
between the Derbyshire Dome and the west of Charnwood
Forest. *J. Geol. Soc. London*, Vol. 138, 653–659.

WHITEHURST, J. 1778. *An enquiry into the original state and
formation of the earth*. (London.)

WILLIES, L. 1976. John Taylor in Derbyshire 1839–1851.
Bull. Peak Dist. Mines Hist. Soc., Vol. 6, No. 3, 146–161.

WILAON, J. L. 1975. *Carbonate facies in geologic history*. (Berlin:
Springer-Verlag.)

WOLFENDEN, E. B. 1958. Paleoecology of the Carboniferous
reef complex and shelf limestones in north-west Derbyshire,
England. *Bull. Geol. Soc. Am.*, Vol. 69, 871–898.

— 1959. New sponges from Lower Carboniferous reefs of
Derbyshire and Yorkshire, England. *J. Palaeontol.*, Vol. 33,
No. 4, 566–568.

WORLEY, N. E. 1975. Geology of the Blende Vein, Magpie
Sough. *Bull. Peak Dist. Mines Hist. Soc.*, Vol. 6, No. 1,
28–32.

— 1976. Lithostratigraphical control of mineralisation in the
Blende Vein, Magpie Mine, Sheldon, near Bakewell,
Derbyshire. *Proc. Yorkshire Geol. Soc.*, Vol. 41, Pt. 1, 95–106.

— 1977. The geology of the Wills Founder Shaft, Winster.
Bull. Peak Dist. Mines Hist. Soc., Vol. 6, No. 6, 257–262.

— and FORD, T. D. 1976. Mandale Forefield Shaft. *Bull.
Peak Dist. Mines Hist. Soc.*, Vol. 6, No. 3, 141–143.

— — 1977. Mississippi Valley type orefields in Britain. *Bull.
Peak Dist. Mines Hist. Soc.*, Vol. 6, No. 5, 201–208.

— WORTHINGTON, T. and RILEY, L. 1978. The geology and
exploration of the Hubbadale mines, Taddington. *Bull. Peak
Dist. Mines Hist. Soc.*, Vol. 7, No. 1, 31–38.

WORSLEY, P. 1970. The Cheshire–Shropshire lowlands. Pp.
83–106 *in* LEWIS, C. A. (Ed.) *The glaciations of Wales and
adjoining regions*. (London: Longman.)

YORKE, C. 1954. *The pocket deposits of Derbyshire*. (Birkenhead:
privately published.)

— 1961. *The pocket deposits of Derbyshire*. (2nd edn.) (Birkenhead:
privately published.)

APPENDIX 1

Six-inch Maps

Geological six-inch maps included wholly or in part in 1:50 000 Sheet 111 (Buxton) are listed below, together with the initials of the surveyors and dates of survey; in the case of marginal sheets all surveyors are listed. The officers involved were: N. Aitkenhead, J. I. Chisholm, R. A. Eden, W. N. Edwards, E. A. Francis, D. Price, G. H. Rhys, E. G. Smith, I. P. Stevenson and B. J. Taylor.

Manuscript copies of the maps are deposited for public reference in the libraries of the British Geological Survey. Uncoloured dyeline copies of these maps are available for purchase. Certain marginal sheets are available as printed copies; these are indicated on the list by an asterisk.

SJ 95 NE	Leek	DP, EAF, NA	1958 – 70
SJ 96 SE	Gun Hill	DP, EAF, NA	1958 – 70
SJ 96 NE	Wildboarclough	BJT, DP, EAF, NA	1957 – 72
SJ 97 SE	Macclesfield Forest	EAF, BJT, IPS	1951 – 71
SK 05 NE	Manifold Valley	NA	1971
SK 05 NW	Mixon	NA	1968 – 70
SK 06 SE	Fawfieldhead and Longnor	NA	1968
SK 06 SW	The Roaches	NA, EAF	1964 – 69
SK 06 NE	Earl Sterndale	NA	1967 – 70
SK 06 NW	Flash	NA, EAF	1964 – 69
SK 07 SE	Buxton	IPS	1952 – 68
SK 07 SW	Goyt's Moss	IPS	1952 – 70
SK 15 NE	Biggin and Alsop en le Dale	DP, NA, IPS, JIC	1969 – 76
SK 15 NW	Alstonefield and Wetton	NA	1970 – 71
SK 16 SE	Middleton	JIC, DP	1970 – 71
SK 16 SW	Hartington	NA, JIC	1968 – 72
SK 16 NE	Ashford and Lathkill Dale	JIC	1971 – 72
SK 16 NW	Flagg and Monyash	NA	1969
SK 17 SE	Monsal Dale	IPS, RAE	1953 – 72
SK 17 SW	Miller's Dale	IPS	1958 – 72
SK 25 NE*	Bonsall and Via Gellia	RAE, GHR, JIC	1955 – 76
SK 25 NW	Aldwark	DP, IPS, JIC	1969, 76
SK 26 SE*	Wensley and Darley Dale	EGS, RAE, GHR, JIC	1954 – 76
SK 26 SW	Stanton, Youlgreave and Winster	EGS, RAE, DP, JIC	1955 – 74
SK 26 NE*	Beeley	EGS, RAE, GHR	1947 – 55
SK 26 NW	Bakewell	IPS	1970 – 71
SK 27 SE*	Baslow	RAE, WE	1939 – 54
SK 27 SW	Hassop	IPS, RAE	1951 – 70

APPENDIX 2

List of boreholes and measured sections

This list includes the permanent record number, location, total depth or thickness, and stratigraphical range of the selected boreholes and measured sections that are referred to in this memoir. Copies of these records may be obtained from the British Geological Survey, Keyworth, Nottingham NG12 5GG, at a fixed tariff. Other measured sections and non-confidential borehole data are given either on the six-inch geological maps listed on p. 48 or are held on open file in the Survey's archives.

SJ 95 NE/9   Ballhaye Green Works, Leek [9890 5689]
   Borehole, 152.40 m   Namurian (E_{2a} to ?E_{2c}) strata
SJ 95 NE/10   Belle View Mills, Leek [9799 5671]
   Borehole, 59.74 m   Hawksmoor Formation
SJ 95 NE/11   Churnet Works, Leek [9756 5704]
   Borehole, 79.86 m   Hawksmoor Formation and drift
SJ 95 NE/13   Messrs. T. Whittles' Works, Leek [9808 5637]
   Borehole, 166.68 m   Hawksmoor Formation and drift
SJ 95 NE/15b and /15i   Near Tittesworth Reservoir [9897 5950 and 9871 5961]
   2 of a series of 9 site investigation boreholes, E_2 strata and drift
SJ 95 NE/17B   Abbey Green, Leek [9790 5765]
   Borehole, 184 m   Hawksmoor Formation and drift
SJ 95 NE/21   Tittesworth – east bank [9921 5896 to 9933 5919]
   Section, 41.35 m E_{2c} to H_{1a} mudstones
SJ 95 NE/22   Tittesworth Reservoir Water Treatment Plant [995 585]
   Temporary section in H_{2c} to R_{1a} mudstones
SJ 96 NE/9   Stream near Cumberland Cottage [9935 6984 and 9959 6989 to 9985 6998]
   Composite section, 83.54 m   Roaches Grit to Chatsworth Grit ($R_{2b} - R_{2c}$)
SJ 96 SE/18   Gun Hill [9723 6182]
   Borehole, 1410.90 m   Tournaisian to Pendleian
SK 05 NW/11   New Mixon Hay [0369 5713]
   Borehole (No. 2), 147.7 m   Ecton Limestones and Mixon Limestone-Shales
SK 05 NW/13   Thorncliff [0176 5859 to 0123 5845]
   Composite section, 181.87 m   *Ct. edalensis* Band to *Ht prereticulatus* Band ($E_{2b} - H_{2c}$)
SK 05 NW/14   Wellington Farm [0214 5612 to 0201 5608]
   Section, 55.91 m E_{2a} to E_{2b} strata including Minn Sandstones
SK 05 NW/15   Holly Dale [0200 5594 to 0189 5588]
   Section, 45.70 m   Hurdlow Sandstones (E_{2c})
SK 05 NE/8   Warslow Hall [0942 5928]
   Borehole, 99.81 m   Ecton Limestones and Mixon Limestone-Shales
SK 05 NE/9   River Manifold south of Wettonmill [0988 5509 to 0987 5504]
   Section, 28.40 m   Milldale Limestones
SK 05 NE/10   Bullclough, Grindon Moor [0603 5502 to 0595 5500]
   Section, 15.50 m   Onecote Sandstones and Pendleian strata ($P_2 - E_{1a}$)
SK 05 NE/11   Clough Head [0862 5925 to 0842 5932]
   Section, 32.70 m   Mudstones and siltstones ($E_{1b} - E_{1c}$)
SK 05 NE/12   Dale Quarry, Warslow [0938 5867 to 0932 5877]
   Section, 43.10 m   Ecton Limestones
SK 05 NE/13   Old railway-cutting, Ecton [0954 5820 to 0910 5794]
   Section, 127.31 m   Milldale Limestones and Ecton Limestones

SK 05 NE/14   Swainsley Farm [0931 5756 to 0925 5752]
   Section, 45.48 m   Ecton Limestones
SK 06 NW/4   Smallshaw Farm [0384 6547 to 0425 6439]
   Section, 94.59 m   Pendleian and Arnsbergian ($E_{1c} - E_{2a}$) strata including Minn Sandstones
SK 06 NW/5   Orchard Farm [0180 6872 to 0179 6870, and 0221 6894 to 0231 6911]
   Composite section; 27.66 m   Chatsworth Grit to *G. cumbriense* Marine Band including Yeadonian stratotype section
SK 06 NW/6   Robins Clough near Knar [0042 6763 to 0002 6792]
   Section, 51.53 m   Westphalian A strata including Red Ash Coal and *G. listeri* Marine Band
SK 06 NW/7   Fairthorn [0446 8699 to 0439 6906]
   Section, 7.19 m   Alportian and Kinderscoutian ($H_2 - R_{1a}$) mudstones
SK 06 NW/8   River Manifold, Dun Cow's Grove [0408 6698]
   Section, 14.62 m   Kinderscoutian mudstones ($R_{1a} - R_{1b}$)
SK 06 NW/9   Fairthorn [0433 6912]
   Section, 11.04 m   Kinderscoutian (R_{1b}) mudstones
SK 06 NW/10   Cistern's Clough, Axe Edge [0337 6978 to 0326 6985]
   Section, 50.20 m   Roaches Grit (R_{2b})
SK 06 NW/11   Stream near Greens [0049 6636 to 0092 6676]
   Section, 25.60 m   Rough Rock to *G. subcrenatum* Marine Band
SK 06 NE/3   Limeworks near Hindlow Station [0871 6910]
   Borehole, 213.51 m   Woo Dale Limestones and Bee Low Limestones
SK 06 NE/10   High Edge [0575 6910]
   Borehole, 60.50 m   Woo Dale Limestones and Bee Low Limestones
SK 06 NE/15   Hindlow Quarry [0919 6788]
   Borehole, 142.04 m   Woo Dale Limestones and Bee Low Limestones
SK 06 NE/17   Glutton Bridge [0827 6653]
   Borehole, 81.69 m   Pendleian and Arnsbergian mudstones
SK 06 NE/20   Stoop Farm [0650 6822]
   Borehole, 107.69 m   Monsal Dale Limestones and Namurian ($E_2 - R_{1c}$) strata including Longnor Sandstones
SK 06 NE/23   Golling Gate, Hollinsclough [0513 6710]
   Borehole, 75.00 m   Longnor Sandstones
SK 06 NE/24   High Edge [0580 6913 to 0630 6873]
   Section, 126.70 m   Bee Low Limestones
SK 06 NE/25   Hillhead Quarry [0743 6924 to 0671 6996]
   Section, 133.47 m   Bee Low Limestones
SK 06 NE/26   Buxton Quarry, Hind Low [0764 6927 and 0846 6903 to 0813 6913]
   Composite section, 57.62 m   Bee Low Limestones and Monsal Dale Limestones
SK 06 NE/27   Brierlow Quarry [0897 6908 to 0907 6898]
   Section, 24.72 m   Bee Low Limestones
SK 06 NE/28   Hindlow and Dow Low quarries [0923 6776 to 0992 6753]
   Composite section, 52.51 m   Bee Low Limestones
SK 06 NE/29   Swallow Brook, Hollinsclough [0652 5756 to 0668 6731]
   Section, 103.35 m   Arnsbergian and Chokierian mudstones ($E_{2b} - H_{1a}$)
SK 06 NE/30   River Dove, Hollinsclough [0596 6714]
   Section, 22.25 m   Longnor Sandstones (R_{1c})

SK 06 NE/31   Hollingsclough [0640 6648 to 0624 6636]
Section, 56.50 m   Longnor Sandstones (R_{1c})
SK 06 NE/32   Daisy Knowl Mine, Longnor [0825 6522]
Section, 9.73 m   Longnor Sandstones (R_{1c})
SK 06 NE/33   Dalehead [0507 6981 to 0500 6978]
Section, 62.1 m   Woo Dale Limestones and Bee Low Limestones
SK 06 SW/6   River Churnet and tributaries near Hurdlow [0165 6062 to 0145 6065]
Composite section, 31.65 m   *R. gracile* Band to *R. bilingue* Band ($R_{2a}-R_{2b}$)
SK 06 SW/7   River Churnet and its tributaries, Stake Gutter [0241 6190 to 0243 6298]
Composite section, 70.57 m   Longnor Sandstones to *R. eometabilingue* Band
SK 06 SW/8   Stream near Hurdlow [0266 6088 to 0257 6093]
Section, 38.76 m   Hurdlow Sandstones (E_{2c})
SK 06 SW/9   Stream near Hurdlow [0314 6081 to 0274 6084]
Composite section, 146.66 m   Minn Sandstones and associated faunal bands ($E_{1c}-E_{2b}$)
SK 06 SW/10   Black Brook [0140 6427 to 0124 6427]
Section, 58.54 m   *G. subcrenatum* Marine Band to Woodhead Hill Rock
SK 06 SW/11   Goldsitch House [0091 6634 to 0100 6427]
Section, 19.19 m   Yard coal to Lower Foot Marine Band
SK 06 SW/12   Warslow Brook [0476 6001 to 0467 6038]
Section, 33.80 m   Onecote Sandstones and Pendleian beds (P_2-E_{1a})
SK 06 SW/13   Stream near Hurdlow [0242 6059 to 0219 6019]
Section, 116.56 m   Namurian mudstones with faunal bands ($E_{2c}-R_{1a}$)
SK 06 SW/14   Stream at Upper Hulme [0123 6103 to 0127 6136]
Section, 81.6 m   Five Clouds Sandstones
SK 06 SE/4   Near Brund [0892 6153]
Site investigation borehole, 34.14 m   Kinderscoutian (R_{1a-b}) mudstones
SK 06 SE/5   Near Brund [0911 6148]
Site investigation borehole, 33.22 m   Kinderscoutian (R_{1b}) mudstones
SK 06 SE/8   Near Brund [0963 6178]
Site investigation borehole, 46.02 m   Longnor Sandstones to *R. bilingue* (early form) Band ($R_{1c}-R_{2b}$)
SK 06 SE/12   Near Brund [0970 6165]
Site investigation borehole, 34.75 m   *R. bilingue* (early form) to *R. bilingue* bands (R_{2b})
SK 06 SE/13   Near Brund [0945 6161]
Site investigation borehole, 47.85 m   Kinderscoutian (R_{1b-c}) strata including Longnor Sandstones
SK 06 SE/14   Blake Brook [0551 6106 and 0558 6075 to 0578 6095]
Section, 151.38 m   Minn Sandstones to *Ct. edalensis* Band ($E_{1c}-E_{2b}$)
SK 06 SE/15   Blake Brook [0578 6095 to 0609 6115]
Section, 103.85 m   *Ct. edalensis* Band to lowest *H. subglobosum* Band ($E_{2b}-H_{1a}$)
SK 06 SE/16   Blake Brook [0621 6119 to 0617 6118]
Section, 82.93 m   lowest *H. subglobosum* Band to low Kinderscoutian ($H_{1a}-R_{1a}$) including the Lum Edge Sandstones and the proposed Alportian (H_2) stratotype section
SK 06 SE/17   Blake Brook [0620 6118 to 0673 6133]
Section, 140.14 m R_{1a} to R_{2a} sequence including Blackstone Edge Sandstones and Longnor Sandstones
SK 06 SE/18   Oakenclough Brook near Oakenclough Hall [0501 6369 to 0539 6364]
Section, 130.11 m E_{2b} to R_{1a} sequence including Lum Edge Sandstones

SK 06 SE/19   Wiggenstall [0899 6080 to 0932 6082]
Section, 129.07 m E_{2b} to H_{2c} sequence including Lum Edge Sandstones
SK 07 SW/3   Portobello Bar, near Buxton [0327 7137]
Borehole, 205.44 m   Roaches Grit (R_{2b}) and drift
SK 07 SW/4   Ladmanlow [0498 7154]
Borehole, 177.09 m   Woo Dale Limestones and Bee Low Limestones
SK 07 SW/9   Near Derbyshire Bridge [0168 7155]
Section, 7.65 m   Beds above Yard Coal with Lower Bassy Marine Band
SK 07 SW/10   Stream near Cavendish Golf Course, Burbage [0479 7353 to 0466 7370]
Section, 29.49 m   Kindescoutian
SK 07 SW/11   Stream near Burbage Edge [0283 7338]
Section, 21.65 m   Roaches Grit and *R. superbilingue* Marine Band
SK 07 SE/24   Woo Dale [0985 7248]
Borehole, 312.19 m   Pre-Carboniferous, and Woo Dale Limestones (?Chadian to Holkerian)
SK 07 SE/39   Green Fairfield [0904 7483]
Borehole, 134.17 m   Woo Dale Limestones and Chee Tor Rock
SK 07 SE/51   King Sterndale [0822 7136]
Borehole, 100.00 m   Woo Dale Limestones and Chee Tor Rock
SK 07 SE/52   Topley Pike [0993 7157]
Borehole, 120.26 m   Woo Dale Limestones and Chee Tor Rock
SK 07 SE/53   Woo Dale – Wye Dale [0974 7253 to 0970 7259]
Section, 31.78 m   Woo Dale Dolomites
SK 07 SE/54   Cowdale Quarry [0818 7227 to 0802 7216]
Section, 50.48 m   Chee Tor Rock
SK 07 SE/55   Below Cowdale Quarry [0807 7246]
Section, 38.29 m   Woo Dale Limestones and Chee Tor Rock
SK 07 SE/56   Deep Dale [0978 7144 to 0979 7138]
Section, 42.51 m   Woo Dale Limestones and Chee Tor Rock
SK 07 SE/57   Ashwood Dale Railway-cutting [0675 7293 to 0666 7300]
Section, 35.16 m   Chee Tor Rock, Lower Miller's Dale Lava and Miller's Dale Limestones
SK 07 SE/58   Harpur Hill Quarry [0634 7079 to 0632 7063]
Section, 48.80 m   Bee Low Limestones
SK 07 SE/59   Railway-cutting, Buxton [0621 7321 to 0631 7348]
Section, 20.32 m   Miller's Dale Limestones and Monsal Dale Limestones
SK 07 SE/60   Cunning Dale [0804 7288 to 0794 7290]
Section, 25.50 m   Woo Dale Limestones
SK 07 SE/61   Holker Road, Buxton [0614 7349 to 0616 7343]
Section, 8.08 m   Eyam Limestones
SK 15 NW/8   Wolfscote Hill [1373 5835]
Borehole, 141 m   Bee Low Limestones
SK 15 NW/9   Wetton [1167 5648]
Borehole, 100.60 m   Hopedale Limestones and Mixon Limestone-Shales
SK 15 NW/10   Alstonefield [1371 5563]
Borehole, 100 m   Milldale Limestones (knoll-reef)
SK 15 NW/12   Wolfscote Dale – Biggin Dale [1440 5684 to 1413 5705]
Section, 85.25 m   Woo Dale Limestones and Bee Low Limestones
SK 15 NW/13   Biggin Dale [1422 5741 to 1418 5778]
Section, 156.20 m   Woo Dale Limestones and Bee Low Limestones
SK 15 NW/14   Stream near Field House [1236 5839 to 1215 5834]
Section, 67.30 m   Arnsbergian and Chokierian ($E_{2c}-H_{1b}$) mudstones
SK 15 NW/15   Paddock House [1060 5793 to 1073 5793]
Section, 93.19 m $E_{2a}-H_{1a}$ sequence

SK 15 NW/16   Trial level near Back of Ecton [1021 5833]
Section, 16.21 m   Ecton Limestones and Mixon Limestone-Shales

SK 15 NE/2   Cardlemere Lane, Uppermoor Farm [1763 5835]
Borehole, 61 m   Bee Low Limestones and Monsal Dale Limestones

SK 15 NE/3   Pikehall [1984 5924]
Borehole, 100 m   Bee Low Limestones and Monsal Dale Limestones

SK 15 NE/4   Biggin [1525 5820]
Borehole, 100 m   Bee Low Limestones and Monsal Dale Limestones

SK 15 NE/5   Lees Barn [1581 5674]
Borehole, 100 m   Woo Dale Limestones and Bee Low Limestones

SK 15 NE/6   Parwich [1932 5712]
Borehole, 100 m   Woo Dale Limestones and Bee Low Limestones

SK 15 NE/7   Twodale Barn, Parwich [1942 5587]
Borehole, 101.30 m   Bee Low Limestones

SK 15 NE/11   Coldeaton railway-cutting [1586 5704 to 1619 5800]
Section, 100.34 m   Bee Low Limestones and Monsal Dale Limestones

SK 15 NE/12   Hindlip Quarry, Alsop en le Dale [164 567]
Section, 43.90 m   Bee Low Limestones

SK 16 NW/8   Town Head [1274 6956]
Borehole, 59.60 m   Monsal Dale Limestones including Upper Miller's Dale Lava

SK 16 NW/9   Chelmorton [1120 6902]
Borehole, 25.02 m   Monsal Dale Limestones including Upper Miller's Dale Lava

SK 16 NW/10   Near Sparklow [1199 6704]
Borehole, 58.31 m   Bee Low Limestones and Monsal Dale Limestones including Upper Miller's Dale Lava

SK 16 NW/11   Sparklow [1284 6588]
Borehole, 60.51 m   Monsal Dale Limestones

SK 16 NW/12   The Jarnett [1476 6947]
Borehole, 102.26 m   Monsal Dale Limestones including Upper Miller's Dale Lava

SK 16 NW/13   Chelmorton Thorn [1195 6970]
Borehole, 33.16 m   Monsal Dale Limestones including Upper Miller's Dale Lava

SK 16 NW/14   Great Low [1136 6765]
Borehole, 60.91 m   Bee Low Limestones and Monsal Dale Limestones including Upper Miller's Dale Lava

SK 16 NE/9   High Low [1552 6817]
Borehole, 37.53 m   Monsal Dale Limestones and Eyam Limestones

SK 16 NE/10   Sheldon [1728 6946]
Borehole, 207.30 m   Bee Low Limestones and Monsal Dale Limestones

SK 16 NE/19   Mogshaw [1889 6778]
Borehole, 143.24 m (inclined)   Monsal Dale Limestones including Shacklow Wood Lava

SK 16 NE/20   Cliff above River Wye [1797 6985 to 1793 6981]
Section, 35.1 m   Monsal Dale Limestones including Shacklow Wood Lava

SK 16 NE/21   Rookery Plantation, Ashford [1922 6982 to 1888 6954]
Section, 109.50 m   Monsal Dale Limestones, Eyam Limestones, Longstone Mudstones

SK 16 NE/22   Wye valley near Ashford [2000 6939 to 1900 6938]
Composite section, 75 m   Monsal Dale Limestones, Eyam Limestones, Longstone Mudstones and Namurian mudstones

SK 16 NE/23   Shacklow Wood [182 696 to 170 690]
Section, 158.40 m   Monsal Dale Limestones including Shacklow Wood Lava, and Eyam Limestones

SK 16 NE/24   Little Shacklow Wood [1790 6928]
Section, 24.60 m   Monsal Dale Limestones

SK 16 NE/25   Little Shacklow Wood [1793 6916]
Section, 12.13 m   Monsal Dale Limestones

SK 16 NE/26   Lathkill Dale [1683 6600 to 1683 6595]
Section, 38.5 m   Monsal Dale Limestones and Eyam Limestones

SK 16 NE/27   Haddon Grove, Lathkill Dale [1807 6585 to 1808 6573]
Section, 73.22 m   Monsal Dale Limestones

SK 16 NE/28   Twin Dales, Over Haddon [1957 6623 to 1972 6613]
Section, 48 m   Monsal Dale Limestones

SK 16 SW/5   Parsley Hay [1471 6369]
Borehole, 61.62 m   Monsal Dale Limestones

SK 16 SW/8   Vincent House [1365 6287]
Borehole, 60.88 m   Woo Dale Limestones

SK 16 SW/11   Mosey Low [1292 6447]
Section, 5 m   Bee Low Limestones

SK 16 SW/12   Hartington Dale [1395 6048 to 1399 6066]
Section, 68.20 m   Bee Low Limestones and Monsal Dale Limestones

SK 16 SW/13   Hartington School [1303 6039 to 1309 6046]
Section, 20.80 m   Bee Low Limestones and Monsal Dale Limestones

SK 16 SW/14   River Dove, Bank Top [1260 6160 to 1265 6155]
Section, 18.50 m   Bee Low Limestones and Monsal Dale Limestones

SK 16 SW/15   Parks Barn, Pilsbury [1219 6317 to 1219 6328]
Composite section, 21.8 m   Monsal Dale Limestones

SK 16 SE/4   Oldhams Farm, Middleton [1704 6254]
Borehole, 61 m   Monsal Dale Limestones

SK 16 SE/5   Thorntree [1826 6334]
Borehole, 61.68 m   Monsal Dale Limestones

SK 16 SE/6   Gratton Moor Farm [1975 6089]
Borehole, 244.45 m   (inclined); Bee Low Limestones and Monsal Dale Limestones

SK 16 SE/15   Station Quarry, Hartington [152 613]
Composite section, 135 m   Woo Dale Limestones and Bee Low Limestones

SK 16 SE/16   Middleton [198 633]
Composite section, 33.10 m   Monsal Dale Limestones and Eyam Limestones

SK 17 SW/15   Topley Pike Quarry [1006 7209]
Borehole, 42.67 m   Woo Dale Limestones

SK 17 SW/18   Pillwell Lane, Fivewells Farm [1206 7053]
Borehole, 34.61 m   Monsal Dale Limestones including Upper Miller's Dale Lava

SK 17 SW/55   Taddington [1399 7138]
Borehole, 100 m   Chee Tor Rock, Lower Miller's Dale Lava and Miller's Dale Limestones

SK 17 SW/62   Topley Pike [1093 7232]
Borehole, 90.05 m   Woo Dale Limestones and Chee Tor Rock

SK 17 SW/65   Wye Dale [1125 7277 to 1132 7284]
Section, 57.25 m   Woo Dale Limestones and Chee Tor Rock

SK 17 SW/66   Wye Dale [1083 7256 to 1103 72268]
Section, 38.37 m   Woo Dale Limestones

SK 17 SW/67   Station Quarry, Miller's Dale [1327 7340 to 1329 7347]
Section, 30.48 m   Miller's Dale Limestones, 'Station Quarry Beds' and Upper Miller's Dale Lava

SK 17 SW/68   Miller's Dale Lime Works [1448 7325]
Section, 19.98 m   Miller's Dale Limestones and 'Station Quarry Beds'

SK 17 SW/69   Miller's Dale Lime Works [1398 7304 to 1411 7297]
  Section, 57.3 m   Monsal Dale Limestones
SK 17 SE/2   Longstone Edge [1912 7332]
  Borehole, 179.63 m   Monsal Dale Limestones including Shacklow Wood Lava and Litton Tuff
SK 17 SE/12   Great Longstone [1969 7253]
  Borehole, 100 m   Monsal Dale Limestones
SK 17 SE/13   Brushfield [1545 7192]
  Borehole, 100.03 m   Monsal Dale Limestones with Upper Miller's Dale Lava
SK 17 SE/14   Lees Bottom, Wye valley [1705 7050]
  Borehole, 40.66 m   Bee Low Limestones, Monsal Dale Limestones and alluvium
SK 17 SE/18   Near Litton Mill [1569 7298]
  Section, 24.50 m   Monsal Dale Limestones
SK 17 SE/19   Crossdale Head, Little Longstone [1827 7312]
  Section, 9.91 m   Monsal Dale Limestones
SK 17 SE/20   Upperdale [1782 7215 to 1807 7217]
  Section, 97.59 m   Monsal Dale Limestones including Upperdale and Hob's House coral bands
SK 17 SE/21   White Cliff, Monsal Head [1825 7195]
  Section, 32.97 m   Monsal Dale Limestones including White Cliff Coral Band.
SK 17 SE/22   Headstone Head, Monsal Head [1824 7182 to 1847 7159]
  Section, 78.98 m   Monsal Dale Limestones including White Cliff Coral Band
SK 17 SE/23   Railway-cutting, Little Longstone [1883 7140 to 1899 7135]
  Section, 25.76 m   Monsal Dale Limestones, Eyam Limestones and Longstone Mudstones
SK 17 SE/24   Hob's House, Monsal Dale [1756 7123 to 1757 7119]
  Section, 51.73 m   Monsal Dale Limestones with Hob's House Coral Band
SK 17 SE/25   South of Fin Cop [1725 7066 to 1739 7070]
  Section, 137.10 m   Monsal Dale Limestones
SK 17 SE/26   Above A6 road, Monsal Dale [1710 7022]
  Section, 42.85 m   Monsal Dale Limestones
SK 17 SE/27   By A6 road, Great Shacklow Wood [1745 7015]
  Section, 19.74 m   Monsal Dale Limestones including Lees Bottom Lava
SK 25 NW/18   Longcliffe [2182 5544]
  Borehole, 61.52 m   Woo Dale Limestones (Griffe Grange Member), Bee Low Limestones and drift
SK 25 NW/19   Longcliffe Quarry [2392 5701]
  Borehole, 56.08 m   Woo Dale Limestones (Griffe Grange Member) and Bee Low Limestones
SK 25 NW/21   Elton [2248 5965]
  Borehole, 79.42 m   Bee Low Limestones and Monsal Dale Limestones
SK 25 NW/22   Aldwark [2125 5815]
  Borehole, 100.39 m   Bee Low Limestones and Monsal Dale Limestones including Winstermoor Lava
SK 25 NW/23   Longcliffe [2277 5575]
  Borehole, 66.75 m   Bee Low Limestones and made ground
SK 25 NW/24   Longcliffe [2288 5573]
  Borehole, 55.25 m   Bee Low Limestones and Pocket Deposits
SK 25 NW/33   Hoe Grange Quarry [2228 5597 to 2233 5608]
  Section, 38.22 m   Monsal Dale Limestones
SK 25 NW/34   Ivonbrook Quarry [2329 5830 to 2336 5846]
  Section, 37.21 m   Monsal Dale Limestones
SK 25 NW/35   Longcliffe Quarry [2399 5702 to 2395 5693]
  Section, 28.37 m   Bee Low Limestones

SK 25 NW/36   Grangemill Quarry, Grangemill [2406 5733 to 2415 5728]
  Section, 23.27 m   Bee Low Limestones
SK 25 NW/37   Prospect Quarry, Grangemill [2458 5753]
  Section, 47.25 m   Woo Dale Limestones and Bee Low Limestones
SK 25 NW/38   Valley near Griffe Grange [2477 5675]
  Section, 12.68 m   Woo Dale Limestones and Bee Low Limestones
SK 25 NE/41   Bottom Leys Farm, Bonsall [2587 5795]
  Borehole, 92.27 m   Monsal Dale Limestones including Shothouse Spring Tuff
SK 25 NE/42   Ryder Point, Hopton Wood [2619 5645]
  Borehole, 100.16 m   Woo Dale Limestones
SK 25 NE/48   Ryder Point, Hopton Wood [2620 5642]
  Borehole (No. 3), 150 m   Woo Dale Limestones including the Griffe Grange Member
SK 25 SE/37   Ryder Point Plant, Hopton [2616 5485]
  Borehole (No. 2), 200 m   Woo Dale Limestones (Griffe Grange Member); Bee Low Limestones
SK 26 NW/8   Bakewell [2080 6843]
  Borehole, 18.9 m   tuff in ?vent
SK 26 NW/12   Field House, Bakewell [2077 6897]
  Borehole, 100 m   Monsal Dale Limestones including Conksbury Bridge Lava, Eyam Limestones and Longstone Mudstones
SK 26 NW/13   Conksbury Bridge [2142 6587]
  Borehole, 110.35 m   Monsal Dale Limestones, Eyam Limestones and boulder clay
SK 26 NW/14   Nutseats Quarry, Bakewell [2369 6585]
  Borehole, 100.05 m   Monsal Dale Limestones including Conksbury Bridge Lava, and Eyam Limestones
SK 26 NW/16   Haddon Fields [2279 6553]
  Borehole, 86.40 m (inclined)   Monsal Dale Limestones including Conksbury Bridge Lava, ?Eyam Limestones and boulder clay
SK 26 NW/21   Haddon Fields [2333 6554]
  Borehole, 62.60 m (inclined)   Eyam Limestones, Longstone Mudstones and Namurian mudstones including *C. leion* Band
SK 26 NW/25   Haddon Fields [2372 6577]
  Borehole, 297.24 m (inclined)   Fallgate Volcanic Formation, Monsal Dale Limestones and Eyam Limestones
SK 26 NW/31   Below Endcliff Wood, Bakewell [2141 6898]
  Section, 27.40 m   Monsal Dale Limestones above Conksbury Bridge Lava including *Orionastraea* Band
SK 26 NW/33   Quarry near Bank Top House, Bakewell [2105 6822]
  Section, 23.50 m   Monsal Dale Limestones and Eyam Limestones
SK 26 NW/35   Shining Bank Quarry, Youlgreave [2281 6501 to 2284 6507]
  Section, 58.44 m   Monsal Dale Limestones, Eyam Limestones, Longstone Mudstones and boulder clay
SK 26 NW/36   Old railway-cutting, Bakewell [2302 6758 to 2300 6760]
  Section, 7.63 m   Namurian mudstone ($H_1a - H_1b$)
SK 26 SW/9   Conksbury Quarry, Youlgreave [2091 6486]
  Borehole, 188.90 m (inclined)   Fallgate Volcanic Formation, Monsal Dale Limestones
SK 26 SW/16   Birchover [2413 6233]
  Borehole, 160.26 m   Namurian ($R_1 - R_{2b}$) mudstones, siltstones and sandstones including Ashover Grit
SK 26 SW/17   Cowclose Farm, Birchover [2376 6236]
  Borehole, 67.94 m   Namurian (R_{2b}) mudstones, siltstones and sandstones, and head

SK 26 SW/18   Gratton Moor Farm [2035 6034]
Borehole, 186.61 m (inclined)   Bee Low Limestones and
Monsal Dale Limestones

SK 26 SW/19   Gratton Moor Farm [2027 6025]
Borehole, 166.08 m (inclined)   Bee Low Limestones and
Monsal Dale Limestones (dolomitised)

SK 26 SW/26   Gratton Moor [2011 6094]
Borehole, 74.60 m (inclined)   Monsal Dale Limestones

SK 26 SW/33   North of Eagle Tor [2322 6360]
Borehole, 172.14 m (inclined)   Monsal Dale Limestones, Eyam
Limestones, Longstone Mudstones and Namurian mudstones

SK 26 SW/36   Bowers Hall [2325 6458]
Borehole, 131.87 m   Monsal Dale Limestones, Eyam
Limestones and Longstone Mudstones

SK 26 SW/37   Upper Town [2444 6154]
Borehole, 376.92 m (inclined)   Monsal Dale Limestones
including Lower Matlock Lava, Eyam Limestones, Longstone
Mudstones, Namurian ($E_{1a} - R_{2b}$) mudstones and sandstones

SK 26 SW/46   Bowers Hall [2349 6456]
Borehole, 121.38 m (inclined)   Eyam Limestones, Longstone
Mudstones, Namurian ($E_{1a} - R_{1a}$) mudstones and boulder clay

SK 26 SW/47   Bowers Hall [2349 6456]
Borehole, 77.42 m (inclined)   ?Longstone Mudstones,
Namurian mudstones, sand and boulder clay.

SK 26 SW/49   Upper Town [2455 6183]
Borehole, 272.16 m (inclined)   Eyam Limestones, Longstone
Mudstones, Namurian ($E_{1a} - R_{2b}$) mudstones, siltstones and
sandstones

SK 26 SW/54   Whiteholmes, Winster [2349 6126]
Borehole, 275.59 m (inclined)   Bee Low Limestones, Monsal
Dale Limestones including Lower Matlock Lava, Eyam
Limestones, Longstone Mudstones and Namurian mudstones

SK 26 SW/57   Near Winster [2389 6140]
Borehole, 267.28 m (inclined)   Monsal Dale Limestones with
Lower Matlock Lava, Eyam Limestones, Longstone Mudstones
and Namurian mudstones

SK 26 SW/64   Upper Town [2396 6140]
Borehole, 369.86 m (inclined)   Bee Low Limestones, Monsal
Dale Limestones including Lower and Upper Matlock lavas,
Eyam Limestones, Longstone Mudstones and Namurian
mudstones

SK 26 SW/66   Upper Town [2365 6132]
Borehole, 81.68 m   Monsal Dale Limestones, Eyam
Limestones, Longstone Mudstones and Namurian mudstones

SK 26 SW/88   Moatlow Knob, Bradford Dale [2016 6380 to
2010 6387]
Section, 63.90 m   Monsal Dale Limestones

SK 26 SW/89   Gratton Dale [2007 5965 to 2114 6143]
Composite section, 349.93 m   Bee Low Limestones, Monsal
Dale Limestones including Lower Matlock Lava, and Eyam
Limestones

SK 26 SW/91   Stanton Mill Quarry, Stanton [2321 6400]
Section, 22.10 m   Eyam Limestones and Longstone Mudstones

SK 26 SW/92   River Bradford [2149 6401]
Section, 15.72 m   Eyam Limestones

SK 26 SE/4   Snitterton [2698 6030]
Borehole, 203.71 m   Bee Low Limestones, Monsal Dale
Limestones including Lower and Upper Matlock lavas, and
Eyam Limestones

SK 27 NW/15   Middleton Dale [2096 7603]
Borehole, 1851.05 m   Ordovician mudstones, Tournaisian
evaporites, Woo Dale Limestones, Bee Low Limestones and
Monsal Dale Limestones including Cressbrook Dale Lava and
Litton Tuff

SK 27 SW/1   Longstone Edge [2144 7367]
Borehole, 226.36 m   Monsal Dale Limestones including
Cressbrook Dale Lava and Litton Tuff

SK 27 SW/4   Longstone Edge [2056 7325]
Borehole, 131.06 m   Monsal Dale Limestones including
Cressbrook Dale Lava and Litton Tuff

SK 27 SW/5   Longstone Edge [2088 7335]
Borehole, 177.77 m   Monsal Dale Limestones including
Cressbrook Dale Lava

SK 27 SW/8   Longstone Edge [2099 7345]
Borehole, 116.74 m   Monsal Dale Limestones including
Cressbrook Dale Lava and Litton Tuff

SK 27 SW/10   Longstone Edge [2250 7368]
Borehole, 272.64 m   Monsal Dale Limestones including Cressbrook Dale Lava and Litton Tuff

SK 27 SW/15   Longstone Edge [2347 7400]
Borehole, 233.20 m   Monsal Dale Limestones including Cressbrook Dale Lava and Litton Tuff

SK 27 SW/19   Longstone Edge [2020 7323]
Borehole, 141.35 m   Monsal Dale Limestones including Cressbrook Dale Lava and Litton Tuff

SK 27 SW/20   Longstone Edge [2203 7363]
Borehole, 611.48 m   Holkerian and Asbian limestones, Monsal
Dale Limestones including Cressbrook Dale Lava and Litton
Tuff

SK 27 SW/26   Buskey Cottage, Great Longstone [2103 7142]
Borehole, 39.40 m   Eyam Limestones, Longstone Mudstones
and boulder clay

SK 27 SW/29   By Sallet Hole Mine, Coombs Dale [2192 7408 to
2189 7393]
Section, 54.38 m   Monsal Dale Limestones

APPENDIX 3

List of petrographical samples

E numbers refer to the Sliced Rock Collection of the British Geological Survey. Each number is followed by locality details where these exist; in the case of records in the Survey's archives, the localities are shown by section or borehole reference numbers. (*See* Appendix 2.)

E 7332   Calton Hill, precise location unknown
E 8340   Calton Hill, precise location unknown
E 8341   Calton Hill, precise location unknown
E 9052   Old Quarry at Ible, precise location unknown
E 42631   Ecton Mines, Clayton Level, 230 m from entrance [0984 5808]
E 46369   [SJ 96 SE/18]; depth 731.52 m
E 46763   [SK 06 NE/32]; level below top 6.40 m
E 46764   [SK 06 NE/32]; level below top 6.40 m
E 46766   Summit tor of Sheen Hill [1105 6255]; unlaminated coarse sandstone
E 46767   Townend Quarry, Sheen [1088 6057]; fine-grained unlaminated sandstone
E 46769   Old Quarry in Bollington Wood, Leek [9891 5545]; sandstone in sandstone – shale sequence
E 46770   Old Quarry at Kniveden, Leek [9998 5607]; sandstone showing sole structures
E 46775   Valley near Easing Farm [0139 5731]; channel sandstone – coarse quartzose grit
E 46777   By junction of farm track and road at Upper Hulme [0101 6089]; base of sandstone leaf
E 46778   Roach Quarries [0042 6228]; massive bed in middle of sandstone leaf
E 46679   Ramshaw Rocks [0177 6192]; massive channel facies
E 46782   Right bank of Black Brook [0172 6451]; cross-bedded facies
E 46783   Above right bank of Black Brook [0152 6434]
E 46786   Left bank of Black Brook [0133 6426]; fine-grained ripple-laminated facies
E 47809   Gratton Dale [2082 6080]
E 48015   Roadside crag in Wye valley [1794 6975]
E 48016   [SK 26 SW/9]; depth 48.62 m
E 48017   [SK 26 SW/9]; depth 75.74
E 48018   [SK 26 SW/9]; depth 93.40
E 48019   [SK 26 SW/9]; depth 107.92 m
E 48020   [SK 26 SW/9]; depth 143.26 m
E 51340   [SK 26 SW/16]; depth 65.30 m
E 51341   [SK 26 SW/16]; depth 66.10 m
E 51342   [SK 26 SW/16]; depth 92.70 m
E 52370   Roadside Quarry near Tophill Farm, Grangemill [2442 5783]
E 52549   [SK 26 NW/25]; depth 52.80 m
E 52550   [SK 26 NW/25]; depth 52.83 m
E 52553   [SK 26 NW/25]; depth 105.30 m
E 52557   [SK 26 NW/25]; depth 147.23 m
E 52558   [SK 26 NW/25]; depth 165.60 m
E 52559   [SK 26 NW/25]; depth 170.00 m
E 52560   [SK 26 NW/25]; depth 175.85 m
E 52561   [SK 26 NW/25]; depth 198.47 m
E 52562   [SK 26 NW/25]; depth 198.97 m
E 52563   [SK 26 NW/25]; depth 205.75 m
E 52564   [SK 26 NW/25]; depth 218.85 m
E 52566   [SK 26 NW/25]; depth 259.80 m
E 52567   [SK 26 NW/25]; depth 272.02 m

E 52568   [SK 26 NW/25]; depth 275.95 m
E 52569   [SK 26 NW/25]; depth 292.89 m
E 53778A   [SK 16 NE/19]; depth 100.75 m
E 73778B   [SK 16 NE/19]; depth 100.75 m
E 53780   [SK 16 NE/19]; depth 114.40 m
E 53781   [SK 16 NE/19]; depth 120.90
E 54943   Stream near Ravensdale Cottages [1724 7384]
E 54944   Stream near Ravensdale Cottages [1724 7393]
E 54946   West of Priestcliffe [1353 7184]
E 54947   Old Quarry, Knot Low [1339 7355]; massive lava (lower part of section)
E 54948   Old Quarry, Knot Low; amygdaloidal lava (upper part of section)
E 54949   Calton Hill [1202 7117]
E 54950   Calton Hill [1198 7118]
E 54951   Calton Hill [1188 7114]
E 54952   Calton Hill [1176 7111]
E 54953   Calton Hill [1194 7117]
E 54954   Calton Hill [1195 7118]
E 54955   Calton Hill [1195 7118]
E 54956   Calton Hill [1195 7117]
E 54988   Old Quarry at Ible [2530 5680] about 3 m above quarry floor
E 54989   Old Quarry at Ible about 10 m above quarry floor
E 54990   Old Quarry at Ible about 14 m above quarry floor
E 54993   Roadside cutting, south of Buxton [0640 7229]; amygdaloidal lava
E 54994   Old Quarry, Tideswell Dale [1545 7382]; gully at S end of quarry; about 3.5 m below quarry floor
E 54995   Old Quarry, Tideswell Dale about 0.5 m below quarry floor
E 54996   Old Quarry, Tideswell Dale [1551 7383]; quarry face (lowest outcrop); about 3 m above quarry floor
E 54997   Old Quarry, Tideswell Dale [1553 7386]; near centre of main face; about 10 m above quarry floor
E 54998   Old Quarry, Tideswell Dale near centre of main face; about 12 m above quarry floor
E 56403   Near Grangemill [2438 5782]
E 56429   Calton Hill [1212 7117]
E 56430   Calton Hill [1192 7117]
E 56431   Calton Hill [1194 7115]; marginal apophysis intruding lava
E 56434   Calton Hill amygdaloidal lava wall-rock at intrusion contact
E 56435   Calton Hill, about 2 m from intrusion contact
E 56749   Old Quarry, Tideswell Dale [1547 7378], prismatic clay below dolerite

APPENDIX 4

List of Geological Survey photographs

Copies of these photographs may be seen in the library of the Geological Museum, Exhibition Road, South Kensington, London SW7 2DE, and of the BGS, Keyworth, Nottingham NG12 5GG. All belong to Series L except where indicated, and most are available as prints or transparencies in both black-and-white and colour. The National Grid references refer to the viewpoint. Photography by K. E. Thornton (Series L), J. M. Pulsford and J. Rhodes (Series A).

GENERAL VIEWS

1148　Milldale [146 552]: Crags of Milldale Limestones flanking the gorge of the River Dove
1149　Milldale. As 1148
A9077　Earl Sterndale [080 670]: Chrome Hill and hollow occupied by Namurian shale
1163　Earl Sterndale [090 675]: Asbian apron-reef scenery, Parkhouse Hill and Chrome Hill
1164　Earl Sterndale [072 676]: Scenic contrast between shelf and apron-reef facies of Bee Low Limestones
1165　As 1148 [086 670]
1166　Earl Sterndale [091 674]: Scenery associated with shelf and apron-reef facies of Bee Low Limestones
1167　As 1166
A9076　Glutton and Parkhouse Hill [095 671]: Steep-sided hill in apron-reef limestones
1173　Wolfscote Dale [131 584]: River Dove valley narrowing as it passes from outcrop of Namurian shale to that of Bee Low Limestones
1195　Dove valley, Hollinsclough [066 661]: Scenery at the junction between Asbian apron-reef limestones and Namurian mudstones and sandstones
1196　Dove valley north of Crowdicote [100 652]: Valley in Namurian shales flanked by Asbian apron-reef limestones
1197　As 1196 [104 651] with Namurian sandstone escarpments also in field of view
1198　South of Crowdicote [108 649]: Dove valley in Namurian shales flanked by Sheen Sandstones and Asbian apron-reef limestones
1199　Dove valley south-west of Hartington [117 597]: Scenery at junction of Namurian shales and Asbian shelf and reef limestones
1200　South-west of Hartington [116 598]: Asbian reef limestone hills and lower relief Namurian topography
1201　South-west of Hartington [117 597]: Contrast in relief between Dinantian limestone and Namurian mudstone and sandstone topography
1202　Wetton Hill [108 575]: Chadian–Arundian knoll-reef topography
1203　Hope [122 549]: Hollow occupied by Namurian mudstone outlier among hills of Hopedale Limestones
1217　Goyt Syncline from south [012 592]: Roaches, Hen Cloud and Ramshaw Rocks
1218　Goyt Syncline from SE [037 605]: As 1217; with landslipped dip slope of Minn Mudstones-with-Sandstones in foreground
1219　Near Merryton Low: As 1218; part of panoramic view

1227　Robin Hoods Stride, Birchover [232 614]: Growth-faulted Ashover Grit scenery
1242　Dane valley near Wildboarclough [978 685]: Shutlingsloe capped by Chatsworth Grit, and dip slope of Roaches Grit
1243　Dane valley near Wildboarclough: As 1242; part of panoramic view
2259　Woo Dale [097 726]: Type area for Woo Dale Limestones
A9079　Miller's Dale [165 728]: Gorge in Miller's Dale Limestones overlain by Monsal Dale Limestones
2268　Chelmorton [114 706]: Scarp of Monsal Dale Limestones overlying Upper Millers Dale Lava
2269　Taddington [143 708]: As 2268
2275　Headstone Head (Monsal Head) [184 715]: Monsal Dale cut in Monsal Dale Limestones with landslip of Hob's House
2277　Headstone Head (Monsal Head) [184 715]: Valley of River Wye cut in Monsal Dale Limestones
2286　Above Wildboarclough [991 725]: Headwaters of Clough Brook and features formed by Chatsworth Grit

CARBONIFEROUS – DINANTIAN

1145　Thors Cave, Manifold valley [099 550]: Massive knoll-reef facies of Milldale Limestones
1146　Wetton Mill [0958 5612]: Contrast between knoll-reef and inter-reef beds of Milldale Limestones
1147　Ecton [0955 5805]: Mine adit in Milldale Limestones
1150　Lode House, Dove valley [1453 5536]: Dark lithofacies of Milldale Limestones
1151　Coldeaton Bridge [1462 5606]: Dark facies of Milldale Limestones
1152　Drabber Tor, Wolfscote Dale [1435 5789]: Uppermost part of Woo Dale Limestones
1153　Wolfscote Dale [1435 5689]: Close-jointed micritic Woo Dale Limestones with small cave
1154　Parsley Hay [1456 6245]: Graded bedding in Woo Dale Limestones
1155　Hillhead Quarry [071 696]: Bee Low Limestones showing thick bedding and a clay wayboard
1156　Hillhead Quarry: As 1155
1157　Hartington Station Quarry [1510 6132]: Bee Low Limestones showing thick bedding and clay wayboards
1158　Buxton Quarry, Hind Low [083 690]: Monsal Dale Limestones overlying Bee Low Limestones
1159　Thirkelow [050 689]: Bee Low Limestones steeply dipping at western margin of Dinantian outcrop
1160　Swallow Tor [065 676]: Apron-reef limestones at the limit of their outcrop in the Earl Sterndale reef belt
1161　Abbots Grove [108 661]: Old quarry showing shelf facies passing into apron-reef
1162　Abbots Grove: As 1161
1168　Glutton [087 674]: Bee Low Limestones transition from shelf to apron-reef facies in dry gorge
1169　As 1168
1170　Wolfscote Dale [1325 5800]: Gently dipping Bee Low Limestones and scree

1171 Wolfscote Dale [1310 5843]: Bee Low Limestones, interdigitation of shelf and apron-reef
1172 Wolfscote Dale [132 583]: As 1171
1174 Wolfscote Dale [1310 5843]: Bee Low Limestones, massive back-reef lagoonal facies
1175 Ecton [095 582]: Ecton Limestones type section
1176 As 1175
1177 As 1175
1178 Dale Quarries, Warslow [0933 5874]: Ecton Limestones
1179 Ecton (Apes Tor) [0990 5867]: Subsidiary folding in Ecton Limestones in Ecton Anticline
1180 Ecton (Apes Tor) [0987 5867]: As 1179
1181 Ecton (Apes Tor) [0996 5865]: As 1179
1182 Ecton (Apes Tor) [0991 5867]: As 1179
1183 Lathkill Dale [194 658]: Monsal Dale Limestones, dark lithofacies
1184 Lathkill Dale [183 655]: Monsal Dale Limestones, shell bed with *Gigantoproductus*
1185 Ashford [193 695]: As 1183
1186 Ashford [1908 6939]: As 1183
1187 Ashford [193 695]: Monsal Dale Limestones, laminated beds
1188 Pilsbury [1200 6330]: Monsal Dale Limestones with abundant brachiopod shells
1189 Dove valley (Ludwell) [1237 6249]: Copious spring from Monsal Dale Limestones near junction with Namurian shales
1190 Monyash [1486 6768]: Bedded facies of Eyam Limestones
1191 Monyash [1491 6774]: Interdigitating bedded and reef facies of Eyam Limestones
1192 As 1191
1193 Monyash [1491 6773]: As 1191
1194 As 1191
1206 Ecton [0955 5805]: Milldale Limestones and adit to mine level
2260 Great Rocks Dale (Buxton Central Quarry) [110 730]: Chee Tor Rock overlying Woo Dale Limestones
2261 Deep Dale [096 708]: Valley with fault exposing Chee Tor Rock and Woo Dale Limestones
2262 Chee Tor [123 734]: Type locality of Chee Tor Rock
2264 Miller's Dale [161 728]: Cliffs of Miller's Dale Limestones
A9081 As 2264 [165 729]
2265 Raven's Tor [1503 7328]: Contact of Miller's Dale Limestones on Lower Miller's Dale Lava
2266 Raven's Tor [1503 7328]: Cliff of Miller's Dale Limestones on Lower Miller's Dale Lava
2270 Old railway cutting, Litton Mill [1570 7298]: Upper Miller's Dale Lava overlain by Monsal Dale Limestones
2271 As 2270 [1617 7289]: Monsal Dale Limestones, resting on Miller's Dale Limestones with a clay wayboard at the top
2273 Station Quarry, Miller's Dale [133 734]: Miller's Dale Limestones overlain by Monsal Dale Limestones ('Station Quarry Beds')
2274 Priestcliff Quarry, Miller's Dale [138 730]: Monsal Dale Limestones ('Priestcliff Beds')
2278 Bakewell [2107 6821]: Quarry with disused working for chert in Monsal Dale Limestones
2280 Shining Bank Quarry, Alport [2297 6520]: Boulder Clay on Longstone Mudstones, Eyam Limestones and Monsal Dale Limestones
2281 Old railway cutting, Little Longstone [1890 7138]: Monsal Dale Limestones
2282 As 2281 [1897 7136]: Eyam Limestones overlain by Longstone Mudstones

A9301 Harboro' Rocks [24 55]: Dolomitised Bee Low Limestones
A9302 As A9301

CARBONIFEROUS – NAMURIAN

1207 Hurdlow [0306 6058]: Interbedded calcareous siltstones and shaly mudstones, E_{1c} Zone
1208 Hurdlow [0311 6053]: As 1207 with *Cravenoceras malhamense* Band overlying protoquartzitic turbidites of Minn Sandstones
1209 Thorncliff [0176 5861]: Pair of K-bentonite bands in mudstones, E_{2b} Zone
1210 Lum Edge [0607 6074]: Lum Edge Sandstones, coarse grit with mudflake cavities
1211 Hollinsclough [060 672]: Basal beds of Longnor Sandstones
1212 Ballbank House [062 651]: Longnor Sandstones, mudstones-with-sandstones facies
1213 Fough [0594 6757]: Longnor Sandstones
1214 Hangingstone Farm [9660 6521]: Sandstones and mudstones between Five Clouds Sandstones and Roaches Grit
1215 Five Clouds [003 626]: Five Clouds Sandstones
1216 As 1215 [998 629]
1222 Hen Cloud [000 610]: Roaches Grit
1223 Ramshaw Rocks [020 629]: Roaches Grit
1224 Hen Cloud [0080 6272]: Roaches Grit
1225 Hen Cloud [0085 6260]: Roaches Grit
1226 Ramshaw Rocks [0190 6205]: Roaches Grit
1228 Birchover (Robin Hood's Stride) [225 623]: Ashover Grit
1229 Birchover (Rowtor Rocks) [2348 6212]: Ashover Grit
1231 As 1229
1232 As 1229
1233 Birchover (Stanton Park Quarry) [2437 6250]: Ashover Grit
1234 As 1233 [2433 6252]: Ashover Grit
1235 As 1234
1236 As 1234
1237 As 1234
1238 Birchover [2417 6242]: Quarry in Ashover Grit
1239 Stanton Moor [2456 6344]: As 1238
A9247 Stanton Moor (Endcliffe Quarry) [2540 6363]: Landslipped Ashover Grit
1240 Cumberland Cottage [9974 6993]: Thinly interbedded mudstone, siltstone and sandstone below Chatsworth Grit
1241 Gibb Tor [018 648]: Chatsworth Grit
1244 As 1241
1245 Reeve Edge Quarries [013 697]: Rough Rock

CARBONIFEROUS – WESTPHALIAN

1246 Goldsitch House [0095 6432]: Yard Coal exposed on left bank of stream

IGNEOUS ROCKS

2263 Buxton [0650 7291]: Miller's Dale Limestones overlying Lower Miller's Dale Lava
2272 Winstermoor Farm (Shothouse Spring) [2425 5892]: Type locality, Shothouse Spring Tuff
2283 Grangemill [2442 5783]: Tuff in vent
A9306 Grangemill [244 578]: Tuff with limestone lapilli in vent
A9083 Calton Hill [117 714]: Quarry in dolerite intrusion
A9084 As A9083
2284 Calton Hill [120 712]: Quarry in dolerite with included masses of vesicular lava
2285 Calton Hill [121 712]: Quarry in dolerite intrusion
A9300 Ible [2530 5682]: Quarry in dolerite sill

MINERALISATION

1204   Ecton [096 582]: Spoil tips of old Ecton Copper Mines
1205   Panoramic view with 1204, includes gravel pit in cemented scree deposit
2287   Long Rake, Youlgreave [2105 6504]: Open working for fluorite

POCKET DEPOSITS

1248   Hind Low (Buxton Quarry) [0832 6908]: Pliocene clay and Namurian shale infilling a solution hollow in limestone
1249   As 1248
A9303 Brassington [243 550]: Kilns for silica brick manufacture using local silica sand
A9304 Brassington [240 550]: Green Clay Sand Pit

QUATERNARY

1250   Hollins Hill [055 676]: Large rotational landslip
1251   As 1250 (part of panoramic view)
1252   Lud's Church [9870 6565]: Large chasm at back of landslip
1253   As 1252
1254   As 1252
1255   Ecton [0968 5814]: Old gravel pit worked in cemented scree
1256   Ecton [097 581]: Cemented scree in old gravel pit
2279   Shining Bank Quarry, Alport [2297 6520]: Boulder Clay

INDEX OF FOSSILS

In compiling the following, signs which qualify identification (e.g. aff., cf., ?) have been disregarded.

GENERAL INDEX

Page numbers in italics refer to illustrations

HMSO publications are available from:

HMSO Publications Centre
(Mail and telephone orders only)
PO Box 276, London SW8 5DT
Telephone orders (01) 622 3316
General enquiries (01) 211 5656

HMSO Bookshops
49 High Holborn, London, WC1V 6HB
(01) 211 5656 (Counter service only)

258 Broad Street, Birmingham, B1 2HE
(021) 643 3757

Southey House, 33 Wine Street, Bristol, BS1 2BQ
(0272) 24306/24307

9-21 Princess Street, Manchester, M60 8AS
(061) 834 7201

80 Chichester Street, Belfast, BT1 4JY
(0232) 234488

13a Castle Street, Edinburgh, EH2 3AR
(031) 225 6333

HMSO's Accredited Agents
(see Yellow Pages)

and through good booksellers

BRITISH GEOLOGICAL SURVEY

Keyworth, Nottingham NG12 5GG

Murchison House, West Mains Road,
Edinburgh EH9 3LA

The full range of Survey publications is available
through the Sales Desks at Keyworth and
Murchison House. Selected items are stocked by
the Geological Museum Bookshop, Exhibition
Road, London SW7 2DE; all other items may be
ordered through the BGS Information Desk
adjacent to the Geology Library in the Geological
Museum. All the books are listed in HMSO's
Sectional List 45. Maps are listed in the BGS Map
Catalogue and Ordnance Survey's Trade
Catalogue. They can be bought from Ordnance
Survey Agents as well as from BGS.

*On 1 January 1984 the Institute of Geological Sciences
was renamed the British Geological Survey. It continues to
carry out the geological survey of Great Britain and
Northern Ireland (the latter as an agency service for the
government of Northern Ireland), and of the surrounding
continental shelf, as well as its basic research projects. It
also undertakes programmes of British technical aid in
geology in developing countries as arranged by the Overseas
Development Administration.*

*The British Geological Survey is a component body of the
Natural Environment Research Council.*